Island Biogeography

Island Biogeography

Ecology, Evolution, and Conservation

ROBERT J. WHITTAKER
School of Geography,
University of Oxford

OXFORD NEW YORK TOKYO
OXFORD UNIVERSITY PRESS
1998

Oxford University Press, Great Clarendon Street, Oxford OX2 6DP
Oxford New York
Athens Auckland Bangkok Bogota Bombay Buenos Aires Calcutta
Cape Town Chennai Dar es Salaam Delhi Florence Hong Kong Istanbul
Karachi Kuala Lumpur Madrid Melbourne Mexico City Mumbai
Nairobi Paris São Paolo Singapore Taipei Tokyo Toronto Warsaw
and associated companies in
Berlin Ibadan

Oxford is a trade mark of Oxford University Press

Published in the United States
by Oxford University Press, Inc., New York

A catalogue record for this book is available from the British Library

Library of Congress Cataloging in Publication Data

(Data applied for)

ISBN 0 19 850021 1 (Hbk)
0 19 850020 3 (Pbk)

Typeset by Newgen, India

Printed in Great Britain by Bookcraft Ltd., Midsomer Norton, Avon

Preface and acknowledgements

I decided to write this book for two reasons. First, the main recommended text for my island biogeography course, Mark Williamson's *Island populations*, was published in 1981. A lot of new work has appeared since then, and I felt it demanded a new synthesis. Secondly, I had been working on the recolonization of a particular island system—Krakatau—for some time, and despite its early recognition in the development of island ecological theory, I had come to believe that it did not fit well into the standard frameworks. Writing this book was therefore an excuse to immerse myself more fully into the broader island biogeographical literature and to work out how it all might fit together. Although several new 'island' books have been published while I have been working on this one, they are mostly more specialized than the present volume. My hope is that this book will provide access to students of differing backgrounds and disciplines to the full array of island biogeographical themes and issues.

Island biogeography is an important subject for several reasons. First, it has been and remains a field that feeds ideas, theories, models and tests of the same into mainstream ecology, evolutionary biology, and biogeography. Secondly, some of these theories have had great weight placed upon them (sometimes too much) in applications in nature conservation—this 'habitat island' biogeography is thus given its own chapter in a volume otherwise concentrating mostly on real islands. Thirdly, at a time of increasing concern with global reductions in biodiversity, it is worth emphasizing that a high proportion of species extinctions in historical time have been of island species, and many more are judged to be on the brink. The protection of the unique biological features of island ecosystems presents us with a considerable challenge not only ecologically, but also because of the fragmented nature of the resource, scattered across all parts of the globe and all political systems, and often below the horizons of even global media networks. It is my hope that this book will foster an increased interest in island ecology, evolution, and, conservation, and that it will be of value for students and researchers working in the fields of geographical, environmental, and life sciences.

I came to be fascinated by island biogeography during my schooldays, through introduction to Darwin's formative observations from the Galápagos. My interest was fostered and broadened by my university tutors John Flenley and David Watts. I became permanently hooked when John Flenley initiated our studies (since involving numerous colleagues) of the Krakatau islands. My personal heroes are those founding fathers of Krakatau ecology, Melchior Treub, Alfred Ernst, Willem M. Docters van Leeuwen, and Karl W. Dammerman, together with H. N. Ridley for his wonderful book *The dispersal of plants throughout the world*. It is salutary how often one finds that an idea seemingly innovative can be discovered in the writings of these and other early twentieth-century authors, or in the great works of Darwin or Wallace, and I have attempted to indicate this at relevant points within the text. This book is fairly heavily referenced for a textbook, and I make no apology for this: I regard it as an essential part of good scholarship.

I am extremely grateful to those who reviewed all or part of the manuscript; most especially Mark Williamson, and also Mark Bush, Heather Viles, Adrian Parker, Jared Diamond, Angela Whittaker, Philip Stott, Steve Compton, Stephen Jones, and Stuart Franklin. I am also indebted to my former tutors, my students, and numerous other colleagues, notably Mark Bush, John Flenley, Graham Wragg, Clive Hambler, and Steve Compton, for discussions of island biogeography. All mistakes remaining are my own. I was assisted in preparing this book by a one-year Special Lectureship awarded by the University of Oxford in 1995/96. The figures were drawn by Ailsa Allen, with the assistance of Peter Haywood. I am indebted to Linda Birch, Mary Eaves, and their colleagues in the zoology libraries and to Linda Atkinson and her colleagues in the geography library, Oxford University. I am grateful to those individuals and organizations who granted permission for the reproduction of copyright material: the derivation of which is indicated in the relevant figure and table legends, supported by the bibliography. Although every effort has been made to trace and contact copyright holders, in a few instances this has not been possible. If notified the publishers will be pleased to rectify any omission in future editions. I would like to extend my appreciation to The Press for continual encouragement during the gestation of this project, and to Linda Antoniw for correcting the text. Finally, the writing of this book was only possible with the tolerance, love, and support of Angela, Mark, and Claire.

Oxford
September 1997 R. J. W.

Contents

1

The natural laboratory paradigm

> ...it is not too much to say that when we have mastered the difficulties presented by the peculiarities of island life we shall find it comparatively easy to deal with the more complex and less clearly defined problems of continental distribution...
>
> (Wallace 1902, p. 242.)

These words taken from Alfred Russel Wallace's *Island life* encapsulate an over-arching idea, which could be termed the central paradigm of island biogeography. It is that islands, being discrete, internally quantifiable, numerous, and varied entities, provide us with a suite of natural laboratories, from which the discerning natural scientist can make a selection that simplifies the complexity of the natural world, enabling theories of general importance to be developed and tested. Under this umbrella, a number of distinctive traditions have developed, each of which is a form of island biogeography, but only some of which are actually about the biogeography of islands. They span a broad continuum within ecology and biogeography, the endpoints of which have little apparently in common. This book explores these differing traditions and the links between them within the covers of a single volume.

This introductory chapter serves to outline these different subject areas and to explain the organization of material in the book, which basically adopts a diminishing temporal focus, progressing from the patterns and processes operating over the longer (geological) time-scales, through to the shorter (ecological) time-scales, leading to a consideration of applied aspects of island biogeography. Table 1.1 picks out some of the dominant island theories or themes and the typical island configurations that, as will become clear later, appear to match these themes (cf. Haila 1990). The first two, Adaptive Radiation and the Taxon Cycle, are each forms of evolutionary change; typically, the best examples of island evolution are from very isolated, high islands. The last three themes fall within what may be termed island ecology, although the distinction between evolutionary and ecological change is, of course, an arbitrary one. The ecological theories have applications within continental land masses as well as to real islands and, indeed, form an important part of the applied biogeography or conservation biology literature.

We cannot begin our investigation of the longer-term island biogeographies until we have explored something of the origins, environments, and geological histories of the platforms on which the action takes place. This forms the subject matter of the second chapter, *Island environments*. The biogeographical data from

Table 1.1
Some prominent island theories and the geographical configurations of islands for which they hold greatest relevance

Type of archipelago	Prominent theories
Large, very distant	Adaptive Radiation
Large, distant	Taxon Cycle
Medium, mid-distance	Assembly Rules
Small, near	Equilibrium Theory of Island Biogeography
Small, very near	Metapopulations

islands form part of a complex jigsaw of different sources of information that help us reconstruct past changes in the configuration of continents, ocean basins, and climate. In certain cases, islands can provide simplified reconstructions of phenomena, such as eustatic sea-level changes, that prove difficult to analyse from continental coastlines, with their complexities of isostatic readjustments. Islands thus hold a wider scientific value than merely to enthuse island biogeographers.

The third chapter, *Biodiversity hot-spots*, concentrates on the biogeographical affinities and peculiarities of island biotas—a necessary step before the processes of evolution on islands are tackled. The chapter thus begins in the territory of the geological biogeographers, who are concerned with tracing the largest scales from the space–time plot (Fig. 1.1) and with working out how particular groups and lineages came to be distributed as they are. This territory has been fought over by the opposing schools of dispersalist and vicariance biogeography. Island studies have been caught up in this debate in part because they seemingly provide such remarkable evidence for the powers of long-distance dispersal, whereas rejection of this interpretation requires alternative (vicariance) hypotheses for the affinities of island species, invoking plate movements and/or lost land bridges, to account for the breaking up of formerly contiguous ranges. Some of the postulated land-bridge connections now appear highly improbable; nevertheless the changing degrees of isolation of islands over time remains of central importance to understanding island biogeography. As will be seen, the vicariance and dispersalist hypotheses have been put into too stark an opposition; both processes have patently had their part to play (Stace 1989; Keast and Miller 1996).

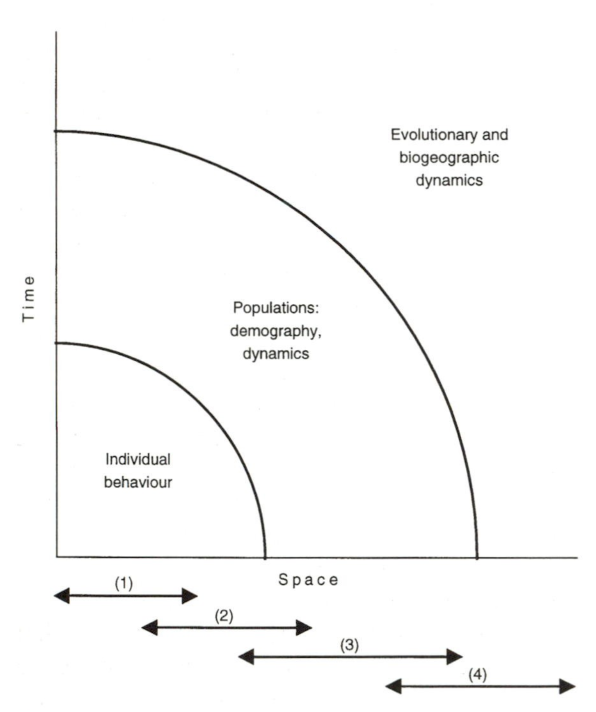

Fig. 1.1. A scheme of different time–space scales of ecological processes and criteria that define corresponding scales of insularity. (1) Individual scale; (2) population scale 1: dynamics; (3) population scale 2: differentiation; (4) evolutionary scale. (Redrawn from Haila 1990, Fig. 1.)

Remote island biotas differ from those of continents in a number of ways, being generally species-poor and disharmonic (peculiar in taxonomic composition), yet rich in species found nowhere else, i.e. that are endemic to those islands (there being no simple relation between endemism and richness). While many of these features are well known through the more popular illustrations from island groups such as Hawaii and the Galápagos, the global significance of island biodiversity is not so well appreciated. Data for a number of taxa are available that reveal strikingly that islands, particularly large and remote islands, contribute disproportionately to global biodiversity, that is they are biodiversity ‘hot-spots’. Island biogeographers have, in recent decades, continued to discover new species on islands, some from extant (i.e. living) populations, others from fossils or subfossils. This point is worth emphasizing, not only for the significance of their finds to theories of island biogeography, but also because of the high rate of loss of these forms.

If we are to understand the special contribution of island forms to the science of evolutionary biology, it is first necessary to outline the units of currency in which evolution is typically measured: species or their varieties. This book is not in any sense a primer in genetics, but I do consider it important to set out what we mean by the species unit and its place in phylogeny. Chapter 4, *Speciation and the island condition*, opens with this intent and then develops a number of alternative frameworks for understanding island speciation, namely the distributional, locational, mechanistic, and phylogenetic. I contend that each of these is important and that students should be familiar with them before delving too far into the evolutionary literature.

Chapter 5, *Arrival and change*, and Chapter 6, *Emergent models of island evolution*, develop the theme of island evolution, expanding on the theories that explain the peculiarities of island biotas, working from the micro- to the macroevolutionary. Features such as gigantism, nanism, and loss of dispersiveness are among the more familiar characteristics that we associate with island evolution, but equally valuable insights have come from studies of phenomena such as changes in feeding niche, and founder effects on islands. Within-species change, which may or may not lead ultimately to speciation, is easier to document than is full speciation, and so a considerable body of literature on these themes has built up from studies on islands. Oceanic islands also provide some of the best textbook examples of macroevolutionary change, taking a range of forms that include the adaptive radiation of lineages, speciation with little or no radiation (anagenesis), and the taxon cycle. Each of these patterns involves the formation of new species, which may be achieved by several different routes, most of which are difficult to observe (Rosenzweig 1995). Thus, just as Darwin and Wallace realized, the number and variety of natural experiments provided by islands have provided unique insights into one of the most fundamental of subjects—evolution. It helps us answer questions such as: How necessary is geographical isolation for evolution and speciation to occur? What determines the successes and failures of evolution? How long does speciation take?

The separation of ecology from evolution is, as already noted, a rather arbitrary one: evolution cannot work without ecology to drive it; ecological communities operate within the constraints of evolution. Thus, in order to understand evolution, a substantial body of ecological theory must be considered. None the less, a distinct body of island ecological theory can be distinguished from the evolutionary ecological literature. This 'island ecology' is packaged here in two parts: Chapter 7, *Species numbers games*, and Chapter 8, *Community assembly and dynamics* (species composition). Undoubtedly, the most influential contribution to this literature has been Robert H. MacArthur and Edward O. Wilson's (1967) *The theory of island biogeography*. MacArthur and Wilson were not the first to recognize or theorize about the species–area effect—that plots of species number versus area have a characteristic form, differing between mainland and island biotas—but it was they who developed a fully fledged mathematical theory to explain it: the Equilibrium Theory of Island Biogeography, based on a dynamic relationship between forces accounting for immigration to islands and losses of species from them.

Whether on islands or continents, the form of the relationship between area and species number is of fundamental biogeographical importance. The current interest in biodiversity is intimately related to this topic, as diversity can (or at least should) only be expressed as a function of area. This is because diversity patterns are the product of the differential overlap of species ranges and, in general, the bigger the area under study, the more species ranges will intersect with it. Attempts to analyse the cause of diversity patterning must therefore control for area. This can be done in two ways: by holding area constant (e.g. O'Brien 1993), or by determining empirically the form of the species–area curve (Rosenzweig 1995). Over the past 30 years or so, there have been numerous island biogeographical studies that have set out to describe and understand the form of the species–area curve, and the influence of numerous other variables on species richness. What have these studies shown? Are we any closer to understanding the factors that control species richness in patches and islands, and how they vary in significance geographically? I will show that MacArthur and Wilson's Equilibrium theory has actually proved remarkably difficult to test and that its predictive value appears limited.

While progress has been made, a grand unifying theory remains elusive, and perhaps illusory.

The study of community assembly and dynamics on islands (Chapter 8) has interested natural scientists since at least the mid-nineteenth century, which explains why the botanist Melchior Treub took the opportunity to begin monitoring the recolonization of the Krakatau islands just 3 years after they were sterilized in the eruptions of 1883 (Whittaker *et al.* 1989). It emerges that island biotas typically are not merely a 'random' sample of mainland pools. There is generally some degree of structure in the data. Some important insights into this structure have come from Jared Diamond's work on birds on islands off the coast of New Guinea in the 1970s (e.g. Diamond 1975*a*), although these studies became embroiled in heated controversy, as we shall see. At the species level, so-called 'incidence functions' can be calculated, which express the probability of an island being occupied by a species as a function of island area or richness. Another pattern noticed by Diamond was the 'checkerboard' distribution of species between islands, which led to the idea of 'assembly rules': that certain combinations of species are found far more often than expected under a null model, whereas others appear less often than expected, or not at all. A more recent insight, from Patterson and Atmar (1986), was that island assemblages frequently exhibit 'nestedness', such that successively smaller biotas are subsumed within each larger assemblage, like a set of Russian dolls.

Much of this work has dealt with patterns of species distributions within archipelagos of islands. The dynamics of island biotas have also been the focus of study and debate. For instance, Diamond developed the idea from his New Guinea studies that some species are 'supertramps', effective at colonizing but not at competing in communities at equilibrium, whereas others are poor colonists but effective competitors. This is a form of successional structuring of colonization and extinction. Successional structure can be recognized through other approaches, for instance by analysis of the dispersal characteristics of the flora of the Krakatau islands through time, or by analysis of the habitat requirements of birds and butterflies appearing on and disappearing from the same islands. Krakatau is unusual in having species–time data for a variety of taxa, and to some extent they tell different stories, yet stories linked together by the emergent successional properties of the system.

In order to make sense of this diverging literature on species–area relations, species composition, and turnover, it is necessary to reconsider the theoretical frameworks of island (ecological) biogeography. The island ecology chapters thus close with a reappraisal of these ideas. Are equilibrium frameworks adequate and generally applicable, or should we be developing non-equilibrium theories? Are island communities stochastic, or deterministically structured? Should MacArthur and Wilson's theory be viewed as: not supported by the evidence and of heuristic value only; true but trivial; true but only within a more limited domain than once hoped; or still the belle of the ball? Much, it appears, depends on scale. There has been an increasing realization that ecological phenomena have characteristic spatial and temporal signatures, which tend to be linked (Delcourt and Delcourt 1991). This simple observation helps us reconcile apparently contradictory hypotheses as actually being relevant to different spatio-temporal domains (Fig. 1.1, Haila 1990).

One of the reasons that all of this is important is because of the contribution of island thinking to the applied field of conservation biology. This is explored in Chapter 9, *Island theory and conservation.* As we turn continents into a patchwork quilt of habitats, we create systems of insular populations. What are the effects of fragmentation and area reduction within continents? Not just the short-term changes, but the long-term changes? To some, the obvious place to turn for a rule book was to theory that had been developed for islands, and in particular that body of it which stemmed from MacArthur and Wilson's seminal contribution. One of the early debates in this tradition concerned the design of nature reserve systems, and whether it was a better strategy to place all your eggs in one basket, the single large reserve, or in many smaller ones, each of which might hold far fewer species. This debate reached something of an impasse, but recent years have seen important developments, for example a focus on the roles of edges and of habitat corridors. Distinct from this whole-system approach, on the single-species level, it is theoretically possible to determine the minimum size of population that will ensure long-term survival and, where suitable data exist, to convert this to an estimate of the minimum area needed for a particular species. Unfortunately, for most species, the necessary empirical details are not available and, in any case, population structures are often more complex than is allowed for by this approach. Some of this complexity is recognized by another development, metapopulation theory. The term 'metapopulation' describes a situation in which a series of isolates are actually sufficiently close to one another that their populations are, in effect, linked; the precise requirements of this linkage we can leave until later. Although largely developed in relation to habitat islands, the metapopulation concept can also be applied to real islands (Haila 1990). It is merely the next step down the scale framework, as indicated in Table 1.1. The trick in all of this may well come down to being able to recognize what sort of island you have and what sort of island biogeography fits it.

I hope this book will engender a renewed enthusiasm in islands themselves and in the fates of their biotas. If the catalogue of known extinctions is drawn up for the period since AD 1600, it can be seen that, for animal taxa that are relatively well known (principally mammals, birds, and land snails), the majority of losses have been of island species (Fig. 1.2). Today, some of the greatest showcases for evolution are in peril, with many species on the verge of extinction. Why is this? These issues are explored in Chapter 10, *The human impact on island ecosystems*, which summarizes the mounting evidence of the long-term pattern of island colonization by humans, followed by extinctions of endemic species, which can be traced across the Pacific, the Caribbean and the Mediterranean—indeed, wherever humans have colonized oceanic islands. Are islands fragile, or are island peoples peculiarly good at extinguishing species? In

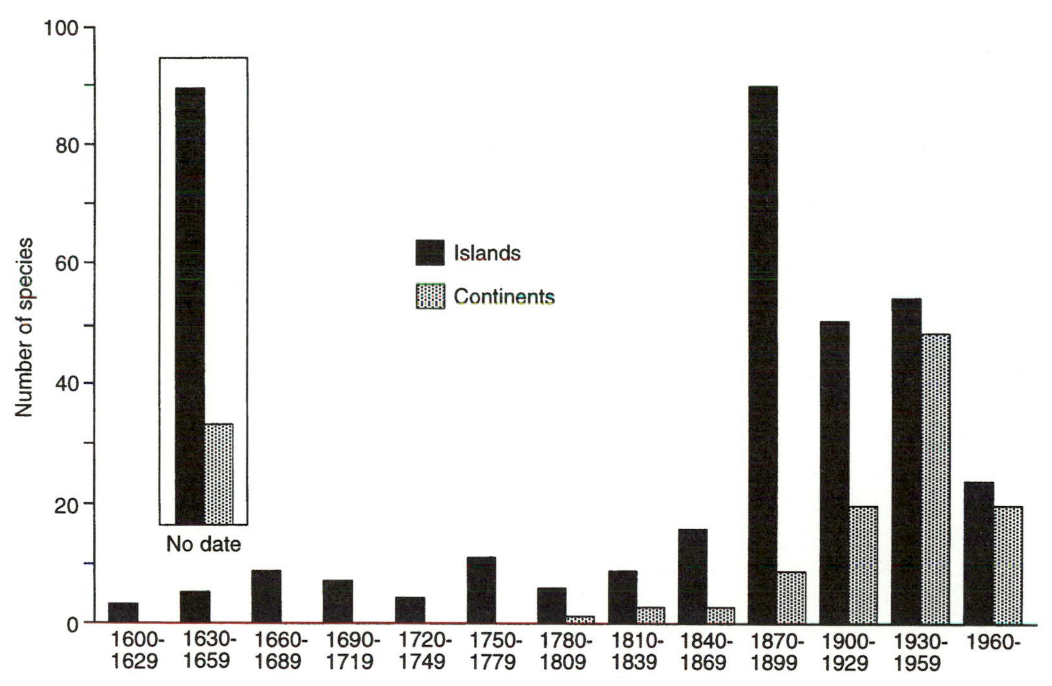

Fig. 1.2. Time series of extinctions of species of molluscs, birds, and mammals from islands and continents since about AD 1600. (Redrawn from Groombridge 1992, Fig. 16.5.)

some senses, island biotas are indeed fragile, but very often the demise of endemic forms can be traced to a series of 'hard knocks' and the so-called synergistic interactions between a number of alien forces, such that the nature of the final demise of the endemic (apparently, in one famous case, the intervention of a lighthouse keeper's cat) may not give a full picture of why the species came to be so reduced in the first place (see Chapter 10, section 4.1). Typically, there have been two major waves of extinctions, one associated with the aboriginal or prehistoric colonizations, and the second following contact with modern European societies. Whereas, on the continents, the increasing insularization of habitats is a prime cause for concern, the problem for oceanic island biotas today is the increasing breakdown of the insularization of theirs. The prime agents of destruction are introduced exotic species (especially mammalian browsers and predators), habitat loss, predation by humans, and the spread of disease. The biological forces involved are now well established, and many of the biological management options are firmly founded in experience, but attaining the twin goals of sustainable development and management for conservation requires attention to fields largely outside the scope of the present volume, concerning the politics, economics, sociology, and culture of island states: continental solutions may be poorly transferable to islands.

What *is* an island, anyhow?

2

Island environments

...from cowpats to South America, it is difficult to see what is not or at some time has not been an island.

(Mabberley 1979, p. 261.)

The islands move horizontally and vertically and thereby grossly modify the environment on and around them. Life forms too must evolve, migrate, or become extinct as the land changes under them.

(Menard 1986, p. 195.)

2.1. Types of islands

Islands come in many shapes and sizes, and their arrangement in space, their geologies, environments, and biotic characteristics are each extremely variable, simply because there are vast numbers of them in the world. If this makes them marvellous experimental laboratories for field ecologists and biogeographers, it also means that attempting to make generalizations about islands holds the danger—almost the certainty—that one will be wrong!

For the purposes of the biogeographer, islands may be divided into those within the sea and those within land masses, and in turn these classes can be split into the following four categories: **oceanic islands**, which are located over oceanic plate and have never been connected to continental land masses; **continental shelf islands**, which are located on the continental shelf and may have been connected to mainland during periods of significantly lower sea levels, such as have occurred several times in the Quaternary (the last *c.* 1.8 million years (Ma)); **habitat islands**, meaning discrete patches of terrestrial habitat surrounded by very different habitats but not by water; and finally, **non-marine islands**, which in their characteristics sit between continental shelf and habitat islands. Both habitat and non-marine islands can, of course, occur within the first two categories of island. Moreover, a more refined classification of marine islands has to recognize that there are, for example, many forms and ages of continental shelf islands, and that there are anomalous islands mid-way in character between oceanic and continental (see Chapter 3). Providing these limitations are understood, this simple classification can serve quite well. Aquatic scientists might wish to develop an equivalent classification, but a line must be drawn somewhere, and I am drawing mine at the low water mark.

Defining the term 'island' is not as straightforward as it seems. Some authors regard land areas that are too small to sustain a supply of fresh water as merely beaches or sand bars rather than islands, the critical size for a supply to be maintained being about 10 ha (Huggett 1995). At the other end of the scale, the distinction between a continent and an island is slightly fuzzy, in that larger islands assume many of the characteristics of continents; indeed Australia is considered an island continent. Where should the line be drawn? If New Guinea is taken as the largest island—somewhat arbitrarily, given that many would consider Greenland an island—then islands in the sea constitute some 3% of the Earth's land area (Mielke 1989). Much of island biogeography—and of this book—is concerned with significantly smaller islands than New Guinea, and commonly treats such substantial islands merely as the 'mainland' source pools. To some extent this reflects a distinction between island evolutionary biogeography, mostly concerned with larger and often oceanic islands, and island ecological biogeography, mostly concerned with the other types of island. At least it can be claimed that islands in the sea have the virtue of having clearly defined limits and thus providing discrete objects for study, in which such variables as area, perimeter, altitude, isolation, and species number can be quantified with some degree of objectivity—the whole surrounded by the aquatic realm. In contrast, habitat islands exist typically within complex landscapes, with which they may share uncertain boundaries and overlapping populations. Because of this, it cannot necessarily be assumed that what goes for real islands also works for habitat islands and vice versa. Yet the literature and theory of island biogeography has been built up from a consideration of all four of the types of island listed in the previous paragraph. This will become clear in later chapters. For the present purposes, there is thus no need to agonize over the precise definition of an island; even thistle heads may count as islands for some purposes. However, as much of this book is concerned almost exclusively with islands in the sea, their biogeographical peculiarities, and their problems, it is important to consider their modes of origin, environmental characteristics, and histories. These topics form the bulk of this chapter. For the sake of clarity, the reader should take it that, throughout the book, unless otherwise stated, the term island refers to a discrete chunk of land surrounded by sea.

2.2. Modes of origin

Are islands geologically distinctive compared with the majority of the land surface of the Earth? If we take this question in two parts, continental shelf islands are a pretty mixed bag, a reflection of their varied modes of origin, whereas oceanic islands are fairly distinctive geologically, being generally composed of volcanic rocks, reef limestone, or both (Darwin 1842; Williamson 1981). The analysis that follows is intentionally a simplified account of an extremely complex reality.

The development of the theories of continental drift and, more recently, **plate tectonics**, has revolutionized our understanding of the Earth's surface and, along with it, our understanding of the distribution and origins of islands. According to the latter theory, the Earth's surface is subdivided into some seven major plates, each larger than a continent, and a number of smaller fragments (Fig. 2.1). The plates themselves are typically made up of two parts, an oceanic part and a continental part. The lighter, **granitic** parts of the plates are of lower density and these form the continental parts, supporting the continents themselves (consisting of a highly varied surface geology), extending to about 200 m below sea level (Fig. 2.2). This zone, from 0 to −200 m, forms the **continental shelf** and supports islands such as the British Isles and the West Frisian Islands (off Germany and the Netherlands), typically involving a mix of rock types and modes of formation, such that any combination of sedimentary, metamorphic, or igneous rocks may be found. As Williamson (1981) noted, about the only generalization that can be made is that the geological structures of continental shelf islands tend to be similar to parts of the nearby continent.

In places, continental plate can be found at much greater depths than 200 m below sea level, and can then be termed **sunken continental shelf**. Islands on these sections of shelf are thus formed of continental rocks, examples including Fiji and New Zealand (Fig. 2.3; Williamson 1981). Typically, however, there is a steeply sloping transition zone from shallow continental shelf down to *c.* 2000 m or more, where the **basaltic** part takes over and the true oceanic islands occur. These are all volcanic in origin, although in certain cases they may be composed of sedimentary material, principally limestones, as the volcanic core has sunk below sea level. True **oceanic plate islands** have never been attached to a greater land mass.

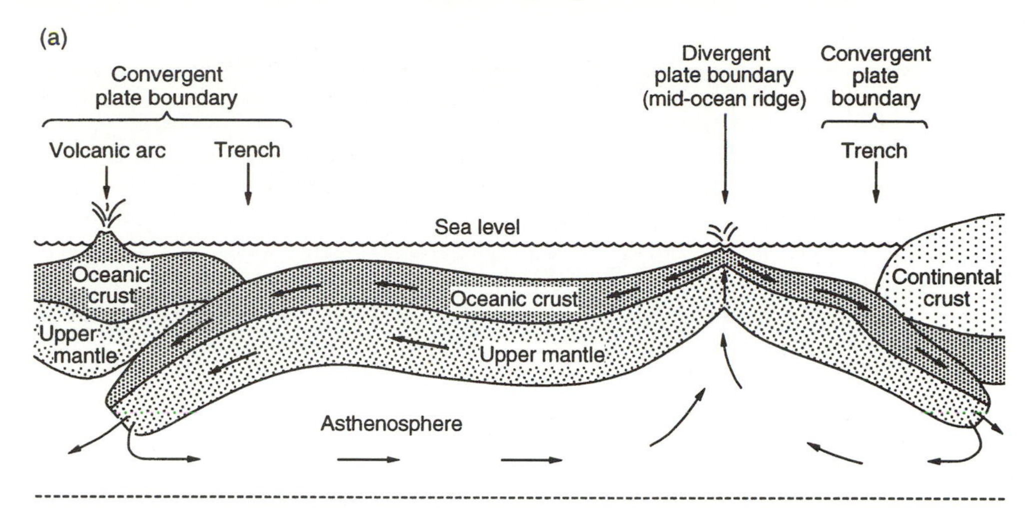

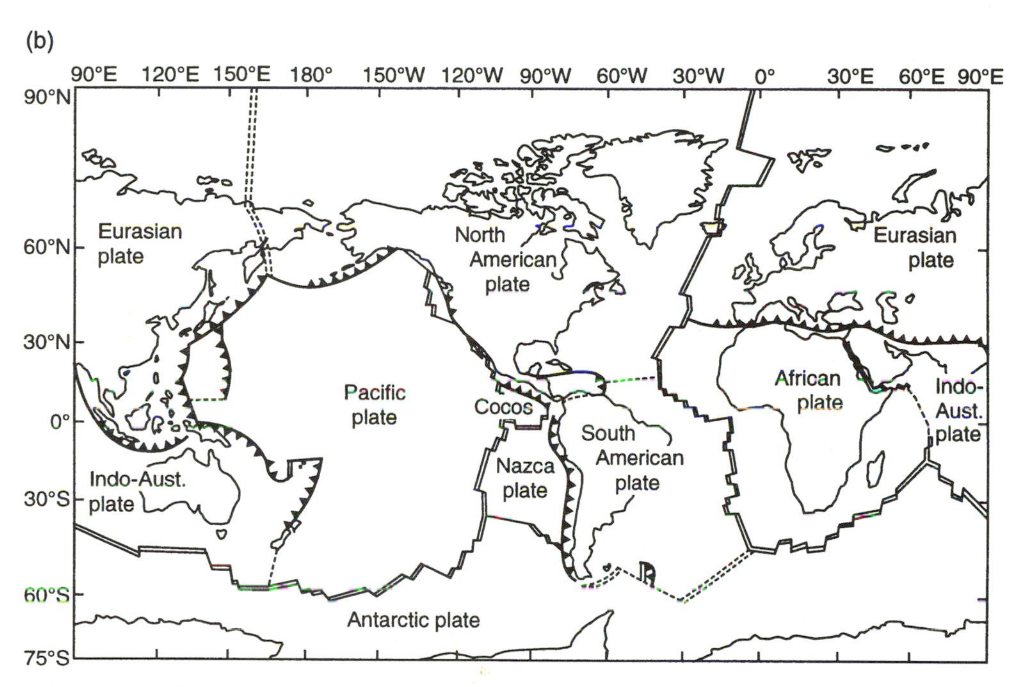

Fig. 2.1. Model and map of plate tectonics. (a) The basic plate tectonics model, exemplified by the South Pacific. New oceanic crust and upper mantle are created along the divergent plate boundary and spread east and west from here. The eastward moving plate (the Nazca plate) is subducted beneath the continental lithosphere of South America. The westward movement of the much larger Pacific plate ends with subduction beneath the oceanic lithosphere of the Indo–Australian plate. At both trenches, the subducted plate begins to melt when it reaches the asthenosphere. (b) The major plates of the world. Divergent plate boundaries (mid-ocean ridges), at which plates move apart, are represented by parallel lines. Convergent plate boundaries (mostly marked by trenches) are represented by lines with teeth along one side: the teeth point from the downward moving plate towards the overriding or buoyant plate. Transverse plate boundaries are shown by solid lines. Broken lines indicate boundaries of an uncertain nature. (Redrawn from Nunn 1994, Fig. 2.4.)

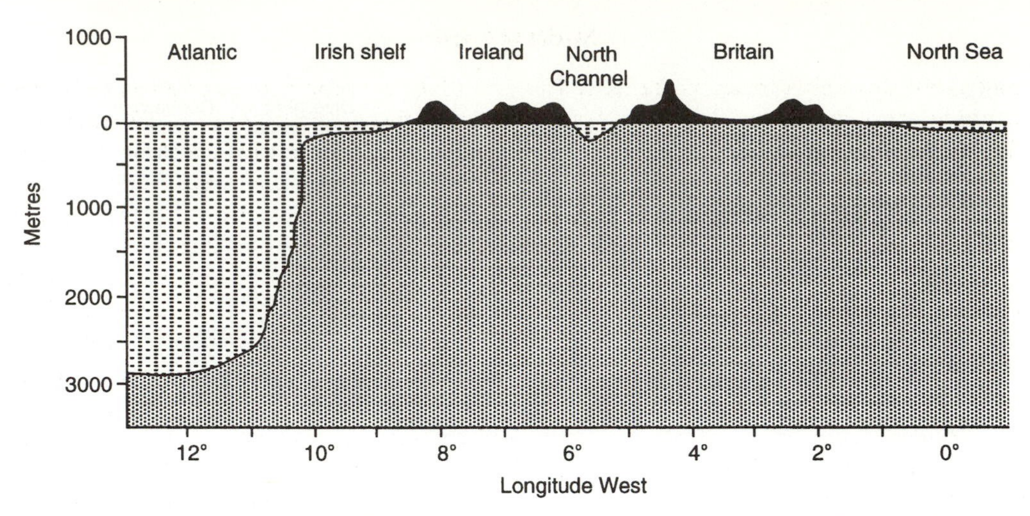

Fig. 2.2. A section across the British Isles at 55°N. The section passes through Londonderry in Northern Ireland, Cairnsmore of Fleet in Scotland, and Newcastle-upon-Tyne in England. (Redrawn from Williamson 1981, Fig. 1.2.)

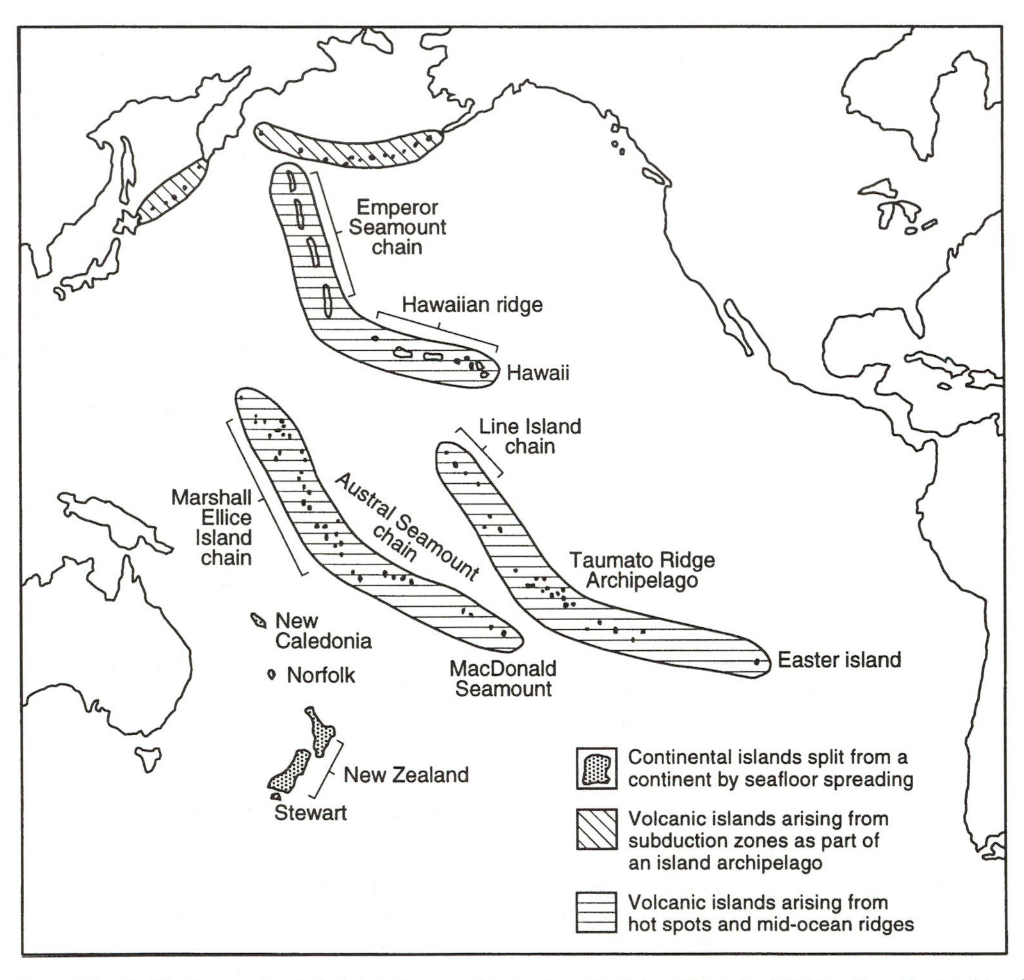

Fig. 2.3. Map of the Pacific Ocean, depicting the three major modes of origin of islands. NB: The New Zealand land mass has been physically isolated from Australia and Antarctica by up to 2000 km for 60 million years, as a result of Late Cretaceous to Late Palaeocene sea-floor spreading (Pole 1994). (Redrawn from Mielke 1989, Fig. 9.1.)

They may grow through further volcanism, or subside, erode, and disappear. In a geological sense, they tend to be transitory: some may last only a few days, others some millennia; relatively few last tens of millions of years. Thus, in addition to the islands that sustain terrestrial biota today, there are also a large number of past and future islands, **seamounts** (a mountain beneath the sea) and **guyots** (a former island planed off below sea level by erosion) that are to be found at varying depths below present sea level.

Islands may originate by existing areas of land becoming separated from other land masses to which they were formerly connected, by erosion, or by changes in relative sea level from a variety of causes. Many islands have originated from volcanism associated with plate movements. The form this volcanism takes and the pattern of island genesis involved depends crucially on the nature of the contact zone between plates, that is whether the plates are moving apart, moving towards each other, or moving past each other laterally. Relatively few islands are ancient biogeographically; some have experienced a complex history of both lateral movements and alternating emergence and submergence, with profound importance to the resulting biogeography, not just of the particular island in question, but also of the wider region (Keast and Miller 1996).

Plate tectonic processes give rise to islands by three main means: first, the breaking away of pieces of continent by sea-floor spreading, as in the case of New Zealand; second, in connection with plate boundaries, volcanic islands may arise to form an archipelago of islands, as in the case of the Greater and Lesser Sundas—the many islands of the Indonesian region; third, volcanic islands may arise from hot-spots and mid-ocean ridges, as is classically the case with the Hawaiian islands and Iceland, respectively.

Setting aside islands of continental origin for the moment, a two-tier classification relating specifically to oceanic islands has been proposed by Nunn (1994). At the first level are the plate boundary and intra-plate island types, each of which may also be subdivided into a number of major types, based on their geographical configuration in relation to the plate boundaries (Table 2.1). The recognition of these distinctive configurations of islands is a helpful starting point. However, Nunn (1994) warns against an over-simplistic use of the classification, pointing out that, whereas many islands with a common origin may be found in proximity to each other, islands of a particular geographical group do not necessarily share a common origin (this is well illustrated by the Mediterranean islands (Schüle 1993)). A brief account (drawn largely from Nunn 1994) of this classification now follows.

Table 2.1

P. D. Nunn's (1994) Genetic classification of islands, with examples

Level 1	Level 2	Examples
Plate boundary islands	Islands at divergent plate boundaries	Iceland, St Paul (Indian Ocean)
	Islands at convergent plate boundaries	Antilles, South Sandwich (Atlantic)
	Islands along transverse plate boundaries	Cikobia and Clipperton (Pacific)
Intra-plate islands	Linear groups of islands	Hawaii, Marquesas, Tuamotu
	Clustered groups of islands	Canary Islands, Galápagos, Cape Verde
	Isolated islands	St Helena, Christmas Island (Indian Ocean), Easter Island

2.2.1. Plate boundary islands

Islands at divergent plate boundaries Divergent plate boundaries produce islands in two rather different circumstances, along mid-ocean ridges and along the axes of back-arc or marginal basins behind island arcs that are themselves associated with convergent plate boundaries. Although divergent plate boundaries are constructive areas involving more magma output than occurs in any other situation, seamounts, rather than islands, tend to form in connection with these boundaries. This is believed to be because of the relative youthfulness of the mounts, the likelihood that the supply of magma decreases as the mounts drift away from the plate boundary, and the increasing depth of the ocean floor away from these boundaries. In some cases, mid-ocean ridge islands may also be associated with hot-spots, providing the seemingly exceptional conditions necessary for their conversion from seamounts to large islands. Iceland is the largest such example (103 106 km^2), being a product of the mid-Atlantic ridge and a hot-spot that may have been active for some 55 million years. Another context in which mid-ocean ridge islands can be found is in association with triple junctions in the plate system, an example being Rodriguez in the Indian Ocean. The second form of divergent plate boundary island, exemplified by the Tongan island of Niuafo'ou, is that sometimes formed in back-arc basins, which develop as a result of plate convergence, but which produce areas of sea-floor spreading (Nunn 1994).

Islands at convergent plate boundaries Where two plates converge, one is subducted below the other. This results in a trench in the ocean floor at the point of subduction. Beyond the trench, a row of volcanic islands develops parallel with the trench axis, often taking the form of an arc on the surface of the upper of the two plates. The majority of such arcs occur in the Pacific and Caribbean. Most commonly, where islands are formed in connection with subduction zones, the edges of the two converging plates are formed exclusively of oceanic crust. The magma involved in island arc volcanism derives from the melting of the subducted ocean crust, complete with its sedimentary load. Its composition thus owes much to the nature of the subducted crust. It is believed that, where basaltic crust and water-bearing sediment are subducted, this leads to andesitic volcanism, which is the more common, explosive type. The Sunda island arc, including the Indonesian islands, is predominantly of this form. Basaltic volcanism in island arcs is more rare, and may be due to a relative paucity of sedimentary load in the subducted crust. The Sandwich arc in the south Atlantic involves both basaltic and andesitic volcanism.

Islands along transverse plate boundaries This is a fairly rare context for island formation as, by definition, less divergence or convergence of plates is involved. However, strike–slip movement, compression, or both, can occur between adjacent parts of plates. An example of an island believed to have been produced by these forces is Cikobia, in the south-west Pacific.

2.2.2. Islands in intra-plate locations

This class of islands contains some of the most biogeographically fascinating islands of all, notably the Hawaii group. They are equally fascinating geologically.

Linear island groups The Hawaiian chain is the classic example of a series of islands with a significant age sequence arranged linearly across a plate (Table 2.2, Fig. 2.3). Among the larger islands, Kauai is the oldest at more than 5 million years, whereas Hawaii itself is less than 1 million years. Islands older than Kauai stretch away to the north-west, but have been worn away to form guyots, or are capped by coral atolls and limestone islands. The chain extends from the Loihi Seamount, which is believed to be closest to the hot-spot centre, through the Hawaiian islands themselves and along the Emperor Seamount chain, a distance of some 6130 km (Keast and Miller 1996). The oldest of the submerged seamounts is more than 70 million years old, so that there must have been a group of islands in this part of the Pacific for far longer than is indicated by the age of the present

Table 2.2

Approximate ages of selected islands and of the Meiji Seamount. This table illustrates the wide variation in the ages of oceanic islands, the simple age sequence of members moving north-west along the Hawaii–Emperor chain, and the lack of a similarly simple sequence moving south in the Austral–Cook cluster. For fuller listings, error margins and original sources see Nunn (1994); for Hawaii, in addition see Wagner and Funk (1995)

Island group	Age (million years)
Hawaii–Emperor chain	
Mauna Kea, Hawaii (The Big Island)	0.38
Kauai	5.1
Laysan	19.9
Midway	27.7
Meiji Seamount	74.0
Austral–Cook cluster	
Aitutaki	0.7
Rarotonga	1.1–2.3
Mitiaro	12.3
Rimatara	4.8–28.6
Rurutu	0.6–12.3
Isolated intra-plate islands	
Ascension	1.5
St Helena	14.6
Christmas (Indian Ocean)	37.5

Hawaiian islands. This may have allowed for colonization of the present archipelago from populations established considerably earlier on islands that have since sunk. However, recent geological findings indicate that there was a pause in island building in the Hawaiian chain, and molecular analyses support the notion that few lineages exceed 10 Ma—that is, the pre-Kauai signal in the present biota is actually rather limited (Wagner and Funk 1995; Keast and Miller 1996).

The building of linear island groups such as the Hawaiian chain is explained by J. T. Wilson's (1963) **hot-spot hypothesis**, which postulates that stationary thermal plumes in the Earth's upper mantle lay beneath the active volcanoes of the island chain (Wagner and Funk 1995). A volcano builds over the hot-spot, then drifts away from it and eventually becomes separated from the magma source, and is subject to erosion by waves and subaerial processes, and to subsidence under its own weight. The change in orientation of the Hawaiian chain has been attributed to past changes in the direction of plate movement about 42 million years ago, with the age of the hot-spot being estimated at 75–80 million years. The Society and Marquesas Island groups, also in the Pacific, provide further examples of hot-spot chains and Nunn (1994) provides a critique of several other postulated cases.

Clustered groups of islands Many island clusters, once regarded as hot-spot island chains that had become slightly less regular than the classic examples, have since been realized to differ significantly from the hot-spot model, such that different groups and different islands within the groups may require distinctive models. The **tectonic-control model**, attributed to Jackson *et al.* (1972), postulates that, instead of lying along a single lineation, the islands lie along shorter lines of crustal weakness (termed ***en echelon* lines**), which are subparallel to each other. This line of reasoning may explain clusters of islands and chains in which there is no neat age–distance relation along the length of the chain, an example being the Line Islands in the central Pacific (Nunn 1994). Two of the largest intra-plate island clusters are the Canary and Cape Verde island groups, both in the central Atlantic and including a number of active volcanoes. The Canary Islands do not appear to conform with the hot-spot model as they lack a simple age sequence and several of the islands have been volcanically active in the historic period. The Cape Verde islands are also of uncertain origin, but Nunn favours the tectonic-control model, on available evidence. The Galápagos islands deserve a mention in this section as being a group of great significance to the development of island evolutionary models. They lie in the western Pacific just south of the divergent plate boundary separating the Nazca and Cocos plates. Again, Nunn regards them as an intra-plate cluster rather than mid-ocean ridge islands.

Isolated islands As knowledge of the bathymetry of the sea floor has improved, a number of islands believed to be isolated have been shown, instead, to be part of a seamount-island chain or cluster. In general, it can be taken that truly isolated intra-plate islands form only at or close to mid-ocean ridges. They result from a single volcano breaking off the ridge with part of the sub-ridge magma chamber beneath it. Nunn (1994) identifies this condition as essential if the island is to continue to grow once it is no longer associated with the ridge crest. Examples include Ascension, Gough, and St Helena in the Atlantic, Christmas Island in the Indian Ocean, and Guadalupe in the Pacific. Isolated intra-plate islands or island clusters may also, in certain cases, be the product of small continental fragments being separated from the main continental mass. The granitic Seychelles provide the best example, where there is good evidence of a continental basement affiliated to the Madagascar–India part of Gondwanaland.

2.3. Environmental changes over long time-scales

> In the case of these islands we see the importance of taking account of past conditions of sea and land and past changes of climate, in order to explain the relations of the peculiar or endemic species of their fauna and flora.
>
> (Wallace 1902, p. 291)

The separation of environmental changes from island origin is artificial. Islands may be built up by volcanism over millions of years; in such cases, active volcanism is thus a major part of the environment within which island biotas develop and evolve. The preceding section may give the impression that all islands have been formed by volcanism, or at least fairly directly by the action of plate-tectonic processes. However, the formation of an island may come about either by land disappearing under the sea to leave disconnected areas of land surrounded by sea, or by land appearing above the water's surface, either by depositional action or by some other process, of which tectonic forms are a large but not exclusive class. Indeed, Nunn (1994) offers the observation that changes of sea level are perhaps the most important reason why islands appear and disappear. This section therefore deals with long-term changes in the environments of islands, focusing particularly on changes in sea level, but also on locational shifts and changes in climate.

2.3.1. Changes in relative sea level—reefs, atolls, and guyots

As already noted, islands may come and go as a consequence of sea-level changes. Some of these sea-level changes are eustatic (i.e. they are due to the changing volume of water in the sea), and others are due to relative adjustment of the elevation of the land surface. This can be brought about by the removal of mass from the land causing uplift, as when an icecap melts, or by tectonic uplift. Subsidence of the lithosphere can be due to increased mass (e.g. increased ice, water, or rock loading) or may be due to the movement of the island away from mid-ocean ridges and other areas that can support anomalous mass. In the right environment, coral reefs build around subsiding volcanoes, eventually forming atolls (Fig. 2.4), an important category of tropical island.

Darwin (1842) distinguished three main reef types: fringing reefs, barrier reefs, and atolls. He explained atolls by invoking a developmental series from one type into the next as a result of subsidence of volcanic islands: fringing reefs are coral reefs around the shore of an island, barrier reefs feature an expanse of water between reef and island, and the final stage is the formation of an atoll, where the island has disappeared leaving only the coral ring (Fig. 2.5). As Ridley (1994) noted, Darwin thought out the outline of his theory on the west coast of South America before he had seen a true coral reef! While the theory may not be globally applicable and requires modification, for instance in the light of contemporary understanding of sea-level change, his basic model remains viable for most oceanic atolls and can account for most of the massive coral reefs (Steers and Stoddart 1977).

Coral reefs are built by small coelenterate animals (corals) that secrete a calcareous skeleton. Within the tissues and calcareous skeleton, numerous algae and small plants are lodged. The algae are symbionts critical to reef formation, providing the food and oxygen supplement necessary to account for the energetics of coral colonies, whilst the algae obtain both growth sites and nutrients from the coral (Mielke 1989). Reef-building corals generally grow in waters less than 100 m deep (exceptionally they can be found to 300 m); they require water temperatures between 23 °C and 29 °C, and are thus found principally in tropical and subtropical areas, notably in the Indo-Pacific and the western Atlantic and Caribbean Oceans (Fig. 2.6).

Coral growth has been found to vary between 0.5 and 2.8 cm/year, with the greatest growth rates occurring in water of less than 45 m depth (Mielke 1989). Historically, these growth rates have been great enough to maintain reefs in shallow water as either the underlying sea floor subsides or the sea level has risen (or both). Drilling evidence from islands with barrier reefs demonstrates this ability, just as predicted by Darwin's (1842) theory. For example, the thicknesses of coral that have accumulated on the relatively young islands of Moorea (1.5–1.6 Ma), Raiatea (2.4–2.6 Ma), and Kosrae (4.0 Ma) vary between 160 and 340 m, whereas the rather

Fig. 2.4. Aldabra Atoll, Indian Ocean, a raised atoll, and home to the giant tortoise, *Geochelone gigantea*. (Photo: Clive Hambler 1983.)

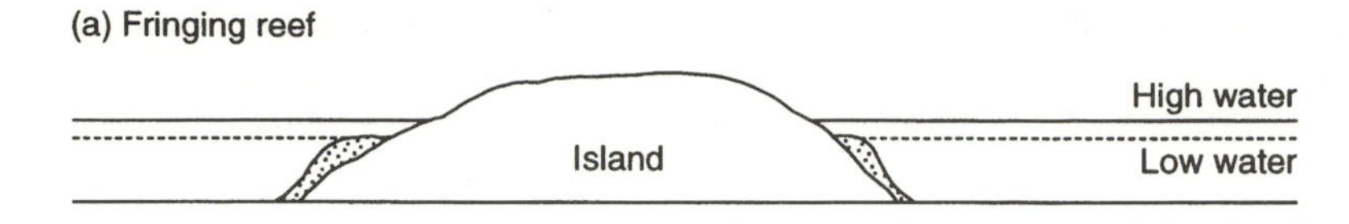

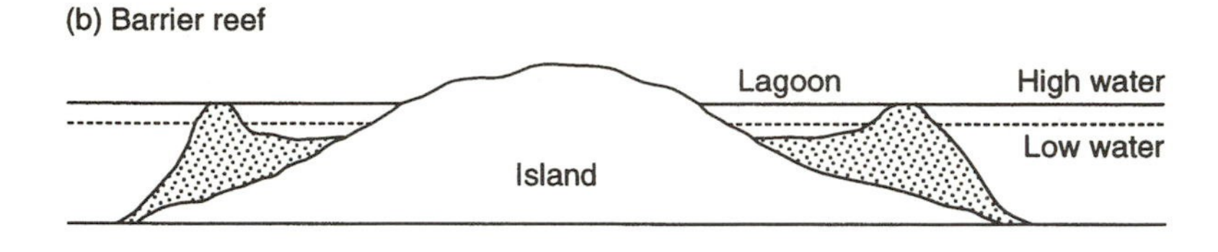

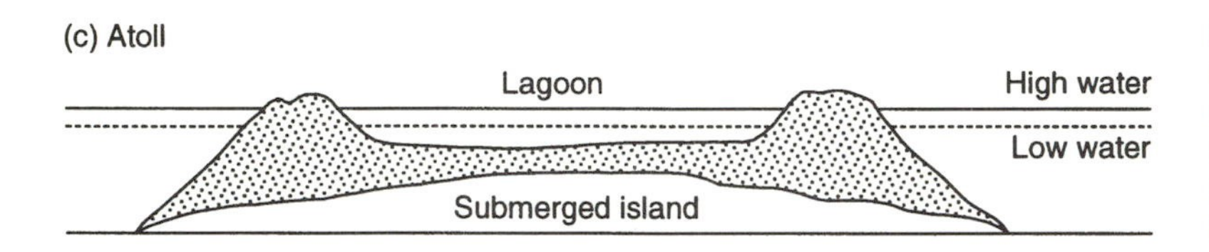

Fig. 2.5. The developmental sequence of coral reefs as a result of subsidence, from (a) to (b) to (c), as hypothesized by Darwin (1842). (Redrawn from Mielke 1989, Fig. 7.10.)

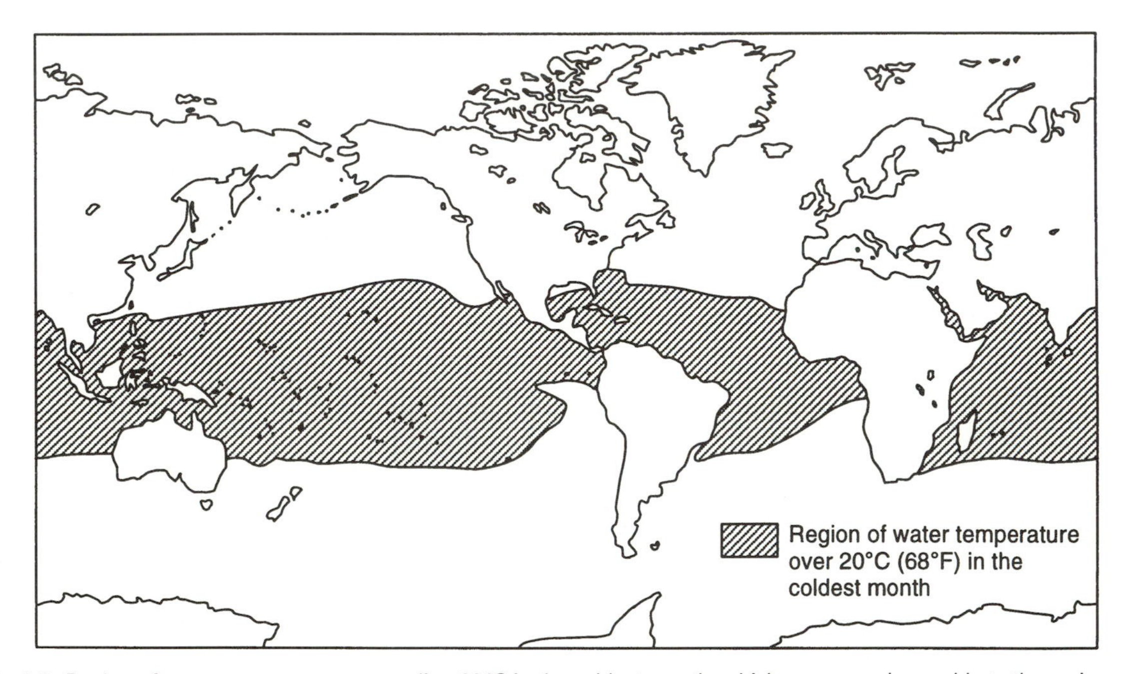

Fig. 2.6. Region of water temperatures exceeding 20 °C in the coldest month, which corresponds roughly to the major reef regions of the world's oceans. (Redrawn from Mielke 1989, Fig. 7.9.)

older islands of Magareva (5.2–7.2 Ma), Ponape (8.0 Ma), and Truk (12.0–14.0 Ma) have accumulated depths of 600–1100 m of coral (Menard 1986). However, it should be noted that massive reef accretion also occurs in the western Atlantic on continental shelves that have not experienced long-term subsidence (Mielke 1989). This lends support to another theory, that reefs were established on wave-cut platforms during periods of lowered sea level associated with glaciation, and that reef development then kept pace with rising sea levels in the interglacial period. This illustrates that several processes—subsidence, eustatic sea-level changes, and wave action—must each be considered with respect to the influence that they have had on the growth of coral reefs.

Given the vagaries of tectonic processes and sea-level changes, coral-capped islands can become elevated. Christmas Island in the Indian Ocean exemplifies this. It has been elevated to form a plateau of 150–250 m above sea level, but in parts reaches 361 m. The general aspect of the island is of coastal limestone cliffs 5–50 m high, which carries with it the implication that the available area for sea-borne propagules to come ashore is considerably less than the island perimeter, a factor in common with many islands of volcanic origin. It is becoming apparent that the relative elevation of particular islands may be subject to influence by the behaviour of others in the vicinity. The loading of the ocean floor by a volcano appears to cause flexure of the lithosphere, producing both near-volcano moating and compensatory uparching some distance from the load; thus one repeated pattern in the South Pacific is the association within island groups of young volcanoes (<2 million years) with raised reef, or **makatea** islands, as exemplified respectively by Pitcairn and Henderson Islands (Benton and Spencer 1995). There are many other examples of islands that contain substantial amounts of limestone as a product of uplift, one that will be considered below being Jamaica.

Over time, sea levels have fluctuated markedly, particularly during the Quaternary glacial–interglacial cycles. These fluctuations, in combination with island subsidence, wave action, and subaerial erosion, have resulted in many islands being levelled off below the current sea level. In nautical parlance, they are known as banks if they are less than 200 m from the surface (Menard 1986). Once they have sunk below about 200 m, these guyots are effectively below the range of eustatic sea level and in this situation will rarely re-emerge as islands. Guyots are generally flat-topped as a result of erosion, although some have become coral-topped (Nunn 1994). The distribution of guyots, and particularly of banks, may be crucial to an understanding of the historical biogeography of contemporary islands (Diamond and Gilpin 1983).

2.3.2. Eustatic changes in sea level

From a biogeographical viewpoint, it does not matter greatly whether the changing configuration of an archipelago is primarily a result of isostatic or of eustatic changes. However, it is important to recognize these complexities because of the need to understand past land–sea configurations (Keast and Miller 1996). That isostatic effects associated with Quaternary glaciations have not been confined to high-latitude continents and their margins, but also extended to low latitudes and ocean basins, was something that became clear only in the 1970s, necessitating a cautious approach to the construction of regional eustatic curves (Nunn 1994). Stratigraphic data from isolated oceanic islands have been of particular value in such analyses (Fig. 2.7).

Given the general emphasis on Quaternary events, it is noteworthy that certain of the sea-level oscillations in the Tertiary appear to have been of greater amplitude, principally as a consequence of high stands at 29, 15, and 4.2 million years (Nunn 1994). At one stage, it was believed that the Quaternary had seen a pattern of glacial episodes in the northern latitudes corresponding with lowered sea levels, and interglacials associated with levels not dissimilar to those of the present day, yet superimposed on a falling trend. In the past 25 years, the idea of the falling trend has been discarded. It has proved difficult to construct regional or global models of sea-level change for the Quaternary (Stoddart and Walsh 1992). Indeed, it appears from the data for the last and present interglacials that sea-level maxima have varied in magnitude and timing across the Earth's surface. One intriguing, if controversial, idea in explanation for some of the irregularities in sea-level changes relates to the configuration of the oceanic geoid surface (i.e. the sea surface itself), which, rather than being perfectly ellipsoid, is actually rather irregular, with a vertical amplitude of about

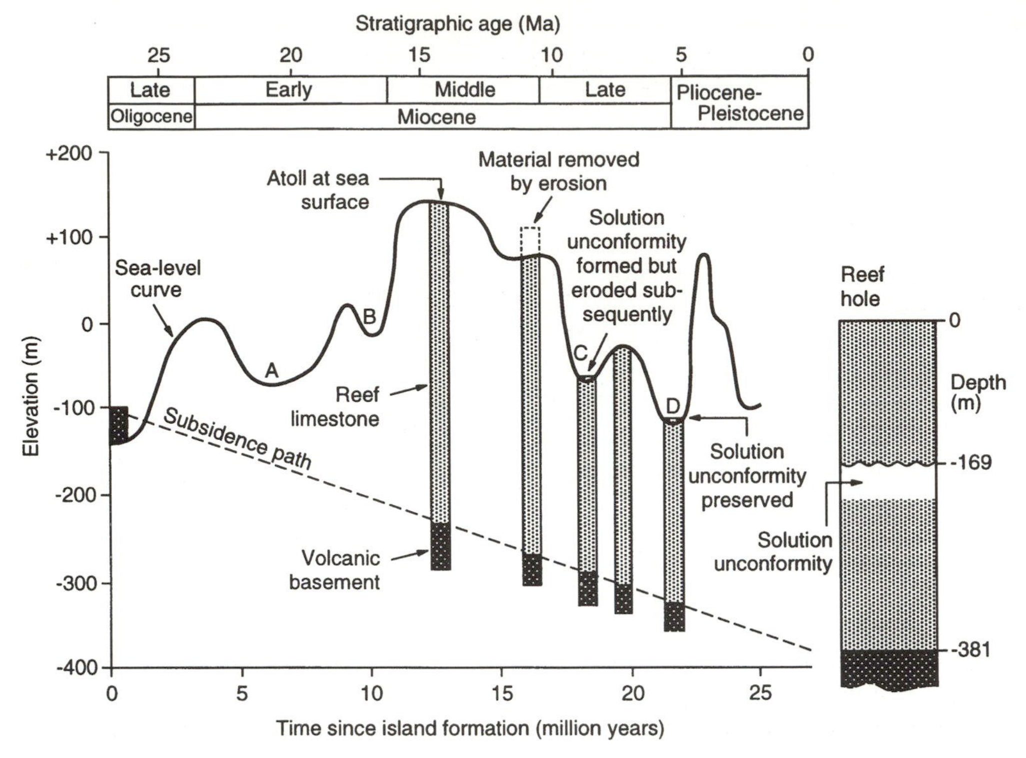

Fig. 2.7. The complex sequence of sea-level changes proposed for Midway Island during the last 25 million years. Solution unconformities formed during low stages A–C were all removed during lowering of the atoll surface by subaerial processes during low sea-level stage D. The solution unconformity formed at this stage has remained preserved because subsidence carried it below the reach of subsequent periods of atoll surface lowering. (Redrawn from Nunn 1994; see for original sources.)

180 m relative to the Earth's centre. Relatively minor shifts in the configuration of the geoid surface, which might be produced by underlying tectonic movements, would be sufficient to cause large amounts of noise in the glacio-eustatic picture (Nunn 1994; Benton and Spencer 1995). However, there is some measure of agreement that stand levels have not exceeded present levels by more than a few metres within the past 340 000 years. The most widely accepted figure for Pleistocene minima is of the order of −130 m, although in places it may have been greater than this (Bell and Walker 1992; Nunn 1994). This order of sea-level depression, given present lithospheric configuration, is sufficient to connect many present-day islands such as mainland Britain to continents, thus allowing biotic exchange between land areas that are now disconnected. Equally important, many islands existed that are now below sea level. However, as will be clear, simply drawing lines on maps in accord with present-day −130 m contour lines is not a sufficient basis for reconstructing past island–mainland configurations.

The rise from the late glacial minima to present levels was not achieved overnight, nor was it a steady or uniform pattern. As a broad generalization, at around 14 500 BP (Before Present), sea levels stood at about −100 m, rising by some 40 m over the next millennium (Bell and Walker 1992). A second major phase of glacial melting around 11 000 BP caused an eustatic rise to about −40 m by the beginning of the Holocene, when ice volumes had been reduced by more than 50%. The pattern for the British Isles for the past 9000 years can be seen in Fig. 2.8. Both eustatic and isostatic elements were involved, leaving a legacy in raised shorelines and drowned valleys. In the North Sea, the

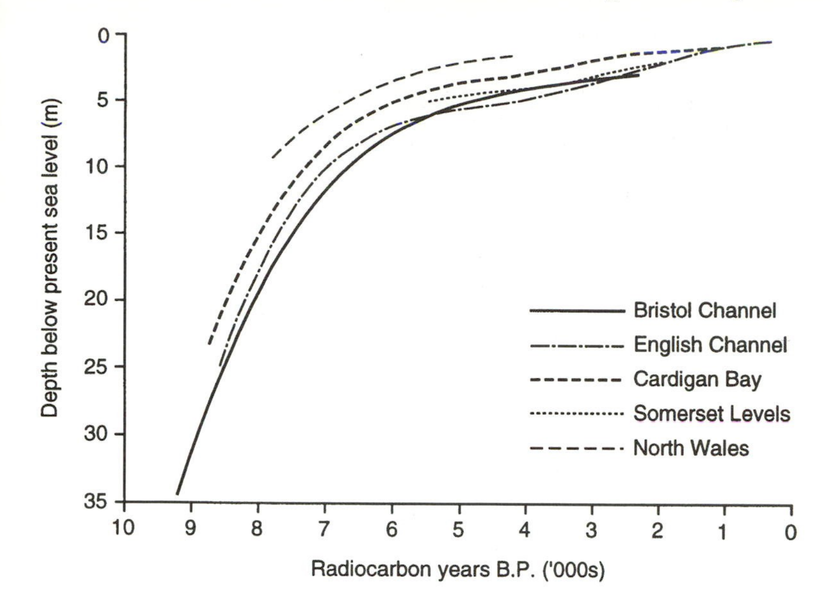

Fig. 2.8. Holocene sea-level rise in southern Britain. (Redrawn from Bell and Walker 1992, Fig. 4.10, after Shennan 1983.)

Dogger Bank was breached by 8700 BP and the Straits of Dover by 8000 BP. The present configuration of the southern North Sea coastlines was more or less established by 7500–7800 BP, and that of the British coastline by *c.* 6000 BP, although slow adjustments continue today. The severing of Britain from Ireland took place some 2000–3000 years before that of Britain and the rest of Europe. In short, the switch from glacial to interglacial conditions took place some 2000–3000 years before 'mainland' Britain became an island once again, with different parts of the British isles being separated from one another at quite different times. These events left a legacy in the biotic composition of the islands that has long been noted by biogeographers (Williamson 1981).

2.3.3. Climate change on islands

An important point made by Mark Williamson (1981) in *Island populations*, is that short-term variations in climate have a lower amplitude than long-term variation, that is the variance within decades is less than the variance with centuries, which is again less than the variance in millennia. This can be summed up in the phrase: climatic variation has a reddened spectrum. The significance of this for oceanic island biota is that species have to be adaptable if they are to survive.

Over the past several million years, the global climate system has undergone continual change, most dramatically illustrated by the glacial–interglacial cycles of the Pleistocene (e.g. Bell and Walker 1992). For much of the last glacial period, for instance, large areas of the British Isles resembled arctic tundra, while the northern and western regions supported extensive glacial ice. Subpolar islands such as the Aleutians (north Pacific) and Marion Islands (south-west Indian Ocean) also supported extensive icecaps at the last glacial maximum, and there is evidence that the Pleistocene cold phases caused extinction of plant species on high-latitude oceanic islands, such as the sub-Antarctic Kerguelen (Moore 1979). The classical model of four major Pleistocene glaciations has long been replaced by an appreciation that there have been multiple changes between glacial and interglacial conditions over the past 2 million years (Bell and Walker 1992). While high-latitude islands have been the most dramatically affected by these climatic oscillations, it would be erroneous to assume that low-latitude oceanic islands have been so effectively buffered that their climates have been essentially stable.

The Galápagos islands in their lower regions are desert-like, whereas in the highlands moist forests occur; however, palaeo-environmental data from lake sediments demonstrate that in the last glacial period the highlands were dry. The moist conditions returned to the highlands about 10 000 BP, but the pollen data for El Junco Lake on Isla San Cristobel indicate a lag of some 500–1000 years before vegetation similar to that of the present day occupied the moist high ground (Colinvaux 1972). This delay may reflect the slow progress of primary succession after expansion from relict populations in limited refugia in more moist valleys, or the necessity of many plants having to disperse over wide sea gaps (the group is 2000 km from the mainland) to reach the site. Analyses of pollen cores from subtropical Easter Island also show the local effects of global climate change. The data indicate fluctuating climate between 38 000 and 26 000 BP, cooler and drier conditions than those of the present day between 26 000 and 12 000 BP, and the Holocene being generally warm and moist, but with some drier phases (Flenley *et al.* 1991). Contrary to the evidence from Easter Island and the Galápagos, it appears from studies of snowline changes on tropical Hawaii that conditions were both cooler and, quite possibly, wetter during the last glaciation (Vitousek *et al.* 1995). Given the dominant influence of ocean and atmospheric current systems on the climates of oceanic islands (below), it is unwise to assume a straightforward relationship between continental and island climate change over the Quaternary.

2.3.4. Case study of an island at sea: the environmental history of Jamaica over the past 50 million years

Several of the themes of this section are well illustrated by Ruth Buskirk's (1985) analysis of the history of Jamaica in relation to the rest of the Caribbean. The changing configurations of Antillean land masses during the Cenozoic, as suggested by a variety of authors, are presented in Fig. 2.9. Beginning in the Eocene, left-lateral faulting along the northern Caribbean plate margin slowly moved Jamaica eastward relative to the North American plate, the island varying in size and becoming increasingly distant from the North American continent in the process. More crucially still, from a biogeographical perspective, in the Upper Eocene, it was entirely submerged, with extensive marine deposits

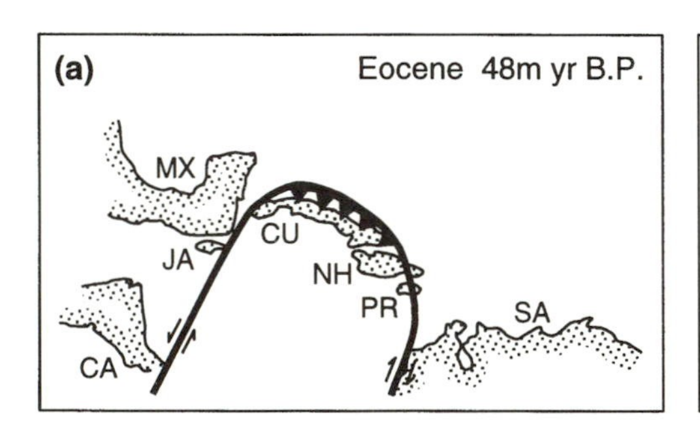

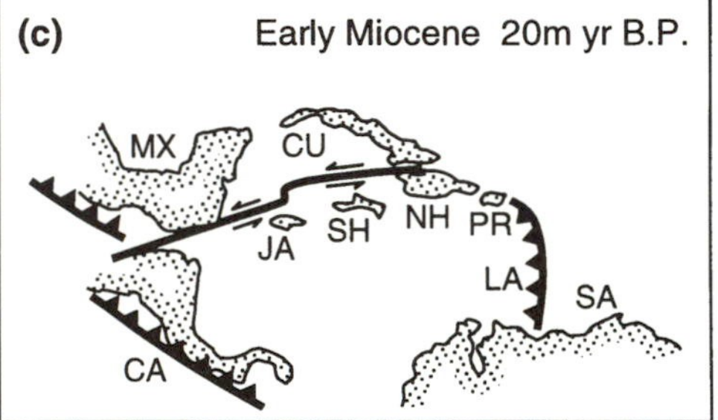

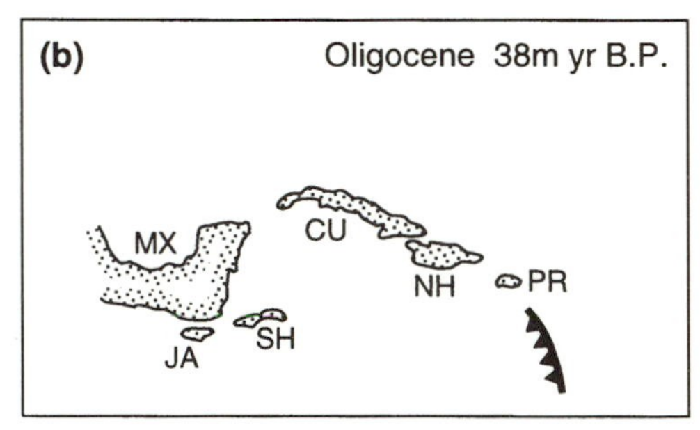

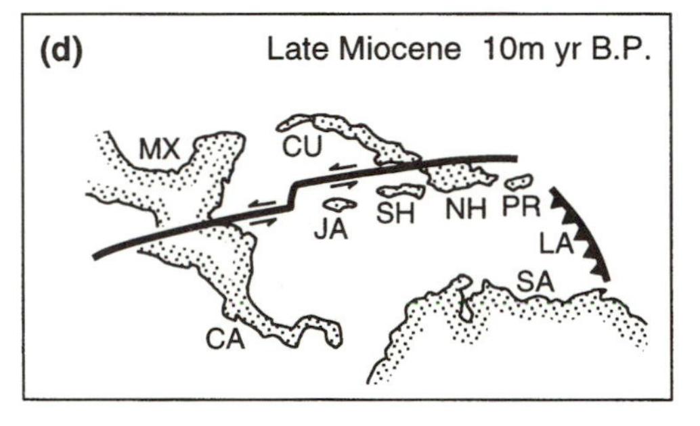

Fig. 2.9. (a–d)

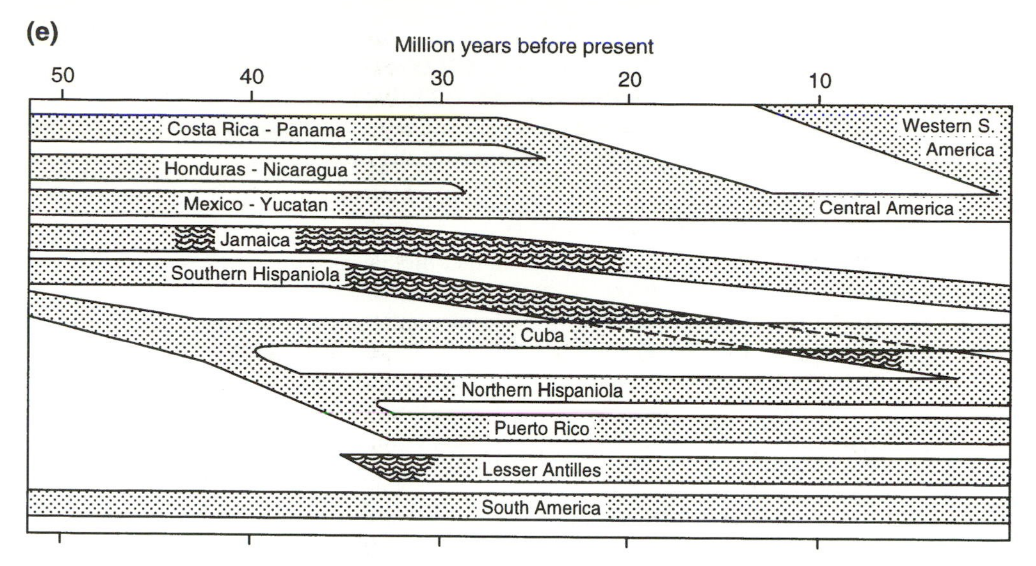

Fig. 2.9. Hypothesized configurations of the Caribbean area during the past 50 million years. (a–d) Maps of changing positions. Present-day land outlines are used for ease of recognition and do not indicate shorelines of the time. CA = Central America, MX = Mexico, JA = Jamaica, CU = Cuba, NH = northern Hispaniola, SH = southern Hispaniola, PR = Puerto Rico, LA = Lesser Antilles, SA = South America. (e) Simplified scheme showing the relative positions of the land areas, increased distance apart indicating increased barriers to dispersal. Areas that were largely inundated are designated with water wave symbols. (Redrawn from Buskirk 1985, Figs 2 and 4.)

being laid down and total subsidence in Jamaica amounting to some 2800 m, before its eventual re-emergence as an island in the early Miocene. The uplift began in the northern and north-eastern coastal region, with general uplift beginning in the middle Miocene. Maximum uplift and faulting occurred in the Pliocene (the final stage of the Tertiary), with as much as 1000 m of uplift in the Blue Mountains area since the middle Pliocene. Even after the major period of emergence began, the island continued to move away from Central America, possibly by as much as 200 km in the past 5 million years.

The message that must be drawn from this and the foregoing sections is that such basic environmental features as island elevation, area, geology, location, isolation, and climate can all be subject to significant change in the long term, and thus, in considering the evolutionary ecology and biogeography of islands, it is unwise to start with the premise that history can be discounted (cf. Stoddart and Walsh 1992; Wagner and Funk 1995). The complexity of Caribbean geology means that Buskirk's (1985) framework may well require updating as new evidence becomes available, yet the illustrative value of her study will surely remain, showing the need to involve a blend of historical and ecological parameters in understanding Caribbean biogeography (see section 3.4).

2.4. Physical environment of islands

2.4.1. Topographic characteristics

Topographic characteristics for a selection of islands in the New Zealand area are provided by Mielke (1989), who notes that, although the largest islands have the highest peaks, smaller islands display no consistent pattern (Table 2.3). In fact, topographic characteristics depend on the type of island. Volcanic islands tend to be steep and relatively high for their area and, through time, become highly dissected. In 1927, Chester K. Wentworth calculated the age of the Hawaiian volcanoes as a function of the degree of dissection. Menard (1986) has updated this original study and compared the values resulting with the ages determined by

Table 2.3
Area and relief of New Zealand and neighbouring islands (from Mielke 1989, Table 9.1)

Island	**Area** (km^2)	**General relief**
South Island	148 700	50 peaks over 2750 m
North Island	113 400	Three peaks over 2000 m
Stewart	1720	Granite; three peaks, up to 987 m
Chatham	950	Cliffs to 286 m
Auckland	600	Volcanic; one peak 615 m
Macquarie	118	Volcanic; 436 m glacial lakes
Campbell	113	574 m glaciated
Antipodes	60	Volcanic; peak of 406 m
Kermadec	30	Volcanic; peak of 542 m
Three Kings	8	Volcanic
Snares	2.6	Granite; cliffs to 197 m

Fig. 2.10. Oceanic islands tend to be a mix of volcanic and coralline rocks. This low-lying coral rag from Mangrove Cay, Andros, in the Bahamas, is useless for agriculture and extremely hazardous to move around on when covered with the usual tangle of hardwood or pineyard woodland. (Photo: RJW 1988.)

potassium–argon dating (up to an age of 1.6 million years), thus reversing the aim of the original study and demonstrating that dissection is indeed a function of age of the volcano. Height of island can be important in respect to changes in sea level, receipt of rainfall, and to other climatic characteristics.

Coral or limestone islands and atolls tend to be very low-lying and flat. There are clear implications here with respect to sea-level changes and resources for island biota. Those which have been uplifted to more than a few feet above sea level are termed **makatea** islands, examples including Makatea itself (Tuamotu Archipelago), Atiu (Cook Islands), and most of the inhabited islands of Tonga. This is an important class of islands. They are characterized by rocky coralline substrates, but some are partly volcanic (Fig. 2.10). Many have had commercially exploitable deposits of phosphate formed through the ages by seabird colonies, and the gradual drying of their central lagoons. For a discussion of the complexities of volcanic, limestone, and makatea (composite) island landscapes, see Nunn (1994).

2.4.2. Climatic characteristics

Island climates have, self-evidently, a strong oceanic influence, and quite often are considered anomalous for their latitude, as a consequence of their location in the path of major ocean or atmospheric current systems (e.g. the Galápagos (Darwin 1845)). Low islands tend to have relatively dry climates. High islands tend to generate heavy rainfall, although they may also have extensive dry areas in the rain shadows, providing a considerable environmental range in a relatively small space. Even an island of only moderate height, such as Christmas Island, Indian Ocean, with a peak of 360 m and general plateau elevation of only 150–250 m, benefits from orographic rainfall, allowing rain forest to be sustained through a pronounced dry season (Renvoize 1979). Islands may also be anticipated to receive rainfall that has a generally different chemical content than that experienced over continental interiors (cf. Waterloo *et al.* 1997).

As noted by J. D. Hooker in his lecture to the British Association in 1866, the climate and biota of islands tend to be more polar than those of continents in the same latitude, while intra-annual temperature fluctuation is reduced (Williamson 1984). Islands near the equator frequently exhibit annual average temperature ranges of less than 1 °C, and even in temperature latitudes, the annual average ranges are less than 10 °C: for example, 8 °C in the case both of Valentia (Iceland) and the Scilly Isles. Some islands do, however, experience quite large interannual variations in other features of their weather. Variability in rainfall, for example related to El Niño–Southern Oscillation (ENSO) phenomena, can be of considerable ecological importance. Islands may also experience periodic extremes such as hurricanes (below) (Stoddart and Walsh 1992).

Contraction, or 'telescoping' of altitudinal zones is another marked feature of smaller islands. For example, on Krakatau, Indonesia, the plentiful atmospheric moisture and the cooling influence of the seas result in the near permanent presence of cloud in the upper parts, further lowering temperatures and allowing the development of a montane mossy forest at around 600 m above sea level, a much lower altitude than the continental norm (Whittaker *et al.* 1989). Taylor (1957) suggested 2000 m as the height at which the vegetational transition occurs in interior Papua New Guinea, and Whitmore (1984) observed that clouds normally form around tropical peaks in excess of 1200 m, but that this altitudinal limit is reduced on offshore islands. Bruijnzeel *et al.* (1993) have pointed out that the effect is not, however, observed on all coastal mountains in the tropics, and that it seems as though average atmospheric humidity levels, rather than proximity to the sea, are decisive. In a recent review of upper limits of forests on tropical and warm-temperate oceanic islands, Leuschner (1996) cited a range of 1000–2000 m for the lowering of the forest line compared with that on continental areas. Factors that have been invoked for these reductions include steepened lapse rates, droughts on trade-wind-exposed island peaks with temperature inversions, the absence of well-adapted high-altitude tree species on some islands, and even the immaturity of volcanic soils (Leuschner 1996). Whatever the precise cause(s) of the 'telescoping', it is a common feature of tropical islands, and it has the effect of (potentially) increasing the number of species that an island can support, by compressing habitats and allowing a relatively low island to 'sample' additional upland species pools (Fig. 2.11).

The range of climatic conditions of an island is determined in large measure by the elevation of the highest mountain peaks. Thus regression studies often find altitude to be an important variable in explaining species numbers on islands, in some cases ranking only second to, or even ahead of, island area. Tenerife, in the Canary Islands, is an excellent example of a volcanic island with a steep central ridge system and a pronounced rain shadow. The Canary Islands lie on the subsiding eastern side of the semi-permanent Azores anticyclone at about 28°N, 100 km off Africa. The subsidence produces a warm, dry atmosphere aloft that is separated at 1500–1800 m from a lower layer of moist, southward-streaming air, producing what is known as a trade-wind inversion

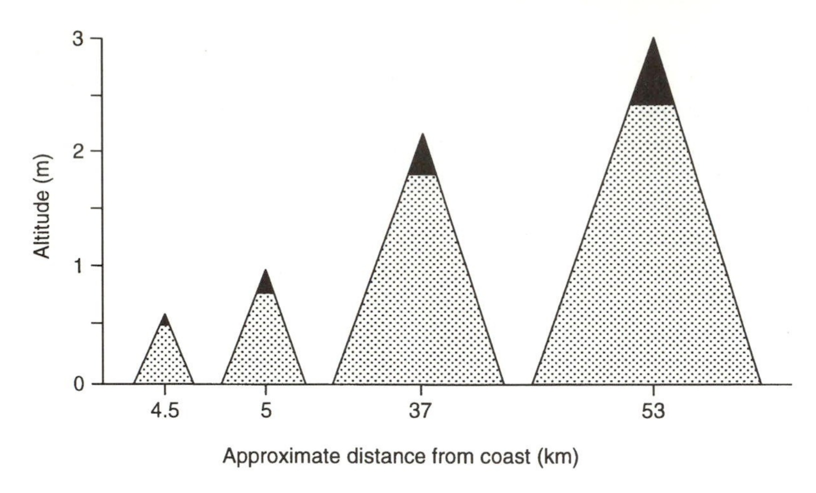

Fig. 2.11. The transition to mossy forest on mountains in Indonesia occurs at lower elevations on small mountains near the sea than on larger mountains inland. From left to right: Mt Tinggi (Bawean), Mt Ranai (Natuna Island), Mt Salak (West Java), and Mt Pangerango (West Java). (After van Steenis 1972.)

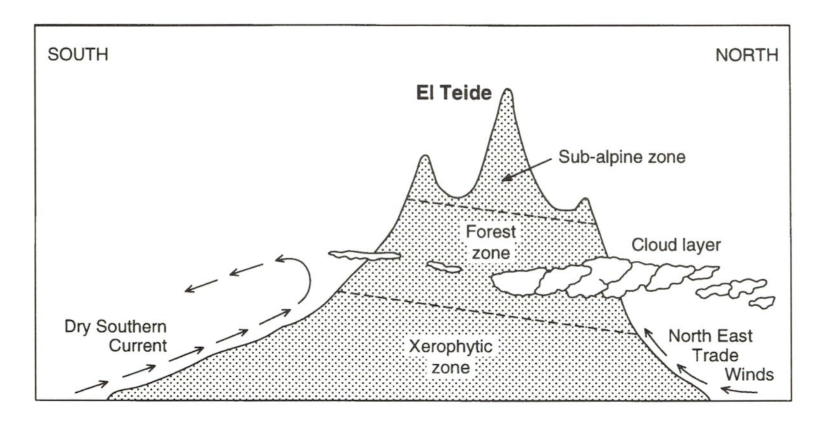

Fig. 2.12. Climatic/vegetation zones of Tenerife. (Redrawn from Bramwell and Bramwell 1974.)

Fig. 2.13. On Tenerife (Canary Islands) the north-east trades bring in moisture from the sea, and the air is forced to rise by the mountain backbone. It cools and this produces a zone of precipitation between about 600 and 1700 m and typically 300 m thick, on the north side of the island. Tenerife contains a dramatic range of habitats, from semi-desert to cloud forest, to alpine desert, and a large percentage of endemic plants and animals. The foreground shows the zone of endemic Canary Island pine (*Pinus canariensis*). Tenerife is a true oceanic island of volcanic origin. (Photo: RJW 1992.)

(Fernandopullé 1976). Tenerife's climate is one of hot dry summers and warm wet winters: in essence, the island experiences a Mediterranean-type climate, despite its latitudinal proximity to the Sahara. The north-east trades bring in moisture from the sea, and the mountain backbone forces the air to rise. It cools, and this produces a zone of precipitation between about 600 and 1700 m, typically 300 m thick, on the north side of the island—a feature that builds up more or less daily (Figs 2.12, 2.13). The southern sector, in the rain shadow, receives far less precipitation. In consequence of the differences in precipitation and temperature, the lower and upper limits of forest growth are higher on the southern side. Indeed, the southern sector lacks a dense forest zone at mid-altitude and is much more xerophytic.

2.4.3. Water resources

Availability of water shapes the ecology and human use of islands (Whitehead and Jones 1969; Ecker 1976; Menard 1986). Most oceanic islands, whether high volcano or atoll, contain large reservoirs of fresh water. Fresh lava flows are highly permeable, but over time the permeability and porosity of the rock decreases as a result of weathering and subsurface depositional processes. The residence time of groundwater in the fractured aquifers of large volcanic islands may be from decades to centuries. Groundwater compartments develop in a tortuous chain of compartments, as shown in Fig. 2.14; they may be subdivided into the vadose zone and the saturated or basal water zone. The vadose zone contains groundwater compartments, often linked into chains, interspersed with dry zones. The basal zone is characterized by many closely placed groundwater compartments and by a high percentage of saturated secondary fractures. Both fresh and sea water occur in this zone, with tides influencing the water level up to 4–5 km from the coast (Ecker 1976). In the absence of rain, the sea water within islands would be at

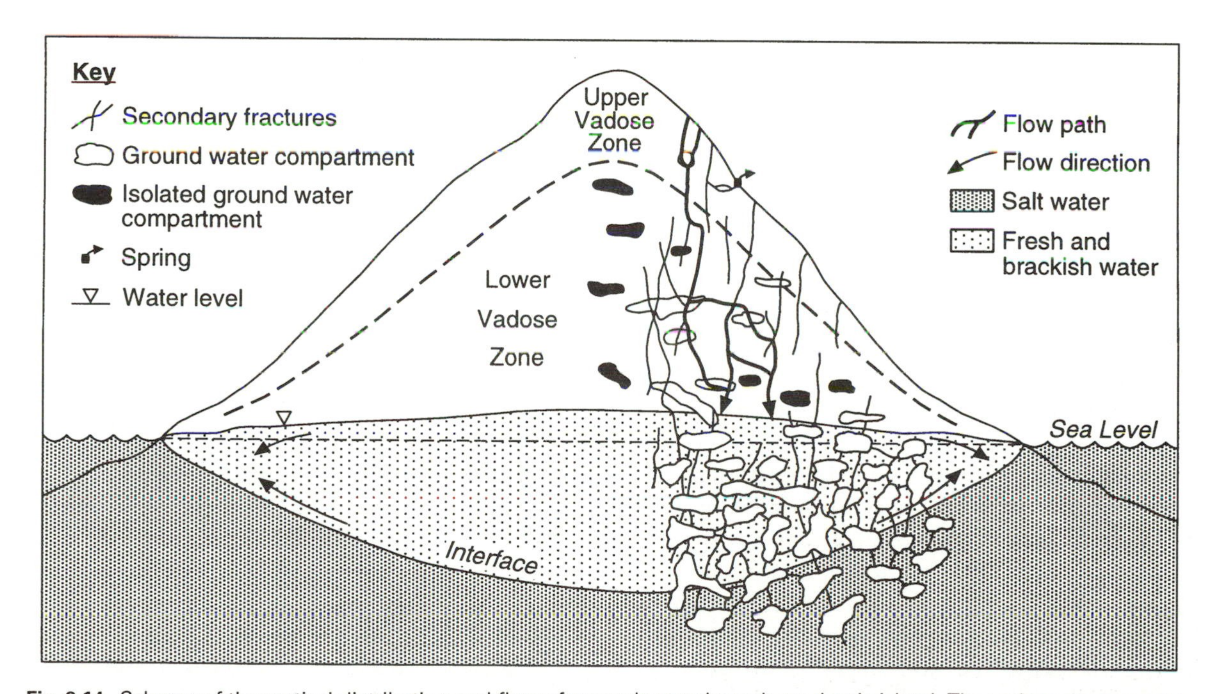

Fig. 2.14. Scheme of the vertical distribution and flow of groundwater through a volcanic island. The vadose zones contain groundwater compartments and chains, interspersed with dry zones. The Ghyben–Herzberg lens of fresh water under an oceanic island rests on the denser salt/brackish water that permeates the base of the island. (Modified from Ecker 1976, Fig. 4, with kind permission of Elsevier Science—NL, Sara Burgerhartstraat 25, 1055 KV Amsterdam, The Netherlands.)

sea level; however, rainwater percolating through an island floats on the denser salt or brackish water that permeates the base of the island, in what is termed a **Ghyben–Herzberg lens** (Menard 1986).

As indicated above, the average annual rainfall varies greatly within the island of Tenerife, from over 800 mm in the highest parts of the Anaga peninsular, to less than 100 mm in the extreme south of the island. The rain shadow thus has profound effects both on the ecology and on the potential human use of the island. The steep west side, which benefits from the majority of the precipitation, is intensively cultivated, whereas much of the flatter east coastal zone is essentially useless for cultivation without irrigation. Thus, even on this large, high island, groundwater (which increasingly has been tapped from aquifers deep in the volcano) is the most important source of water for the human inhabitants, and water constitutes a key limiting resource for development. Even low-lying atolls maintain a freshwater lens, but, as noted in the introduction to this chapter, very small islets (below the order of about 10 ha) can lack a permanent lens. Such habitats are liable to be hostile to plants other than those of strand-line habitats, thus limiting the variety of plant species that can survive on them (Whitehead and Jones 1969).

2.4.4. Tracks in the ocean

One of the intriguing features of the island biogeographical literature is that isolation is, in general, rendered merely as distance to mainland. This ignores the geography of the oceans. Figure 2.15 displays the pattern of surface drifts and ocean currents in January. This pattern is, of course, variable on an interannual and an intra-annual basis. However, in some areas, the ocean currents and wind currents are strongly directional and persistent. In some cases, knowledge of current systems provides a parsimonious explanation of biogeographical patterning, as in the case of the butterflies of the island of Mona in the Antilles. The island is equidistant between Hispaniola and Puerto Rico and is 62 km^2 in area. Its 46 species of butterflies feature nine subspecies in common with Puerto Rico and none with the larger Hispaniola, whereas the ratio of source island areas is 9 : 1 in favour of Hispaniola. The explanation appears to lie in a remarkably constant bias in wind direction, which is from Puerto Rico towards Hispaniola (Spencer-Smith *et al.* 1988). However, it should be recalled that palaeoenvironmental data from islands such as Hawaii and the Galápagos indicate that significant changes in ocean and atmospheric circulation patterns occur over time (Vitousek *et al.* 1995).

2.5. Natural disturbance on islands

One of the themes that I will develop later is that the significance of naturally occurring disturbances has not been given due recognition in the development of island ecological theory (sections 7.4 and 8.6). We have already seen that any ancient island will, over the course of thousands of years, have experienced substantial environmental changes. This section is concerned with shorter-term changes in environment, such as the individual volcanic eruptions that may intermittently add to the bulk of an island while, at least temporarily, reducing its plant and animal populations (e.g. Partomihardjo *et al.* 1992).

Ecologists have defined disturbance as ‘any relatively discrete event in time that removes organisms and opens up space which can be colonized by individuals of the same or different species’ (Begon *et al.* 1990, p. 754). Such a definition is rather inclusive and, in the context of island biogeography, it will be recognized that different scales of perturbation are relevant in differing contexts (Fig. 2.16). Increasingly, ecologists are recognizing that natural systems are largely structured by disturbance (Pickett and White 1985; Huggett 1995). Often, a single event such as a hurricane (tropical cyclone), will impact on both mainland and island systems; however, because of the geographical and geological idiosyncrasies of small islands and their location in oceans, their disturbance regimes

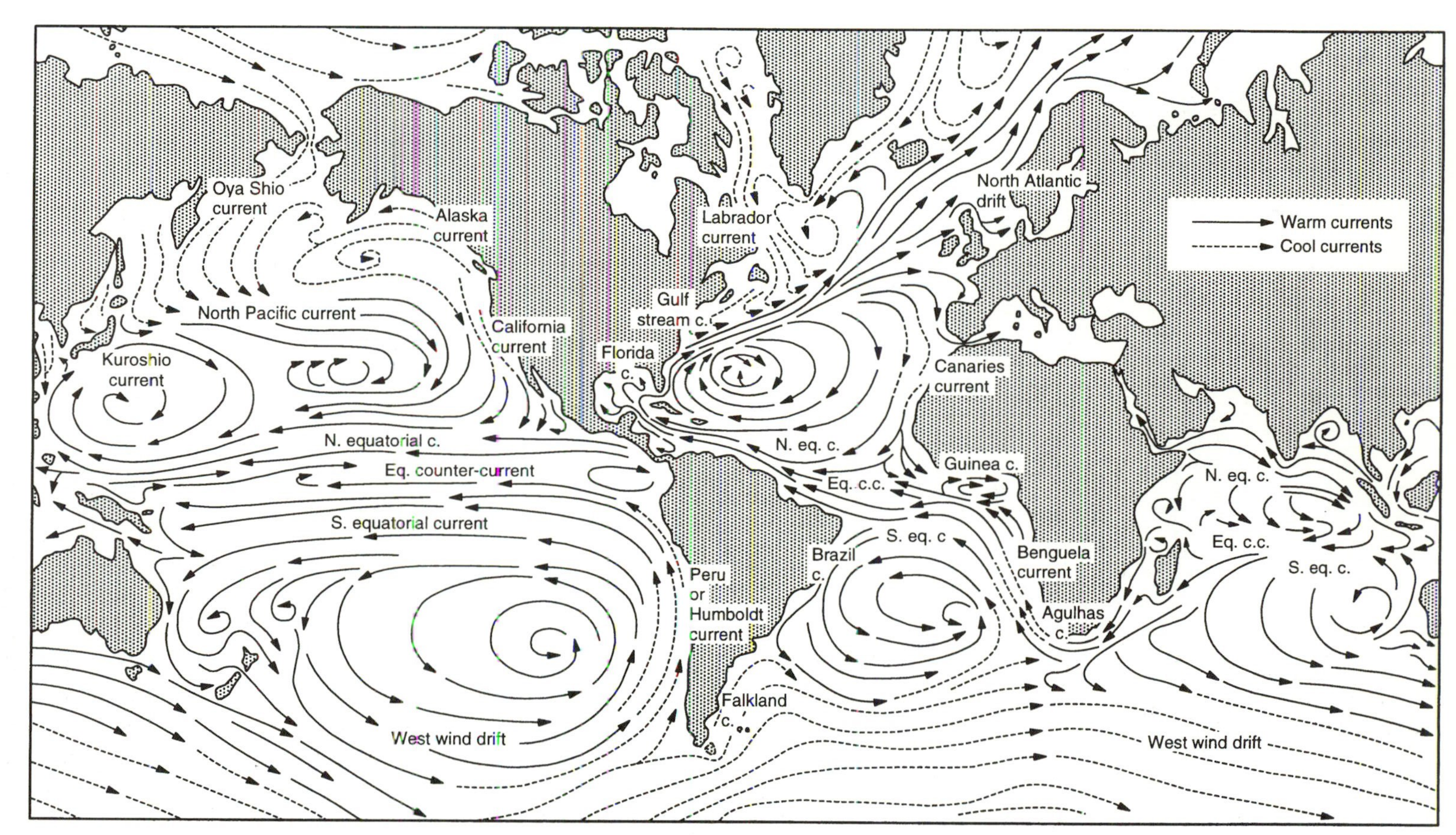

Fig. 2.15. Surface drifts and ocean currents in January. (Redrawn from Nunn 1994, Fig. 4.)

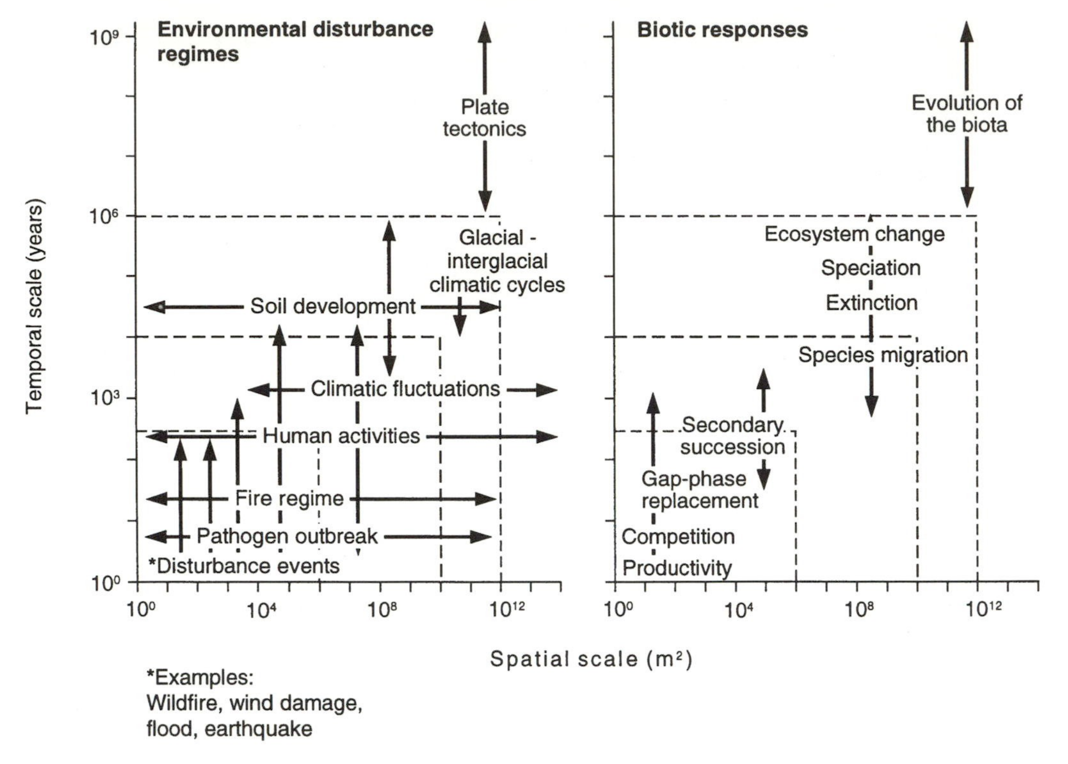

Fig. 2.16. Environmental disturbance regimes and biotic responses, viewed in the context of four space–time domains (shown here bounded by dashed lines), named micro-, meso-, macro-, and mega-scales by the scheme's authors. (Redrawn from Delcourt and Delcourt 1991, Fig. 1.6; from an original in Delcourt and Delcourt 1988, with kind permission from Kluwer Academic Publishers.)

when scaled to island size are atypical of continental land masses.

In order to incorporate the disturbance regime into island biogeography, some form of classificatory framework is required (Whittaker 1995). Figure 2.16 represents an example of one attempt to develop such a framework. Another scheme, devised for the Caribbean Islands and reproduced in Table 2.4, indicates the principal types of disturbance phenomena, the area liable to be affected, the primary nature of the impact, its duration, and the recurrence interval. In an approach owing much to H. T. Odum, Lugo (1988) has classified the disturbance phenomena into five types, each characterized by their impact on energy transfers. Type 1 events change the nature or magnitude of the island's energy signature before the energy can be 'used' by the island; for example, shifts of weather systems through ENSO events, leading to droughts or heavy rainfall (Stoddart and Walsh 1992; Benton and Spencer 1995). Type 2 events are those acting on the major biogeochemical pathways of an island, for example as a result of changes caused by an earthquake. Type 3 events are those which remove structure from an island ecosystem, but without altering the basic energy signature, so that recovery can proceed rapidly after the event. Hurricanes are Type 3 events, a well-researched example being Hurricane Hugo, which caused complete defoliation of a large part of the Luquillo forest on Puerto Rico in 1989, although one might note that, despite a rapid 'greening-up' of the forest, the signature of such an event will be evident in the unfolding vegetation mosaic for many decades (Walker *et al.* 1996). Type 4 events alter the 'normal' rate of material exchange between the island and either the ocean or atmosphere. For instance, if the trade winds are inhibited by changes in atmospheric pressure, exchanges may be reduced. Type 5 events are those which destroy

Table 2.4
Disturbance phenomena affecting Caribbean islands (after Lugo 1988)

Disturbance phenomena	Type	Area affected	Primary impact	Duration	Recurrence
Hurricanes	3, 5	Large	Mechanical	Hours–days	20–30 years
High winds	3, 4, 5	Large	Mechanical	Hours	Annual
High rainfall	4	Large	Physiological	Hours	Decade
High pressure systems	1	Large	Physiological	Days–weeks	Decades
Earthquakes	2, 5	Small	Mechanical	Minutes	10^2 years
Volcanism	all	Small	Mechanical	Months–years	10^3 years
Tsunamis	3, 4, 5	Small	Mechanical	Days	10^2 years
Extreme low tides	1	Small	Physiological	Hours–days	Decades
Extreme high tides	3, 4, 5	Small	Mechanical	Days–weeks	1–10 years
Exotic genetic material	2, 3	Large	Biotic	10^2 years	Decades
Human, e.g. energy	1	Small	Biotic	Years	1–10 years
Human, e.g. war	5	Small	Mechanical	Months–years	?

consumer systems, by which is meant human systems, possibly with subsequent repercussions for the rest of the island's ecology; examples again include hurricanes and earthquakes.

2.5.1. Magnitude and frequency

Between 1871 and 1964, an average of 4.6 hurricanes per year were recorded in the Caribbean, resulting in, for example, a return time of 21 years for the island of Puerto Rico (Walker *et al.* 1991*a*; Fig. 2.17). With minimum wind speeds of 120 km/h and paths of tens of kilometres width, hurricanes can have a profound impact, fundamental to an understanding of the structure of natural ecosystems in the region. However, the Caribbean by no means corners the market in tropical storms. Hurricanes develop in all tropical oceanic areas where sea surface temperatures exceed 27–28 °C, although they are generally absent within 5° north and south of the equator, where persistent high pressure tends to prevent their development (Nunn 1994). Wind defoliation and large blow-downs are important and frequent disturbances for forested tropical islands throughout the hurricane belts, centred 10–20° north and south of the equator (e.g. Elmqvist *et al.* 1994). Recent studies on high islands, such as the impact of tropical cyclone Ofa on Upolu (West Samoa), emphasize the destructive potential of such phenomena. Although storm damage to lower islands can also be severe, storms can be important to island growth by throwing up rubble ramparts (Stoddart and Walsh 1992). Thus some very small islands, such as the so-called *motu* (sand islands), may be in a kind of dynamic equilibrium with both extreme and normal climatic phenomena (Nunn 1994). A comprehensive analysis of disturbance regimes requires quantification of both magnitude and frequency of events (Stoddart and Walsh 1992). Lugo (1988) attempted this for a subset of the Caribbean disturbance phenomena given in Table 2.4. From the figures he derived, he argued that, whereas hurricanes have the highest frequency of recurrence among his major 'stressors', the susceptibility of Caribbean islands to hurricanes is intermediate, partly because they do not change the base energy signature over extensive periods. Also, with the larger Caribbean islands, hurricane damage is unlikely to impact catastrophically on the entire island area. Given their recurrence interval, these island ecosystems should be evolutionarily conditioned by such storm events (Whittaker 1995; Walker *et al.* 1996).

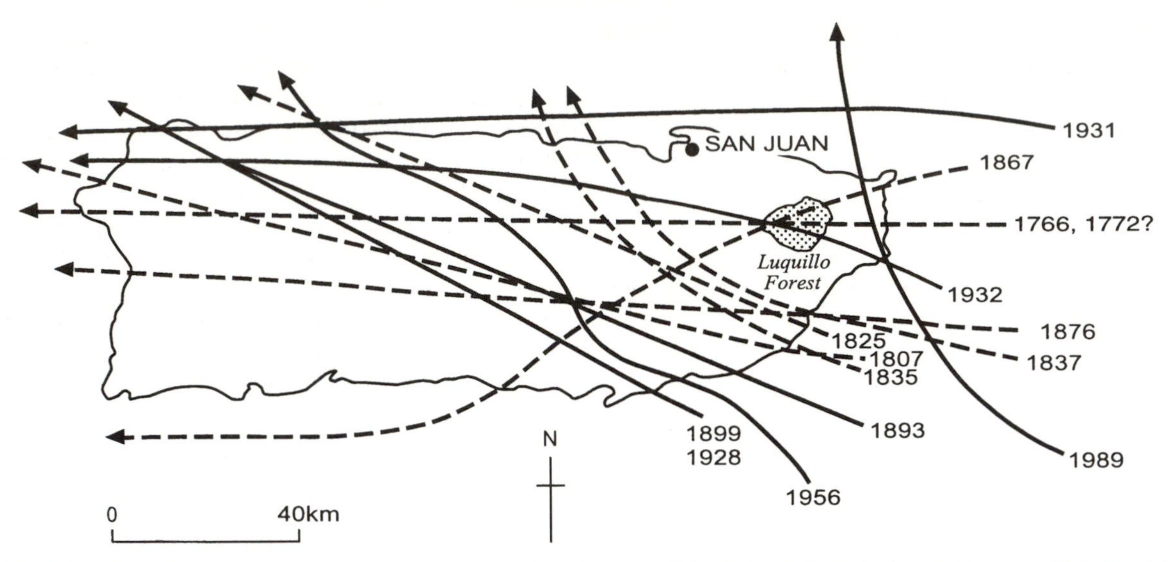

Fig. 2.17. Paths of hurricanes that have crossed Puerto Rico since AD 1700. (Redrawn from Scatena and Larsen 1991, Fig. 1.)

Not all extreme weather phenomena take the form of storms. ENSO phenomena are large-scale interannual events that are the product of variations in air pressure and associated rainfall patterns between the Indonesian and South Pacific regions, coupled with ocean-current and temperature variation. In the Asia–Pacific region, ENSO events are associated with a heightened amplitude of interannual climatic variation—that is, more intense droughts and wet periods. In illustration, the island of Aututaki (Cook Islands) received 3258 mm rainfall in December 1937, which is in excess of the average annual total (Nunn 1994). That events of the magnitude of ENSO phenomena have a detectable ecological impact, even on populations conditioned evolutionarily to the island way of life, was shown by the significant increase in numbers of four Galápagos endemic bird species (three finches and a mockingbird) as a consequence of the increased food supply resulting from the exceptionally high rainfall during 1982–83 (Gibbs and Grant 1987). Analyses of historical data for hurricanes show that the frequency of climatic phenomena such as hurricanes and ENSO events varies over decadal time scales, further complicating analysis. This goes to illustrate that, despite the reddened spectrum of climatic variation, short-term fluctuations deserve the attention of ecologists and biogeographers (Stoddart and Walsh 1992; Benton and Spencer 1995; Schmitt and Whittaker 1998).

There is a premise that, because of the comparative simplicity of geology and climate of many oceanic islands, their *landscapes* should attain a condition of dynamic equilibrium. Nunn (1994) argues that this notion cannot be supported as a generality:

> ... as is evident from the often catastrophic impacts of certain irregular climatic phenomena on oceanic island landscapes, the degree to which particular environments can accommodate the impacts of particular uncommon or extreme events without fundamental alteration, while variable, is generally low. For those oceanic island landscapes in a state of dynamic equilibrium, the effects of such events may be to cause a threshold of landscape development to be crossed... . The effects of irregular climatic phenomena are so variable, so site-specific, that it is pointless to attempt a broad generalization. (pp. 159–60)

These observations might prompt us to ask ourselves, first, with what portion of the magnitude–frequency spectrum of climatic variation might oceanic island biotas best be considered to be in dynamic equilibrium and, second, might not extreme climatic events have caused ecological or evolutionary thresholds to be crossed in the same manner as described above for their landscapes?

2.5.2. Continued volcanism

It has been established that volcanic islands are typically active over an extensive period, often spanning many millions of years, and that there are a variety of distinctive tectonic situations in which volcanic islands are built (above). A corollary of this is that a number of major types of volcanic eruption can be identified. In increasing degrees of explosiveness these are: Icelandic, Hawaiian, Strombolian, Vulcanian, Peléan, and Plinian (Decker and Decker 1991). Icelandic volcanic eruptions are fluid outpourings from lengthy fissures, and they build flat plateaux of lava, such as typify much of Iceland itself. Hawaiian eruptions are similar, but occur more as summit eruptions than as rift eruptions, thereby building **shield volcanoes**. Strombolian eruptions take their name from a small Italian island that produces small explosions of bursting gas that throw clots of incandescent lava into the air. Vulcanian eruptions, named after the nearby island of Volcano, involve the output of dark ash clouds preceding the extrusion of viscous lava flows, thus building a **stratovolcano** or **composite cone**. Peléan eruptions produce pyroclastic flows termed **nuées ardentes**, high-speed avalanches of hot ash mobilized by expanding gases and travelling at speeds in excess of 100 km/h. Plinian eruptions are extremely explosive, involving the sustained projection of volcanic ash into a high cloud. They can be so violent, and involve so much movement of magma from beneath the volcano, that the summit area collapses, forming a great circular basin, termed a **caldera**. Like all generalizations, this classification is over-simplistic; for example, calderas have formed by collapse in Iceland and Hawaii as well as at major explosive volcanoes. However, it serves to illustrate that there are great differences in the nature of volcanism, both between islands and within a single island over time. The eruptive action within Hawaii over the past few centuries has been both more consistent and less ecologically destructive than that of Krakatau, which, in 1883, not only sterilized itself but, through caldera collapse, created a tsunami (giant wave) killing an estimated 36 000 inhabitants of the coastlines of Java and Sumatra. Even fairly small volcanic eruptions can have very important consequences for island ecosystems. For instance, eruptions on Tristan de Cunha in 1962 covered only a few hectares in ejecta, but toxic gases affected one-quarter of the island area. The evacuation of the human population left behind uncontrolled domestic and feral animals, which transformed the effects of the eruption and produced a lasting ecological impact (Stoddart and Walsh 1992).

The dynamism of island environments suggested by this review holds relevance, in my view, to their biogeography. We might wish to ask, if the habitats are shaped and reshaped in this way, what of the inhabitants?

2.6. Summary

Four categories of island are identified: from the high seas, the oceanic and continental shelf islands; within land masses, the habitat and non-marine islands. This chapter considers only islands in the sea. These islands are rarely ancient in geological terms, and in many cases are significantly less ancient biologically than geologically. Oceanic islands have volcanic foundations, are concentrated in a number of distinctive inter- and intra-plate settings and have commonly experienced a dynamic history involving varying degrees of lateral and vertical displacement, the latter confounded by, but often in excess of, eustatic sea-level changes. In the tropics and subtropics, upward growth of reef-forming corals at times of relative or actual subsidence has led to the formation of numerous islands of only a few metres elevation, contrasting with the generally steep topography of the volcanic high islands, to a large degree according with Darwin's 1842 theories on coral reef formation.

Some of the more obviously important environmental features of islands are directly related to these geological factors and include characteristic topographic, climatic, and hydrological phenomena. Subsurface water storage is a particularly important feature of both volcanic and low, sedimentary islands: the possession of a subsurface freshwater 'lens' is a characteristic of all but the smallest islets. Island environments might be thought to have been shielded from the full amplitude of continental climatic fluctuations, but the palaeo-ecological record indicates significant climatic changes within the late Pleistocene and Holocene on a range of islands, including examples from the tropics and subtropics. With limited scope for range adjustments within the island setting, the biogeographical significance of such environmental changes should not be ignored. Shorter-term environmental variation, perturbation, or disturbances (broadly overlapping categories) characterize many island environments, not least for those islands which lie in the tropical cyclone belt and for those which remain volcanically active. Some appreciation of the distinctive nature of the origins, environmental characteristics, and history of islands is almost self-evidently important to an understanding of the biogeography of islands, although, in this scene-setting chapter, the biogeographical content has intentionally been kept within limits.

3

Biodiversity hot-spots

...the scarcity of kinds—the richness in endemic forms in particular classes or sections of classes,—the absence of whole groups, as of batrachians, and of terrestrial mammals notwithstanding the presence of aerial bats,—the singular proportions of certain orders of plants,—herbaceous forms having developed into trees, &c.,—seem to me to accord better with the view of occasional means of transport having been largely efficient in the long course of time, than with the view of all our oceanic islands having been formerly connected by contiguous land with the nearest continent...

(Darwin 1859, p. 384.)

3.1. Introduction: the global significance of island biodiversity

The term **biodiversity** is a contraction of biological diversity, and is most commonly taken to be synonymous with the species richness of an ecosystem. It may also be applied at higher or lower points in the taxonomic hierarchy, such that familial, generic, subspecific, or even gene frequency data may be analysed under the header 'biodiversity'. But the main interest here centres around the species level. As a first-order generalization, islands are species poor for their size but rich in forms found nowhere else, i.e. **endemic** to that island or archipelago. It is in this sense that certain islands warrant the description 'biodiversity hot-spots'. Given that the total number of living species on the planet is not yet known to within an order of magnitude (Groombridge 1992), it is difficult to provide a quantitative measure of the relative contribution of islands to global biodiversity. Yet, for particular taxa there are sufficient data to demonstrate that, *taken collectively*, islands contribute disproportionately for their area to global species totals.

In order to appreciate the special significance of island biotas it is important to consider some of their general properties, and in what ways they are peculiar. This chapter sets out to do that, distinguishing the notion of disharmony from simple species poverty. It also emphasises the historical biogeographical context and points to the importance of dispersal limitations in conditioning the balance of island assemblages. Once this context has been established, the distribution of island endemics can be examined. Before moving on to the theories that may explain island evolution, however, it is important to note that island biotas have contributed disproportionately to species extinction in historic and prehistoric times, such that in places much of the original biogeographical structure has already been lost or distorted.

3.2. The split between continental and oceanic islands revisited

In Chapter 2, islands were classified into four types, with those in the sea split between those built from oceanic crust and those founded on continental crust (continental shelf islands). This classification is imperfect both geologically and biogeographically, but it is a good first approximation, and a useful way to divide up what should be seen as a continuum of island types. The key distinctions were summed up by Wallace (e.g. 1902) as follows. **Oceanic islands** are built over the oceanic plate, are of volcanic or coralline formation, they are remote and have never been connected to mainland areas, from which they are separated by deep sea, and they lack indigenous land mammals and amphibians (but typically have a fair number of birds and insects and usually some reptiles). **Continental (shelf) islands** are more varied geologically, containing both ancient and recent stratified rocks, they are rarely remote from a continent and always contain some land mammals and amphibians as well as representatives of the other classes and orders in considerable variety. We may add to this that if they are of any antiquity there is a good chance that they will have been connected to larger land masses as a result of changes in relative sea level (Chapter 2).

Problems may arise with this simple division because the biogeographical distinction depends greatly on the ability of potential inhabitants of an island to disperse to it, whether over land or sea. As a simplification in biogeographical terms, an oceanic island is one for which evolution is faster than immigration, a continental island is one where immigration is faster than evolution (Williamson 1981, p. 167). Thus, a particular island may be thought of as essentially continental for some highly dispersive groups of organisms (e.g. ferns (Bramwell 1979) and some types of birds), yet essentially oceanic in character for organisms with poor powers of dispersal through or across seas (e.g. conifers, terrestrial mammals, and freshwater fish). Also, as Wallace recognized, there are a few islands, such as New Zealand, which are far more remote and represent ancient disjunctions in the continental plates. History is clearly important in distinguishing the oceanic from the continental. This is exemplified by the avifauna of the Philippines, which is intermediate in character between an oceanic and a continental island type (Diamond and Gilpin 1983), including 16 genera that appear to be unable to cross water—far more than an oceanic island of similar size and far fewer than a typical land-bridge island of similar size. **Land-bridge islands** are those formerly connected to the mainland during Quaternary sea-level minima. Of the 265 bird genera of the Malaysian region, 88 lack a water-crossing ability, so the Philippines have 16 of 88 poorly dispersing genera. Diamond and Gilpin (1983) explain this in terms of the geometry of a former land-bridge connection, which may have taken the form of a long, narrow umbilicus: either a continuous isthmus, or a chain of stepping-stone islands. This umbilicus would have been a strong enough filter to prevent more than a few of the 88 genera making it.

3.3. Species poverty

Islands typically have fewer species per unit area than mainland, and this distinction is more marked the smaller the area of the island, i.e. inter-archipelago species–area curves are steeper than curves constructed by subdividing a large mainland area (Rosenzweig 1995). This will be examined more fully in Chapter 7, and a couple of examples should suffice here. Figure 3.1 illustrates that whether the Californian mainland plant data are built up in spatially nested sets or non-nested data are considered, the points lie above the regression line for the Californian island data. A second example (after Williamson 1981) is provided by Jarak Island in the Straits of Malacca (96 km from Sumatra, 64 km from Malaya, 51 km from the nearest island). It is a 40-ha island. It is forested but lacks dipterocarps, the family that provides the most common dominant species in the forests of Malaya, but whose members have poor long-distance dispersal ability. Within a 0.4-ha plot on the island, Wyatt-Smith (1953) recorded 34 species of trees having a trunk diameter greater than 10 cm, compared with figures of 94 and 102 species, respectively, for two Malayan mainland plots. These mainland figures actually exceeded the total spermatophyte (not just tree!) flora noted by Wyatt-Smith in his survey of Jarak Island, and while it would be a surprise if this was found to be a complete inventory of all the species on the island (a notoriously difficult feat to achieve),

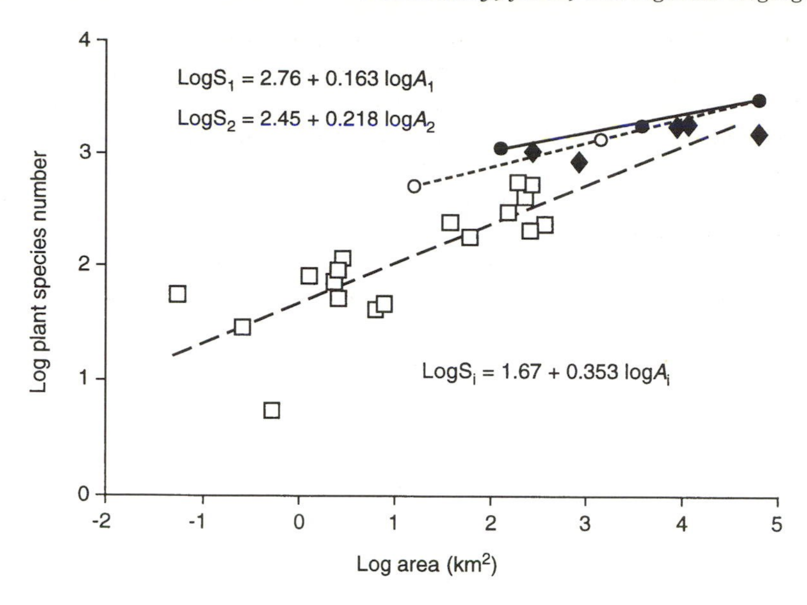

Fig. 3.1. Species–area curves for Californian plants, showing the steeper slope for islands ($\log S_i$, open squares), compared with two alternative nested sets from within the mainland ($\log S_1$, dots, San Francisco area; $\log S_2$, open circles, Marin County). The diamonds represent other areas from the mainland that lie in neither nested set. (Redrawn from Rosenzweig 1995, Fig. 2.9.)

none the less it gives an indication of the extent of impoverishment of a not particularly remote island.

St Helena (15°56' S, 5°42' N) is a remote island of about 122 km^2, and 14.5 million years old, which has a known indigenous flora of just 59 species, a fairly remarkable degree of impoverishment (Cronk 1989). Of the vascular plant species, 40 are endemic. The successful spread of species transported to such islands by people in recent centuries would appear to indicate that such islands are undersaturated with species and could support more in total (D'Antonio and Dudley 1995). However, some turnover is often involved, with endemic species becoming extinct as part of this process. Moreover, as will become clear in Chapter 10, the introductions of new plant species are tied up with all the other human influences—introductions of animals, forest clearance, horticultural and agricultural activities—which may well have caused the extinction of other endemic species on St Helena before any botanical investigations were undertaken (Cronk 1989).

3.4. Disharmony, filters, and regional biogeography

This section deals with the regional biogeographical setting of islands, beginning with some fairly simple patterns, and progressing through increasingly complex scenarios.

Islands tend to have a different balance of species compared to equivalent patches of mainland: they are thus said to be **disharmonic**. There are two aspects to this disharmony (Williamson 1981). First, as Hooker noted, both climate *and biota* of islands tend to be more polar than those of nearby continents (Williamson 1984). Thus, the Canary Islands, located off Africa, have a generally Mediterranean flora, Kerguelen in the Indian Ocean is bleak and Antarctic-like for its latitude, and the Galápagos, although equatorial, are closer to desert (i.e. subtropical) islands, being influenced by upwellings of cool subsurface waters. Secondly, islands are disharmonic in that effectively they sample only from the dispersive portion of the mainland pool. This effect has, of course, to be distinguished from simple impoverishment, i.e. it should not be merely a random subset of a potential mainland pool that is missing. Amongst vertebrates, there is considerable variation in the largest span of ocean crossed without human assistance: for freshwater fishes the limit is about 5 km; for elephants and other large mammals it is about 50 km; tortoises, snakes, and rodents have each made it as far as the Galápagos, a span of about 1100 km; while bats and land birds have reached as far as Hawaii, some 3600 km from the nearest continental land mass (Menard 1986). Present-day distances may not reflect distances at the time of colonization (cf. Myers 1991) but these data provide at least a rough guide. The partial sampling of mainland pools is also exemplified by the flora of Hawaii, which, relative to other tropical floras has few orchids, only a single genus of palms, and lacks altogether gymnosperms and primitive flowering plant families. The largest eight families, Campanulaceae, Asteraceae, Rutaceae, Rubiaceae, Lamiaceae, Gesneriaceae, Poaceae, and Cyperaceae together constitute over half of the native species (Davis *et al.* 1995).

Many other examples could be given. For instance, the Azores and the islands of Tristan da Cuhna possess no land mammals or amphibians. In this case, this disharmony is clearly related to dispersal ability. Tristan da Cuhna also lacks birds of prey: with very few species of land birds, no land mammals, and a small land area, it is doubtful whether a predator population could maintain itself (Williamson 1981). This interpretation suggests that impoverishment among one group may lead through food-chain links to disharmony in another.

The notion of disharmony has, however, been criticised by Berry (e.g. 1992) who has written:

> It is a somewhat imprecise concept indicating that the representation of different taxonomic groups on islands tends to be different to that on the nearest continent, and carrying the implication that there is a proper—or harmonious—composition of any biological community. This is an idea closer to the 'Principle of Plenitude' of mediaeval theology than to modern ecology.
>
> (Berry 1992, pp. 4–5)

Berry has a point: that island environments have 'sampled' distinctive subsets of the mainland pool does not necessarily imply that there is something wrong with island assemblages and that they are further from balance with their environment than those of continents (this is a question worth posing, but it can be distinguished as a separate question). None the less, consideration of remote oceanic islands shows that they are distinctive, and there seems little point in developing a longer phrase to capture the same idea as the term disharmony: it is offered up here with no theological overtones!

Another classic example of the filtering out of particular groups of animals with increased isolation is provided by Fig. 3.2, which shows the distributional limits of different families and subfamilies of birds with distance into the Pacific from New Guinea. Such patterns have sometimes been termed **sweepstake dispersal**. The data in Fig. 3.2 seem to be interpretable broadly in terms of the differing dispersal powers of the groups of birds concerned, illustrating that insular disharmony can be identified not only at the level of orders but also below this at the familial level (cf. Whittaker *et al.* 1997). However, as pointed out by Williamson (1981), within Fig. 3.2 there is a general thinning out of islands and decline in size to the east. This raises the question of whether dispersal difficulties alone have caused the filter effects—a problem that, using presently available data, appears insoluble (but see Keast and Miller 1996).

The New Guinea–Pacific study provides a basically unidirectional pattern, but in other cases a two-way filter appears to have operated (Carlquist 1974). This sort of filter effect can best be seen by examining the composition of certain linear archipelagos. A classic example is shown in Fig. 3.3, which quantifies the decline in reptiles and birds of oriental affinity and the increase in Australian species going from west to east along the Lesser Sunda islands. A similar sort of two-way pattern can also be detected in certain circumstances on mainlands, particularly on peninsulas (e.g. Florida; Brown and Opler 1990), but island archipelagos generally provide the clearest filter effects.

These two-way filter effects evidence a merging of faunas or floras between two ancestral regions. They indicate the biogeographical imprint of extremely distant events in the Earth's history, intimately connected with the

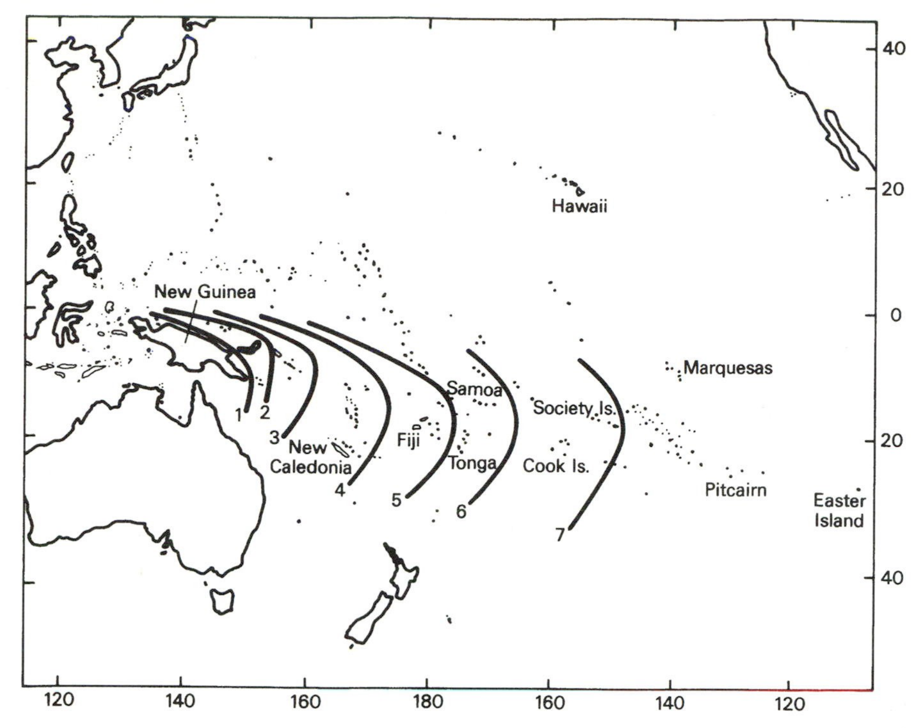

Fig. 3.2. Eastern limits of families and subfamilies of land and freshwater breeding birds found in New Guinea. The decline in taxa is fairly smooth, and shows both differences in dispersal ability and that there is a general decline in island size to the east. Not beyond: 1, New Guinea (14 taxa), pelicans, snakebirds, storks, larks, pipits, logrunners, shrikes, orioles, mudnesters, butcherbirds, birds of paradise, bowerbirds, Australian nuthatches, Australian tree-creepers; 2, New Britain and Bismarck islands (2 taxa), cassowaries, quails, and pheasants; 3, Solomon Islands (10 taxa), owls, frogmouths, crested swifts, bee-eaters, rollers, hornbills, pittas, drongos, sunbirds, flower-peckers; 4, Vanuatu and New Caledonia (7 taxa), grebes, cormorants, ospreys, button-quails, nightjars, wren warblers, crows; 5, Fiji and Niuafo'ou (4 taxa), hawks, falcons, brush turkeys, wood swallows; 6, Tonga and Samoa (7 taxa), ducks, cuckoo-shrikes, thrushes, whistlers, honeyeaters, white-eyes, and waxbills; 7, Cook and Society islands (3 taxa), barn owls, swallows, starlings; 8 (beyond 7), Marquesas and Pitcairn group, herons, rails, pigeons, parrots, cuckoos, swifts, kingfishers, warblers, and flycatchers. Others: owlet-nightjars, one species from New Caledonia, otherwise limit 1; ibises, one species from the Solomon Islands, otherwise limit 1; kagu, endemic family of one species from New Caledonia. (From Williamson 1981, Fig. 2.5. I have not attempted to update this figure in the light of more recent fossil finds.)

plate tectonic processes that have seen the break-up of supercontinents, parts of which have come back into contact with one another many millions of years later. Early students of biogeography, notably Sclater and Wallace, recognized the discontinuities and, on the basis of the distribution patterns of particular taxa, were able to divide the world into a number of major world zoogeographic or phytogeographic regions. One such scheme is shown in Fig. 3.4.

The boundary between Oriental and Australian zoogeographic regions is marked by what has long been known as Wallace's line (Fig. 3.5), which marks a discontinuity in the distribution of mammals, and divides the Sunda islands between Bali and Lombok—not an altogether obvious place to draw the line from the point of view of natural gaps in the present-day configuration of islands. Modifications, and additions in the form of Weber's line, distinguishing a barrier for Australian mammals like Wallace's line for Oriental mammals, have also been proposed (Fig. 3.5). However, the precise placing of lines is problematic, in part because the area between the Asian and Australian continental shelves (sometimes referred to as 'Wallacea') actually contains

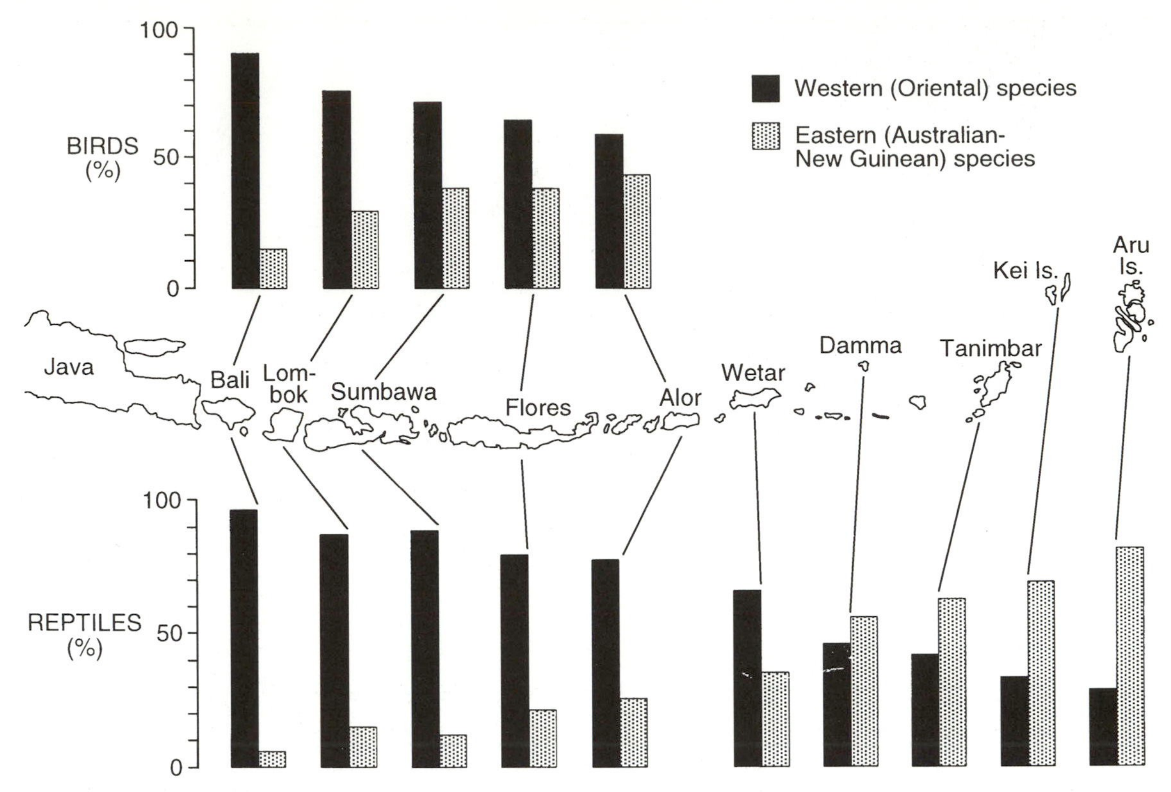

Fig. 3.3. The two-way filter effect, showing trends in the proportion of Oriental and Australian species of birds and reptiles across the Sunda islands, South-East Asia. In contrast to the gradual changes shown here, mammals 'obey' Wallace's line between Bali and Lombok (see Fig. 3.5). (Modified with permission, from Carlquist, S. (1965). *Island life: a natural history of the islands of the world.* Natural History Press, New York, Fig. 3.7.)

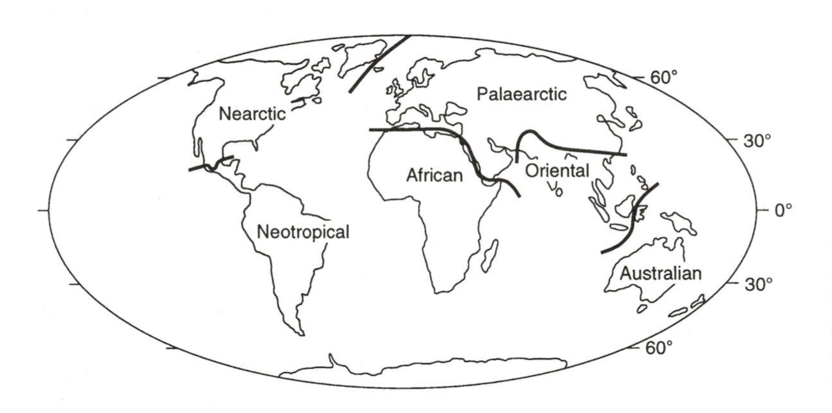

Fig. 3.4. Based on the distribution of extant mammals, A. R. Wallace in 1876 divided the world into zoogeographical regions. (The map shown here has been redrawn from Cox and Moore 1993, Fig. 8.1.)

relatively few Asian or Australian mammals (Cox and Moore 1993), and in part because other groups of more dispersive animals, such as birds, butterflies, and reptiles, show the filter effect (see Fig. 3.3) rather than an abrupt line. Another problem for the drawers of lines is that the insects of New Guinea are mainly of Asian origin. There are several such enigmas of distribution in relation to these lines: for a fuller discussion see Whitmore (1987) and Keast and Miller (1996).

The placing of distant oceanic islands into this sort of traditional biogeographical framework has drawn criticism from Carlquist (1974, p. 61) who has argued that because of the isolation of

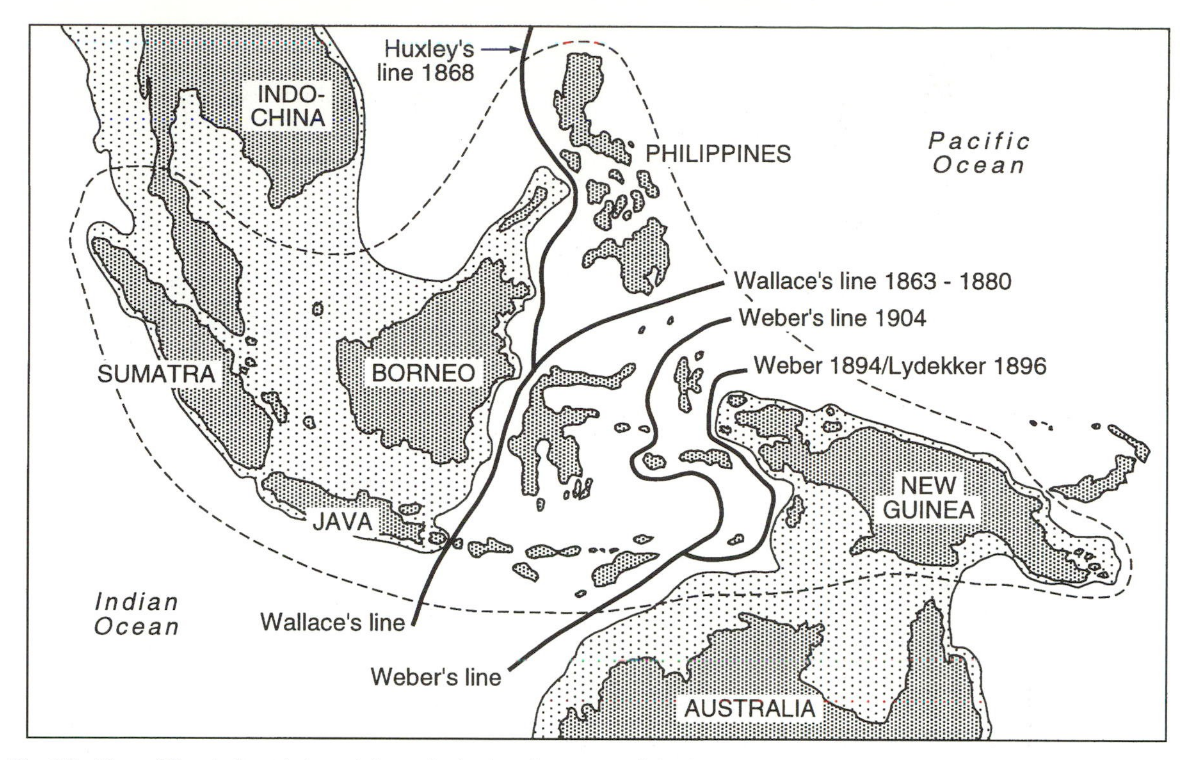

Fig. 3.5. Map of South-East Asia and Australasia showing some of the boundaries proposed by A. R. Wallace and others, dividing the two faunal regions. The continental shelves are shown lightly shaded. In contrast to the sharp discontinuity first noted by Wallace in the distribution of mammals between Bali and Lombok—marked by the original version of Wallace's line—birds and reptiles show a more gradual change between the two regions (see Fig. 3.3). (From various sources.)

islands, and their differing histories from continents, their biotas provide very poor fits with regional biogeographical divisions. Commonly, oceanic islands will have gained significant proportions of their species from more than one land mass and more than one biogeographical province. For instance, Fosberg (1948) calculated that the flowering plants of Hawaii could have descended from some 272 original colonists, of which 40% were believed to be of Indo-Pacific origin; 16–17% were from the south (Austral affinity); 18% were from the American continent; less than 3% were from the north; 12–13% were cosmopolitan or pantropical; and the remainder could not be assigned with confidence. Can the Hawaiian islands logically be put into an Oriental province when appreciable portions of their flora and fauna are American in origin? Most phytogeographers, in fact, assign the flora to its own floristic region (Davis *et al.* 1995).

Renvoize's (1979) analysis of the dominant regional phytogeographical influences within the Indian Ocean (Fig. 3.6) tells a similar story. His paper illustrates that many island biotas are not the product of simple one- or two-way filter effects, nor do they enjoy full affinity to a particular continental region, but instead they consist of a mix of species of varying sources and origins. Christmas Island, close to Indonesia, is dominated by colonists from South-East Asia, whereas the Farquhar Group has an African component, a reflection of the location of these low-lying islands north-east of the northern tip of Madagascar. Renvoize's (1979) analysis of Christmas Island was based on a list of 145 indigenous vascular plants from the end of the nineteenth century. The breakdown of this flora was: 31 pantropical, 21 palaeotropical, 76 tropical Asian (Madagascar to Polynesia and Australia), and 17 endemic species. This contrasts with the indigenous flora of the Farquhar

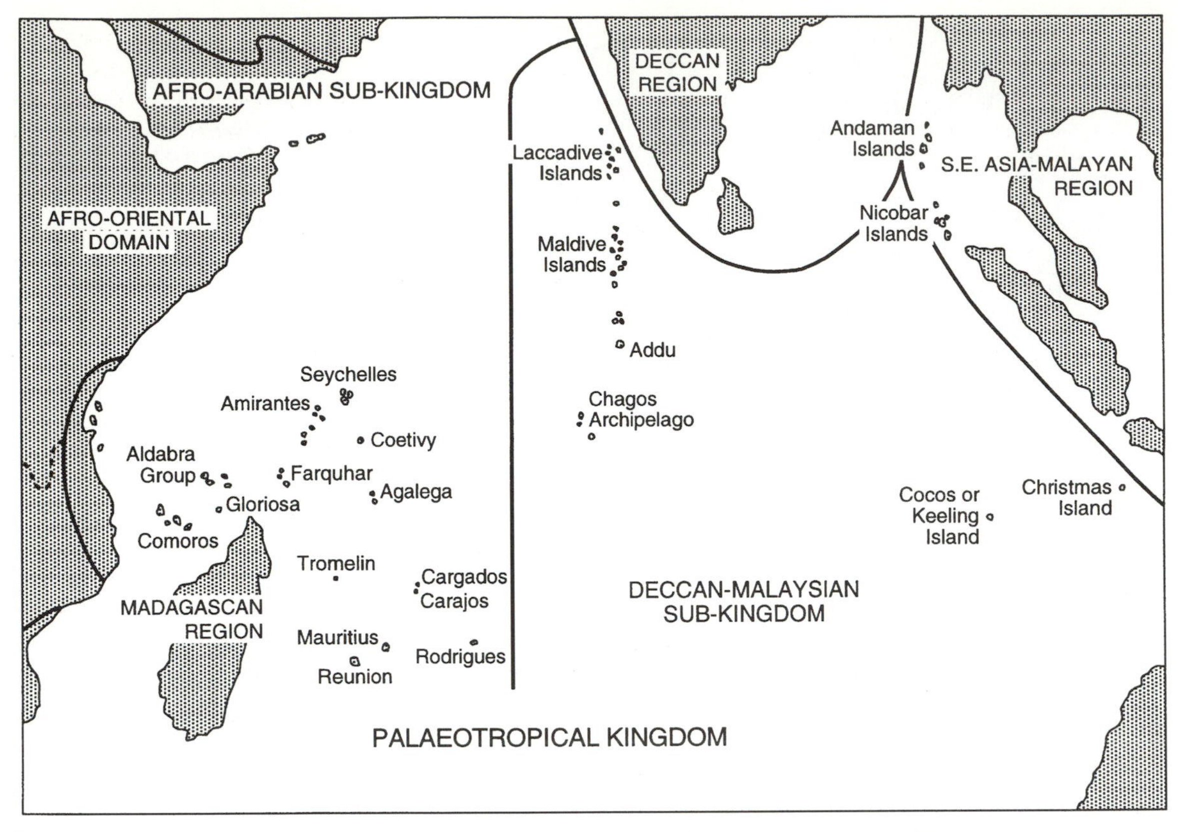

Fig. 3.6. Phytogeographical classification of the Indian Ocean islands. (Redrawn from Renvoize 1979, Fig. 1.)

Group, which consisted of 23 pantropical, 18 Indo-Pacific, and 5 African species—which seems to be a fairly typical sort of mix for the low islands of the Indian Ocean. Leaving aside introduced species, the pantropical species constitute a far greater proportion of the flora of the Farquhar Group than of Christmas Island (respectively 50% and 21%). Although several factors may be involved in this comparison, the most obvious differences are location and in the range of habitats available on the high as opposed to low-lying islands (Renvoize 1979). Christmas Island is an elevated coral-capped island, having a plateau at 150–250 m above sea level (ASL), a peak of 361 m, and a land area of 135 km^2. On low islands there are few areas beyond the direct influence of the sea. Strandline species thus dominate the floras and, since they are nearly all sea-dispersed and since the most constant agent of plant dispersal is the sea, the coastal habitats receive a steady flow of possible colonizers and a steady gene flow, not conducive to the evolution of new forms. High islands do have coastal habitats but they and their characteristically widespread sea-dispersed species do not dominate the floras as they do on the low islands. Those oceanic islands with large proportions of endemic plants are typically high islands with varied habitats (Humphries 1979).

Renvoize (1979) declined to attribute the low islands of the Indian Ocean below the level of palaeotropical kingdom because they contain too high a proportion of very widespread species. This restrained approach to demarcation is realistic, and although the categorization of islands into particular biogeographical provinces or regions is clearly problematic, such phytogeographical and zoogeographical analyses have some value. They establish the biogeographical context within which the compositional structure of a particular island or archipelago needs to be placed (cf. van Balgooy *et al.* 1996).

Distance from a larger source pool can be a poor indicator of biogeographical affinity, as introduced in section 2.4.4. First, there may be historical factors, as in the cases of Madagascar and the granitic islands of the Seychelles, which are ancient continental fragments from Gondwanaland and so began their existence as islands millions of years ago, complete with a full complement of species (Fig. 3.7). In the case of Madagascar the split from Africa may have occurred as long ago as 165 million years (Davis *et al.* 1994). Secondly, ocean or wind currents (or even migration routes) may have determined colonization biases, such as apparently in the case of the butterflies of the Antillean island of Mona (section 2.4.4). Another classic example is that of the flora of Gough Island (Moore 1979). The predominant wind circulation system around the South Pole provides a plausible explanation for the affinities with New Zealand and South America (Fig. 3.8). Changes in relative sea level, land-mass positions, and in oceanic and atmospheric circulation over geological time (Chapter 2) can mean that such analyses are difficult to bring to a definite conclusion.

Fig. 3.7. Curieuse Island, one of the granitic Seychelles—ancient continental islands which contain some poorly dispersing groups of endemics, including a frog, indicating an ancient biogeographic legacy. The Seychelles are rich in endemics. The palm featured here, *Phoenicophorium borsigianum*, is one of six endemic palm species. (Photo: Clive Hambler 1990.)

Historical biogeography has lately been rather polarized into two supposedly opposing camps, the **dispersalist** and **vicariance biogeographers**, each concerned with how disjunct distributions arise. In essence, dispersal hypotheses have it that descendent forms are the product of chance, long-distance dispersal across a pre-existing barrier, whereas vicariance biogeographers envisage species ranges being split up by physical barriers, often followed by speciation in the now separate populations (Myers and Giller 1988). However, according to available data concerning island origins and environmental histories, this opposition can be seen to be in some senses false: both types of process operate, and both contribute to the complexities of biogeographical patterning (Stace 1989; Keast and Miller 1996).

This point can be exemplified with reference to the geological history of the Caribbean, introduced in section 2.3.4. Buskirk (1985) set out to employ this framework in interpreting the phylogenetic relationships of modern beetles, reptiles, amphibians, and other terrestrial animals in the Greater Antilles. The evidence indicates that the Antilles have been islands throughout most of the Cainozoic and that therefore most of their faunas were derived from over-water colonists and not from the fragmentation of original populations on a large land mass. However, Buskirk suggests that tectonic events have affected dispersal and colonization distances. The shear along the northern Caribbean plate boundary has moved Jamaica further away from Central America and thus modern Jamaican species that are relicts or

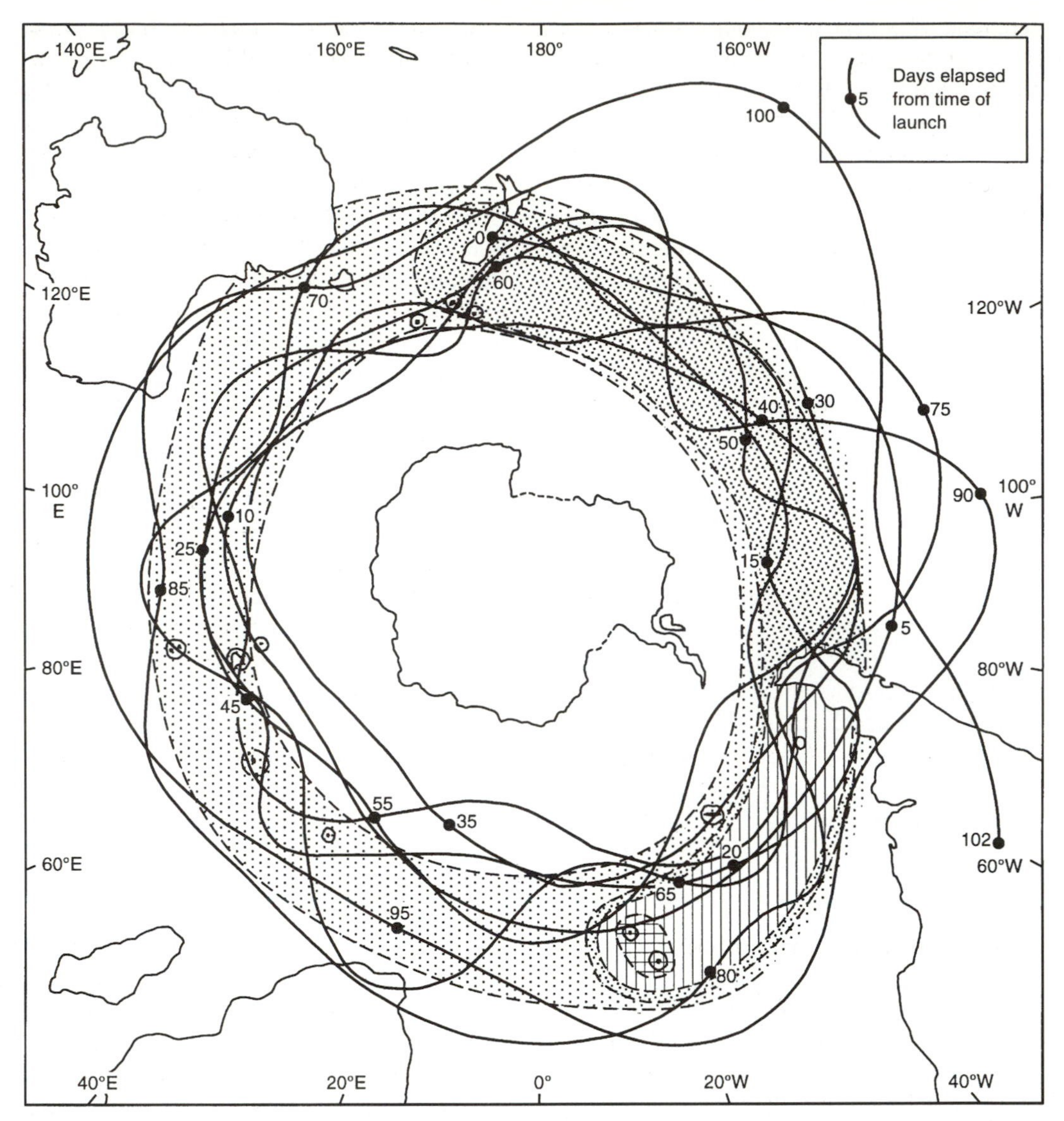

Fig. 3.8. Floristic affinities of Gough Island (Tristan da Cunha) in relation to predominant wind direction as revealed by the flight of a balloon: areas sharing 20–25 species, cross hatch; 15–20 species, single hatching; 10–15 species, heavy stipple; 5–10 species, light stipple. (After Moore 1979, Figs 1 and 2.)

radiations from relatively early colonists have more Central American affinities than do the endemic species of neighbouring Antillean islands. Jamaica was largely submerged in the early to mid-Tertiary, and its modern endemic fauna lacks the groups that invaded the other Antilles at about that time. As well as the change in sea level, the accompanying climatic changes have been important, and in Jamaica have apparently selected against members of some animal groups which required more mesic habitats (Buskirk 1985). Thus, Buskirk shows that to explain the uniqueness of Jamaica's fauna, it is necessary to refer to tectonic events (horizontal movement and extensive late Tertiary uplift), combined with Pleistocene cycles of climate and sea-level change. Both dispersal and vicariance processes are involved in this history.

It is probably true to say that over the past few million years, vertical changes of land and sea have been of more general significance to Caribbean biogeography than horizontal land movements. Another classic example is provided by Williams' (1972) studies of the *Anolis* genus of lizards. *Anolis* are small green or brown lizards which are more or less the only diurnal arboreal lizards found in most of the region. Much of the distributional pattern within this group can be explained most parsimoniously in relation to the submarine banks on which today's islands stand. Each bank has many islands on it, but as recently as about 7000 BP, as a generalization, many of these islands were joined, as they would have been for lengthy periods during the Pleistocene. The present-day distributions of *Anolis* thus owe much to events during the low sea-level stands (Williams 1972).

The unravelling of such movements and a determination of whether a particular species could reach newly suitable territory, such as the British Isles, before the sea rose to isolate it is problematic, as much depends on assumptions concerning the powers of spread of particular species. Many questions in island biogeography come back to the issue of dispersal, as it is inherently extremely difficult to determine the effective long-distance range of propagules (potentially viable units) (but see Hughes *et al.* 1994; Whittaker *et al.* 1997). We will return to this theme again, suffice to note for the moment that although dispersal limitations may provide parsimonious interpretations of data such as the *Anolis* distributions, it is unwise to base interpretations on fixed notions of dispersal limits. This point has been most entertainingly driven home by Johnson's (1980) documentation, complete with photographs, of the ability of elephants to swim across ocean gaps, using the trunk as a sort of snorkel. It had been assumed in some studies that this behaviour was either beyond the ability of elephants, or would not have been indulged in even were they capable of it. Thus, the existence of elephants on an island was formerly taken to be proof of a former land connection. Once they have arrived on an island, the dispersal capabilities of elephants, or indeed of other animals and plants, may subsequently alter (section 5.3.1; Schüle 1993).

3.5. Endemicity

3.5.1. Neo- and palaeo-endemicity

A fair number of island endemics belong to groups that formerly had a more extensive, continental distribution. But, since colonizing the island in the remote past (either over land or sea), they have been replaced in the rest of their range for one reason or another—perhaps outcompeted by newly arisen forms. The implication is that not all species now endemic to an island or island group have actually evolved *in situ*. Cronk (1992) refers to such relict or stranded forms as **palaeo-endemics**, whereas species that have evolved *in situ* are **neo-endemics**. As with all such distinctions, this is an over-simplification, as it implies a lack of change in either island or mainland forms, respectively, which may not strictly be the case. It should be noted that the significance of relict forms or relictual ground plans as against *in situ* evolution is a topic that generates considerable heat, and indeed some of the more keenly contested issues in biogeography (cf. Heads 1990; Pole 1994; Carlquist 1995). Cronk (1992) argues that ancient, taxonomically isolated, relict species should hold a special importance in conservation terms, in that their extinction would cause a greater loss of unique gene sequences and morphological diversity than the extinction of a species with close relations.

There is no doubt as to the importance of endemic forms on particular islands, although as taxonomic work continues, estimates of the degree of endemicity for an island group vary. Figure 3.9 demonstrates this for the flora of the Galápagos. Since fairly complete inventories became available, the number of known endemics has changed relatively little: the refinements come from improved knowledge of mainland source areas and as a result of taxonomic revisions. Although the proportion of

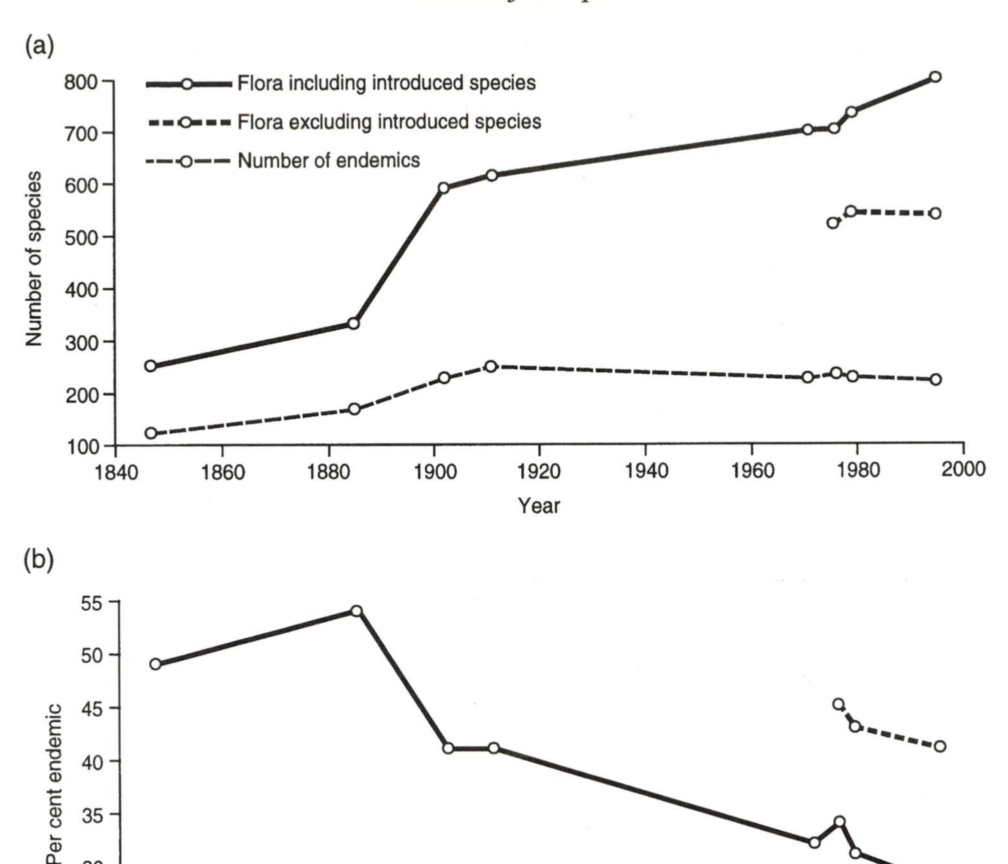

Fig. 3.9. Changing estimates of the Galápagos flora through time: (a) the size of the flora; and (b) the proportion of endemics. (Data from Porter 1979, and from Davis *et al.* 1995).

endemics appears to have declined (Fig. 3.9b), this is mostly the result of increases in numbers of introduced species; if these were these removed from the count, the proportion of endemics in the flora would actually have remained reasonably constant. The basic pattern established by Hooker and Darwin (Ridley 1994) has thus proved to be robust. As a proportion of the indigenous flora of the Galápagos today, the endemics represent 43% of the indigenous taxa (species, subspecies, or varieties); allowing for recent extinctions and uncertainties in taxonomy, the figure could fall to 37% (Porter 1979). Recently published estimates for Hawaii are that the following proportions of native taxa are endemic: birds, 81%; angiosperms, 91%; molluscs, 99%; and insects, 99% (Sohmer and Gustafson 1993). The following sections provide a few estimates which put such figures into the broader context of regional or global diversity.

3.5.2. Endemic plants

Groombridge (1992) provides the following 'best estimates' for the number of higher plant species in the world: 12 000 pteridophytes (ferns and their allies); 766 gymnosperms (conifers, cycads, and their allies); and 250 000 angiosperms (flowering plants). A grand total of about 263 000 species. It will be recalled that if Papua New Guinea is taken as the world's largest, islands constitute about 3% of the land surface area of the world. The small selection of 13 islands and island archipelagos in Table 3.1 provides a minimum estimate of about 36 500 species endemic to those same islands, amounting to some 13.8% of the world's higher plant species. It has been calculated that the Pacific Ocean islands contain about 22 000 vascular plant species, nearly half of which may be introduced. Of the 11 000–12 000 native species, about 7000 are considered to be endemic, most being found within a single island or island archipelago (Davis *et al.* 1995). All such figures come with large error margins. Indeed, estimates of the number of flowering plant species in the world have varied between 240 000 and 750 000 (Groombridge 1992, p. 65). However, as island species are as much a part of these uncertainties as are continental species (see data for New Guinea and Borneo, Table 3.1), their relative importance is unlikely to be seriously diminished by further refinements of the estimates. It is a reasonable estimate that about one in six of the Earth's plant species grows on oceanic islands, while one in three of all known threatened plants are island forms (Groombridge 1992, p. 244). Clearly, in terms of plant biodiversity, the islands of the world make a disproportionate contribution for their land area and are also suffering further disproportionate pressure in terms of the maintenance of that biodiversity.

The percentage of endemics varies greatly among islands. The highest proportions are often associated with ancient continental islands, such as Madagascar and New Zealand. Large numbers of endemics are also associated with the larger, higher oceanic islands in tropical and warm-temperate latitudes. Hawaii, to provide a classic example, has about 1200 native vascular plants, and of the angiosperms (flowering plants) about 91% are endemic. The conservative estimate for their numbers is 850 endemics, although the figure could conceivably be as high as 1000 species (Groombridge 1992; Davis *et al.* 1995). Some lineages have radiated spectacularly. The silversword alliance comprises

Table 3.1

Higher plant species richness and endemism of selected islands (from Groombridge 1992; Davis et al. *1995.) These figures should be treated cautiously as different values for some islands can be found, in cases even within a single publication (see Groombridge 1992, Tables 8.3 and 14.1)*

Island or archipelago	Species number	Endemics	% Endemic
Borneo	20 000–25 000	6000–7500	30
New Guinea	15 000–20 000	10 500–16 000	70–80
Madagascar	8000–10 000	5000–8000	68.4
Cuba	6514	3229	49.6
Japan	5372	2000	37.2
Jamaica	3308	906	27.4
New Caledonia	3094	2480	80.2
New Zealand	2371	1942	81.9
Seychelles	1640	250	15.2
Fiji	1628	812	49.9
Mauritius, including Réunion	878	329	37.5
Cook Islands	284	3	1.1
St Helena	74	59	79.7

28 endemic species, apparently of monophyletic origin (i.e. a single common ancestor), but placed in three genera (*Argyroxiphium*, *Dubautia*, and *Wilkesia*) (Wagner and Funk 1995). The genus *Cyrtandra* (Gesneriaceae) is represented by over 100 species, and although they may not be monophyletic, their radiation is no less impressive (Otte 1989; Wagner and Funk 1995). Smaller oceanic islands, even in the tropics (e.g. the Cook Islands), have much smaller floras and typically lower proportions of endemics. This is due in part to the reduced variety of habitats and to the wide mixing of the typically sea-dispersed strand-line species that dominate their floras. Another feature of the data which must be borne in mind when examining statistics for island endemicity is that in some areas there are numerous species which are endemic to islands, but which are shared between different geographical entities. Thus, in the Lesser Antilles (the Leeward and Windward Islands) the shared or **regional island endemics** form a considerable proportion of the endemic flora as a whole.

3.5.3. Endemic animals

The following constitutes a selection of some of the animal groups that are sufficiently well known to be placed into either a regional or global context with some degree of confidence.

Land snails There may be as many as 30 000–35 000 land snail species in the world (Groombridge 1992), of which a substantial proportion occur on islands (Wallace 1902). Some of the better known of island snail faunas are quantified in Table 3.2. Although only eight archipelagos are listed in the table, they account for approximately 7.7–9.0% of the world's land snails by present estimates (some regions of the world remain poorly known and the current picture could change) (Groombridge 1992). Studies in the Pacific suggest that, once again, it is the larger, higher oceanic islands (15–40 km^2, >400 m ASL) that typically have both most species and most endemics, whereas low-lying atolls have neither high richness nor high degrees of endemism. On most islands with high snail diversity, the snails appear to be concentrated in the interiors, especially mountainous regions with 'primary' forest cover. For those islands with good data, a marked positive correlation exists between numbers of endemic plant species and endemic molluscs, but not between molluscs and birds (Groombridge 1992).

Insects No attempt will be made to cover the insects in a systematic fashion, but the following statistics from Hawaii give an idea. In the family Drosophilidae (fruit flies), 511 species are currently named and described for Hawaii, with another 250–300 awaiting description (Wagner and Funk 1995). Given the pattern of discovery, it is estimated that there may be as many as 1000 endemic species. The radiation of tree crickets on the archipelago has resulted in three

Table 3.2

Land snail species richness and endemism for selected islands for which the data are complete enough for the proportion of endemics to be estimated (from Groombridge 1992, Table 14.3)

Island or archipelago	Species number	Endemics	% Endemic
Hawaiian Islands	*c.*1000	*c.*1000	99.9
Japan	492	487	99
Madagascar	380	361	95
New Caledonia	300	*c.*299	99
Madeira	237	171	88
Canary Islands	181	141	77.9
Mascarene Islands	145	127	87.6
Rapa	>105	>105	100?

endemic genera and some 68 species. The orthotyline plant bug genus *Sarona* is another endemic Hawaiian genus, with 40 known species. Excepting the biogeographers' hot-spot archipelago, few groups of insects are sufficiently well known for broad biogeographic patterns to be established. The Lepidoptera provide an exception.

The tropical Pacific Ocean butterfly fauna consists of 285 species, of which 157 (55%) occur on continental land masses, i.e. they have not speciated after arrival on the islands. One hundred are endemic to a single island/archipelago and 28 are regional endemics, i.e. they are found on more than one archipelago but not on the mainlands (Adler and Dudley 1994). Interestingly, butterfly speciation in the Pacific archipelagos is primarily a result of limited inter-archipelago speciation. Intra-archipelago speciation appears important only in the Bismarcks and Solomons, which contain, respectively, 36 and 35 species endemic at the archipelago level. Of the other 24 archipelagos in the survey, only New Caledonia makes double figures, with 11 endemic species, and the others contribute just 18 species between them. The Bismarcks and Solomons are the two largest archipelagos in land area and the closest to continental source areas. Butterflies are generally specialized herbivores, their larvae feeding only on a narrow range, or even a single species, of plant. On the smaller and more remote archipelagos, the related host plants simply may not be available for the evolution of new plant associations, thereby impeding the formation of new butterfly species (Adler and Dudley 1994).

Lizards In their survey of the 27 oceanic archipelagos and isolated islands in the tropical Pacific Ocean for which detailed herpetological data exist, Adler *et al.* (1995) record 100 species of skinks in 23 genera, of which 66 species were endemic to a single island/archipelago and a further 13 were regional island endemics. Of the 23 genera, nine are endemic to the islands of the tropical Pacific. A few species are widespread; for example *Emoia cyanura*, which occurs on 24 of 27 archipelagos in the survey. The Bismarcks, Solomons, and New Caledonia appear particularly rich in skinks, but most archipelagos contain fewer than 10 species. Hawaii has only three species, none of which is endemic, in stark contrast to the birds, insects, and plants.

Another important group of island lizards are the Caribbean anoles, which are typically small, arboreal insectivores. This is one of the largest and better studied vertebrate genera: about 300 species of *Anolis* have been described, half of which occur on Caribbean islands (Losos 1994).

Birds The examples of adaptive radiation best known to every student of biology from their high school days are probably the Galápagos finches and the Hawaiian honeycreepers. Therefore it will come as no surprise that islands are important for bird biodiversity. In excess of 1750 species are confined to islands, representing about 17% of the world's species; of these 402, or 23%, are classified as threatened, considerably in excess of the 11% of birds worldwide (Johnson and Stattersfield 1990; Groombridge 1992, p. 245).

Adler (1994) has examined the pattern of bird species diversity and endemism for 14 tropical Indian Ocean archipelagos. Probably the most famous island endemics of all, the dodos of Mauritius, Rodriguez, and (possibly) Réunion, although extinct (section 10.4.2), live on as members of this data set, which includes species known only from subfossils. Using stepwise linear regression it was found that numbers of each of the following were related positively to the number of large islands and total land area: total species number; number of continental species; and number of regional endemics. In addition, less remote and low-lying islands tended to have more continental species (of the low islands, only Aldabra had its own endemics), and higher islands tended to have more local endemics and proportionately fewer continental species. Similar results were found by Adler (1992) for the birds of the tropical Pacific, but the Indian Ocean avifauna contained fewer species, had lower proportions of endemics, and had several families that were much more poorly represented, despite being diverse in the mainland source areas. The generally small size of the Indian Ocean islands may be significant in these differences. The total land area is only 7767 km^2, compared to 165 975 km^2 for the Pacific study, and there are also fewer island archipelagos in the Indian Ocean and they tend to be less isolated. The poor representation of certain families, notably the ducks and kingfishers, might be due to a shortage of suitable habitats, such as freshwater streams and lakes. Intra-archipelago speciation has been rare in the Indian Ocean avifauna, possibly occurring once or twice in the Comoros and a few times in the Mascarenes. Adler (1994) suggests that there may not be sufficient numbers of large islands to have promoted intra-archipelago speciation in the Indian Ocean. In contrast, it has been common within several Pacific archipelagos, accounting for most of the endemic avifauna of Hawaii, being important in the Bismarcks and Solomons, and occurring at least once in New Caledonia, Fiji, the Societies, Marquesas, Cooks, Tuamotus, and Carolines. In conclusion, Adler (1994) echoes Williamson's (1981) remark that in bird speciation on islands 'geography is all important'.

Bats Native mammals are not a feature of isolated oceanic islands, with the exception of the bats. While other types of bat occur on islands, flying foxes are arguably the most important group. There are believed to be 161–174 species of flying fox (Chiroptera: Pteropodidae) and they are distributed throughout the Old World Tropics (but not the New World). In number of species, range, and population sizes, flying foxes have been the most successful group of native mammals to colonize the islands of the Pacific. The Pacific land region consists of about 25 000 islands, most of them very small. In addition to occupying small islands, most flying foxes have restricted distributions; 38 of the 55 island species occupy land areas of less than 50 000 km^2 and 22 live on less than 10 000 km^2, while 35 occur on a single island or group of small islands (Wilson and Graham 1992).

Comparisons between taxa at the regional scale Within the figures cited above, those from G. H. Adler and his colleagues provide a common methodological approach and enable comparisons of patterns of endemism of different taxa at the regional scale (Table 3.3). The studies are based on 26–27 Pacific Ocean islands or island archipelagos, plus in one case 14 Indian Ocean islands/archipelagos. The data include all known extant species plus others known only from subfossils (i.e. recently extinct species).

Both birds and butterflies are capable of active dispersal over long distances and have colonized virtually every archipelago and major island within the tropical Pacific. Birds have a higher frequency of endemism than butterflies. Adler and Dudley (1994) consider that birds have a superior dispersal ability, and that this might be anticipated to have maintained higher

Table 3.3

Degree of endemism among selected ocean basin faunas (from Adler 1992, 1994; Adler and Dudley 1994; Adler et al. *1995)*

Group	Total number	Continental	Regional endemics	Local endemics
Pacific Ocean butterflies	285	157 (55%)	28 (10%)	100 (35%)
Pacific Ocean lizards	100	21 (21%)	13 (13%)	66 (66%)
Pacific Ocean birds	592	145 (25%)	59 (10%)	388 (65%)
Indian Ocean birds	139	60 (43%)	10 (7%)	69 (50%)

Continental, species also occurring on continental land masses; regional endemics, species occurring on more than one archipelago within the region but not on continents or elsewhere; local endemics, species restricted to a single archipelago or island.

rates of gene flow than in butterflies, contrary to the higher degree of endemism. One plausible explanation is that the lower rate of endemism in butterflies might be a consequence of the constraints of the co-evolutionary ties with particular host plants required by butterfly larvae. This may limit their potential for rapid evolutionary change on islands (see above). As an aside, the capacity of lepidopterans to reach moderately remote oceanic islands should not be too lightly dismissed, as evidenced by Holloway's (1996) long-term light-trap study on Norfolk Island, which recorded 38 species of non-resident Macrolepidoptera over a 12-year period. Norfolk Island is 676 km from New Caledonia, 772 km from New Zealand, and 1368 km from the source of most of the migrants, Australia. Most of the arrivals appear to be correlated with favourable synoptic situations, such as the passage of frontal systems over the region.

Lizards are also widely distributed in the tropical Pacific. The proportions of skinks in the three categories of endemicity are remarkably similar to the equivalent figures for Pacific birds (Table 3.3). In general, in each of skinks, geckos, birds, and butterflies, the proportion of species endemic to an archipelago is best explained by reference to the number of large (high) islands and to total land area. However, scrutiny of the data on an archipelago by archipelago basis reveals greater differences. For instance, 100% of New Caledonia's skinks are endemic to the Pacific Ocean islands, and as many as 93% are endemic to the New Caledonian islands themselves. The equivalent figures for birds (including subfossils) are 47% and 33%, respectively (Adler *et al.* 1995). These data contrast with those for Hawaii, on which most birds are endemic, but where there are no endemic lizards. Adler *et al.* (1995) suggest that these differences are explicable in relation to the differing dispersal abilities of lizards and birds. Birds, being better dispersers, reached Hawaii relatively early and have radiated spectacularly, whereas the three species of skink may only have colonized fairly recently and possibly with human assistance. New Caledonia, in contrast, may not be sufficiently isolated to have allowed such a degree of avifaunal endemism to have developed.

Numbers of birds and butterflies are remarkably similar on the less remote Pacific archipelagos east of New Guinea, but on the more remote archipelagos the numbers of bird species are far greater than those of butterflies (Adler *et al.* 1995). This is reflected statistically by archipelago area being the most important geographical variable in explaining bird species richness, whereas for butterflies isolation is the more important variable (supporting the point made by Adler and Dudley (1994) about relative dispersal abilities of the two groups). Another pattern noted by Adler *et al.* (1995) is that in Pacific butterflies, Pacific birds, and Antillean butterflies the number of endemic species increases more steeply with island area than does the number of more widely distributed species. This observation supports the idea that a larger island area provides the persistence and variety of habitats most conducive to radiation on isolated islands.

To summarize, while different taxa have radiated to different degrees on particular islands or archipelagos, it appears to be the case that the greatest degree of endemicity is found towards the extremity of the dispersal range of each taxa. Islands that are large, high, and remote typically have the highest proportion of endemics. In total, islands account for significant slices of the global biodiversity cake.

3.6. Extinct island endemics: a cautionary note

The foregoing account has taken little note of the 'state of health' of island endemics, and assessments have included many highly endangered species and others already believed to be extinct. Before moving on from the geography of endemism to an examination of the theories that may account for the patterns of island evolution, it is important to consider the extent of the losses already suffered. Otherwise, there may be a danger of misinterpreting evolutionary patterns which are just the more resistant fragments of formerly rather different tapestries (Pregill and Olsen 1981; Pregill 1986).

Schüle (1993) summarizes recent studies of vertebrate endemism for the Mediterranean islands, noting that although the small-animal faunas have shown good persistence, the larger terrestrial vertebrates of the early part of the Quaternary appear to have included endemic ungulates, carnivores, giant rodents, tortoises, and flightless swans, not one of which has survived to the present day. He regards it as plausible that some of the losses could be attributed to the arrival of new species driving older endemic species to extinction by competition. Some of this flux may well be attributable to climatic change, sea-level change, and other environmental forcing factors that exposed the existing island assemblages to new colonists. But from the middle Pleistocene onwards, the extinctions—which included species of pigmy elephants, hippos and cervids, flightless swans, tortoises, and rodents—appear to be explained most parsimoniously by invasions of *Homo* sp. The role of humans in bringing about extinctions of island endemics is of sufficient importance that it warrants a separate chapter (Chapter 10) and for the moment these examples can serve to place a health warning on attempts to explain current suites of endemic species on islands. Even in well-studied groups such as birds, new species have been described in very recent times. There have been new discoveries of extant species, e.g. the flightless rail, *Gallirallus rovianae*, endemic to the Solomon Islands and first collected in 1977 (Diamond 1991*a*). But, many are known only from subfossils. For example, estimates of the numbers of Hawaiian birds which were lost following Polynesian colonization 1500 years ago, but before European contact, now stands at a minimum of 40 species (Olson and James 1982). Each of the 16 Polynesian islands that have yielded in excess of 300 prehistoric bird bones approaches or exceeds 20 extirpated species (i.e. lost from those particular islands, though not necessarily globally extinct), pointing to the likelihood of far more extinct species yet to be catalogued. About 85% of bird extinctions during historical times have occurred on islands (Steadman 1997*a*), a rate of loss 40 times greater than for continental species (Johnson and Stattersfield 1990).

The cataloguing of insular losses is clearly far from complete and while we have some basis for estimating how many species of birds may be involved, for most taxa such calculations remain to be made (Milberg and Tyrberg 1993). Morgan and Woods (1986) appraise the problem for West Indian mammals. They calculate an extinction rate for the period since human colonization of one species every 122 years, and they emphasize that this is a minimum estimate. The mammals of the West Indian islands today are thus a highly impoverished subset of species which escaped the general decimation of the fauna. Morgan and Woods caution that biogeographical analyses that fail to incorporate the extinct forms are unlikely to prove reliable. It is unsurprising, but significant biogeographically, that in addition to species disappearing altogether, a number of species have gone extinct from particular islands during the past 20 000 years, while still surviving on others. This is particularly evident in the bats, which supply the majority of persisting species of mammal, but is also a pattern found among lizards and birds. Many such losses occurred due to climatic changes and the associated sea-level changes and habitat alterations at the end of the Pleistocene, and could be considered a 'natural' part of the biogeographical patterning of the region; others were doubtless due to human activities, including habitat alteration and the introduction of exotics. Interpretation of the present-day biogeography of the region thus requires knowledge not only of past faunas but also of the extent to which humans have been involved in their alteration.

The difficulty of this task is illustrated by a study from Henderson Island. In his investigation of 42 213 bird bones, 31% of which were identifiable, Wragg (1995) found that 12% of the fossil bird species recorded were accounted for by just 0.05% of the total number of identifiable bones, indicating them to be of uncertain status, quite possibly vagrants. Consequently, biogeographical studies that rely simply on a list of fossil birds might, exceptionally, assign resident status to temporary inhabitants. A different form of bias probably occurs in other studies because of the large mesh sizes (6 or 13 mm) commonly used in sieving soil samples. Larger-boned species such as seabirds, pigeons, and rails are usually found, but small passerines and hummingbirds are much less likely to be recovered (Milberg and Tyrberg 1993).

As yet, we have only a fragmentary picture of past losses, biased towards groups for which the fossil record has been revealing, notably vertebrates and especially birds. Much is still unknown. What, for instance, has been the impact of the losses from New Zealand of the giant moas on the dynamics of forests they formerly browsed? How might the loss of hundreds of populations and a few entire species of Pacific seabirds have influenced marine food webs, in which seabirds are top consumers? (Steadman 1997*a*). Many island fruit bats and land birds (including many now extinct) have undoubtedly had crucial roles as pollinators and dispersers of plants, and thus their extirpation must be anticipated to have important repercussions for other taxa, many of which may be ongoing. That is, unless there is a greater degree of so-called functional redundancy within oceanic island biotic assemblages than most authors currently recognize (Rainey *et al.* 1995; Vitousek *et al.* 1995).

It has been observed that the rapidity of human-induced extinctions of island birds has been influenced by the nature of the island. The species of high, rugged islands, and/or those with small or impermanent human populations, have probably fared better than those of low relief and dense, permanent settlement. This is exemplified by the finding that a larger percentage of the indigenous avifauna survives on large Melanesian islands than on small Melanesian islands, or on Polynesian or Micronesian islands. This may be due to the buffering effects that steep terrain, cold and wet montane climates, and human diseases have had on human impact (Steadman 1997*a*). Steadman (1997*a*) has pointed out that the Kingdom of Tonga did not qualify in a recent attempt to identify the Endemic Bird Areas of the world because of its depauperate modern avifauna, yet bones from just one of the islands of Tonga, 'Eua, indicate that at least 27 species of land birds lived there prior to human arrival around 3000 BP. Forest frugivores/granivores have declined from 12 to four, insectivores from six to three, nectarivores from four to one, omnivores from three to none, and predators from two to one. As is consistent with predation by humans, rats, dogs, and pigs, the losses have been more complete for ground-dwelling species. It is highly likely that the means of pollination and seed dispersal, not just of the plants of 'Eua but of islands across the Pacific, have over the past few thousand years been diminished greatly as a consequence of this pattern of loss, replicated, as it has been, across the Pacific. That the attrition of endemic species from oceanic islands may have so altered distributional patterns, compositional structure, and ecological processes serves as a caution on our attempts in the following chapters to generalize biogeographic and evolutionary patterns across islands.

3.7. Summary

This chapter set out to establish the peculiarities of island biotas, particularly remote island biotas, and to provide an indication of their broader biogeographical significance. The distinction between oceanic and continental (shelf) islands, which can be made on geological grounds, provides a first-order division of marine islands on biogeographical grounds. In each case, some islands are hard to assign to one side of this binary divide. Biogeographically, much hinges on the dispersal capabilities of the taxa under consideration, as an island may be a remote isolate (oceanic in character) for a taxon with poor ocean-going powers, but be linked by fairly frequent population flows (continental in character) to a source area for a highly dispersive taxon.

Islands are typically species poor for their area in comparison to areas of mainland, and this poverty is accentuated by increasing isolation and decreasing island relief and altitude. Remote islands are typically disharmonic in relation to source areas. This can take the form of a climatic difference and thus a sort of biome shift on the island, but more commonly it is taken to mean that they sample only a biased subset of mainland taxa. Such biases may take the form of the absence of mammals, or the relative lack of representation within a particular family of plants. Such ideas need to be phrased more precisely than they often have been, to allow for critical appraisal. Traditional zoogeographic and phytogeographic analyses have enabled biogeographers to identify unidirectional dispersal filters (so called sweepstake dispersal routes), filter effects between different biogeographic regions, and cases where islands have been subject to the influence of multiple source areas. Such analyses set the scene for island evolutionary change, providing the 'relictual ground plan' from which island endemics develop. In some cases, island forms may have changed less than the mainland lineages from which they sprung, in which case they may be termed palaeo-endemics. However, where novelty has arisen predominantly in the island context, they are called neo-endemics, although once more it should be recognized that this is to divide a continuum.

In global terms, for a variety of taxa, islands make a contribution to biodiversity out of proportion to their land area, and in this sense collectively they can be thought of as 'hot-spots'. Some of course, are considerably hotter in this regard than others. This chapter has purposefully avoided discussion of the mechanisms of evolutionary change on islands, although noting some correlates with high proportions of endemics in island biotas. With increasing isolation, island size, and topographic variety, the number and proportion of endemic species increases. Endemicity within a taxon appears to be at its greatest in regions which are near the edge of the effective dispersal range. Thus, although a high proportion of its plants and birds are endemic, Hawaii has only three skinks, none of which is endemic. Their colonization may perhaps have been human-assisted and fairly recent: Hawaii is thus effectively beyond the effective dispersal range for skinks. In contrast, the less remote islands of the Pacific Ocean evidence a high proportion of endemicity. Ancient 'continental' islands (e.g. Madagascar) can also be rich in endemics. Some lineages have done particularly well out of oceanic islands, land snails and fruit bats among them. Particular genera, such as the Hawaiian *Drosophila* (fruit flies) and *Cyrtandra* (a plant example) have radiated spectacularly. Terrestrial mammals have made a poor showing: their greatest successes have been on islands insufficiently remote to have escaped the attention of *Homo sapiens* at an early stage, and many have thus failed to survive to historical times. New discoveries continue to be made of living (extant) and fossil (extinct) island forms and this serves as a cautionary note, such data must be appraised carefully prior to attempts to explain current biogeographical patterns in terms of evolutionary theory.

4

Speciation and the island condition

...it is the circumstance, that several of the islands possess their own species of the tortoise, mocking-thrush, finches, and numerous plants, these species having the same general habitats, occupying analogous situations, and obviously filling the same place in the natural economy of this archipelago, that strikes me with wonder. It may be suspected that some of these representative species, at least in the case of the tortoise and some of the birds, may hereafter prove to be only well-marked races; but this would be of equally great interest to the philosophical naturalist...

(C. Darwin 1845, writing about the Galápagos archipelago. From Ridley 1994, p. 79.)

4.1. Introduction: first know your species

Islands hold a special place in our understanding of evolutionary change and speciation. Both Darwin and Wallace gained many of their insights and some superb illustrations of evolution from their consideration of island biotas. The full significance of his Galápagos collections from the voyage of the *Beagle* did not immediately strike Darwin. Indeed, his key insights into species transmutation came some time after the voyage, as he was attempting to make sense of his observations and of the taxonomic judgements made of his collections. At the time he collected them he viewed the different kinds of mocking-thrushes and finches as merely variants of a single species (Browne and Neve 1989).

Of course, the processes of evolution on islands are not unique to islands. It is just that in their special circumstances of origin, configuration, and environmental history they provide, as described in Chapter 1, a series of 'natural laboratories' in which certain very striking evolutionary patterns have arisen. In this and the following two chapters, some of the more important of these patterns and the principal explanations that have been put forward for them will be introduced. The basic patterns and early accounts of Darwin and Wallace are beautifully written and should be required reading. More recent developments in the subject have benefited from advances in the fields of systematics, molecular biology, and genetics: understanding such lines of evidence requires a fairly specialized knowledge base. In order to make these recent developments accessible, some of the more technical information will be omitted. Further research into the literature will reward the dedicated reader.

The aim of this chapter is to provide the basic frameworks by which we understand evolutionary change on islands: it is thus concerned with the nature of the species unit, and with distributional, locational, mechanistic, and phylogenetic models of speciation. Evolutionary change does not necessarily imply speciation, and indeed much of Chapter 5 will be concerned with the changes in niche that characterize island forms (whether 'good' species or merely varieties). These include a variety of types of niche shift, and fairly subtle changes in genotype, such as may be brought about by the founder effect and by genetic drift, all of which are part of the driving mechanism of evolutionary change on islands. Only then, will we move on to a detailed examination of the macroevolutionary patterns most strongly associated with island biotas.

4.2. The species concept and its place in phylogeny

No one definition has as yet satisfied all naturalists; yet every naturalist knows vaguely what he means when he speaks of a species.

(Darwin 1859, p. 101)

... it will be seen that I look at the term species, as one arbitrarily given for the sake of convenience to a set of individuals closely resembling each other, and that it does not essentially differ from the term variety, which is given to less distinct and more fluctuating forms.

(Darwin 1859, p. 108)

If we are to study distributions of organisms, let alone speciation, it is a prerequisite that we have a currency, i.e. units that are comparable. Thus the most fundamental units of biogeography are the traditional taxonomic hierarchies by which the plant and animal kingdoms are rendered down to species level (and beyond). So, for example, a particular form of the buttercup can be described by the following nomenclature:

Kingdom	Plantae
Division	Spermatophyta
Order	Ranunculales
Family	Ranunculaceae
Genus	*Ranunculus*
Species	*bulbosus*
Subspecies	*bulbifer*

How is this taxonomy decided? Are the units actually comparable? The following is a fairly typical textbook definition of the species unit: 'groups of actually or potentially interbreeding natural populations which are reproductively isolated from other such populations' (Mayr 1942, p. 120). This is a form of the biological species concept, promoted notably by Ernst Mayr, which has been predominant for the past 60 years (Mallet 1995). Yet the question of 'reproductive isolation' is often an unknown quantity, as populations may be differentiated yet remain capable of interbreeding were they not geographically isolated (i.e. having **allopatric distributions**). Then again, species may be capable of interbreeding in the laboratory and have **sympatric distributions** (i.e. occur in the same geographic area) but remain reproductively isolated in the wild because of behavioural differences. Sometimes a rider is added to restrict the definition to those individuals that can successfully interbreed to produce viable offspring. Yet even this rule can be broken, as species which are recognized as 'good' may interbreed successfully in hybrid zones (Ridley 1996), an example being the native British oaks *Quercus petraea* and *Q. robur*. Studies of the genetics of British populations have shown that both maintain their integrity over extensive areas of sympatry, but form localized 'hybrid swarms' in disturbed or newly colonized habitats (White 1981). Apparently this is a stable situation, as evidence from Miocene fossil leaf impressions suggests that these taxa became differentiated no less than 10 million years BP. Some prefer to treat such partially interbreeding populations as 'semispecies' within 'superspecies' (White 1981).

For many groups, moreover, the taxonomy remains poorly known and subject to revision. For example, recent molecular analyses of the *Bufo margaritifera* complex of toads in Central and South America, suggest that toads recognized by this name actually represent composites of morphologically cryptic species that are genetically unique (Haas *et al.* 1995). If the criterion of being reproductively isolated is a difficult one to employ on living organisms, it is even more so with many fossil organisms. Often the best that can be done is to determine whether the morphological gaps between specimens are comparable with those for living species that are reproductively isolated. Thus even the concept of the species unit, although central to so many biogeography purposes, has blurred edges (see also Otte and Endler 1989; Ridley 1996). The units below the species are a matter of considerable debate and strong opinions are held that they are either important ecologically and therefore matter or that they are arbitrary and the whim of the taxonomist (see discussion in Mallet 1995). The significance of this issue can be shown by reference to the Lepidoptera: butterflies are thought to comprise about 17 500 full species, but the number of currently recognized subspecies approaches

100 000 (Groombridge 1992). Some would argue that lepidopterists are terrible splitters and that in terms of diversity estimates we should stick with the 'full species'. None the less, recognizable variation between populations of species do exist, as can be seen, for example, in Spencer-Smith *et al.*'s (1994) beautifully illustrated guide to Caribbean butterflies. These problems are general to taxonomy and systematics, but have particular force in relation to islands on which there is often no adequate way of testing the potential reproductive isolation of geographically isolated 'species' or 'varieties' on different islands within an archipelago.

In a recent review of the species concept, Mallet (1995) notes the prevalence, even at textbook level, of as many as seven or eight different notions of how to define species. This may be bewildering to many and for the present purposes a way out is needed. This is not the place for a lengthy review of the problems of taxonomy and systematics, the intent has instead been to ensure that all those reading this chapter appreciate that difficulties exist with the species concept and its application, whether to speciation on islands or on continents. The solution Mallet recommends is to return to Darwin's pragmatic usage of species and varieties, but to make use of new knowledge from genetics as well as morphology in determining when the species label is appropriate. Darwin's position is made clear by the quote given above, and in the following passage from *On the origin of species*... 'varieties have the same general characters as species, for they cannot be distinguished from species, except, firstly, by the discovery of intermediate linking forms...; and except, secondly, by a certain amount of difference for two forms, if differing very little, are generally ranked as varieties...'

If species are in essence arbitrary units, or 'well-marked varieties', then it will be apparent that determining the point at which speciation has occurred within a radiating lineage is also problematic. 'That is like asking exactly when a child becomes an adult. I am content to know that initially there was one lineage and now there are more.' (Rosenzweig 1995, p. 87). The debates on this crucial matter will continue, and while some may not like it, the view taken here is a pragmatic one, following Darwin and Wallace: the precise level at which species are defined is arbitrary. As with other currencies, exchange rates can change as a function of geography and of who is doing the dealing.

At higher levels, species are grouped into genera, then families, and so forth. In flowering plants such groupings of species are based principally on the evolutionary affinities of floral structure and so it is possible to find many different growth forms and ecologies within a single family. Thus the Asteraceae (Compositae—the daisy family) all have a composite flower structure, but some are herbs while others are trees. In general, there tends to be less variation in functional characteristics within a genus than between genera (within the same family) and there is presumed to be a closer evolutionary relationship, with all members of a genus ultimately descended from a common ancestor, different from that of other genera within the same family. Some species are so distinct that they may be the only members of a genus or family, while others belong to extremely species-rich genera, e.g. figs (*Ficus*; Moraceae). Insular examples include, at the one extreme, *Lactoris fernandeziana*, the monotypic representative of the endemic family Lactoridaceae from the Juan Fernández Islands (Davis *et al.* 1995); and at the other extreme, on the Hawaiian islands, over 100 species of *Cyrtandra* of the widespread family Gesneriaceae (Otte 1989). In general, the greater the taxonomic distinction between organisms, the more ancient their differentiation, although calibrating the time-scale is difficult, particularly for fossil organisms.

In recent decades the methods of traditional evolutionary systematics have been challenged by the development of phylogenetic systematics, or **cladistics**, which presumes to supply a more objective means of quantifying the relatedness of a taxonomic group. This involves scoring the degree to which different organisms share the same uniquely derived characteristics that other taxa outside the group do not possess. These characteristics are then used to construct a **cladogram**, or **phylogenetic tree**, which is a branching sequence setting out the most parsimonious ('simplest') model for the relationships between taxa, i.e. it provides a hypothesis for the evolutionary development of a lineage. The data employed may be morphological or genetic (i.e. based on DNA sequences). The papers in Wagner and Funk (1995) provide examples of both as applied to Hawaiian plants and animals, as well as an excellent explanation of the methods of cladistics.

While the proponents of phylogenetic systematics are often vociferous in praising the approach and condemning traditional systematics, it should be recognized that the decisions as to which characters to include in the phylogeny, and the rules used in its construction, are the choice of the user. The phylogeny thus constructed is therefore merely one approximation, albeit hopefully a basically reliable one, to the history of events that produced the lineage. Once such a phylogeny has been constructed, it can then be used to explore the likely sequence of inter-island and intra-island speciation events within archipelagos, or between islands and mainlands.

The most exciting opportunities have been opened up by the advances in molecular biology and genetics (e.g. Haas *et al.* 1995) and these methods have been seized upon in island evolutionary studies (e.g. Wagner and Funk 1995). Such studies benefit from use of the 'genetic clock' which, on the assumption of a more or less constant rate of accumulation of mutations and thus of genetic differences between isolates, allows estimates of the dates of lineage divergence. An excellent illustration is provided by the studies of Thorpe *et al.* (1994) on the colonization sequence of the western Canary Island lizard, *Gallotia galloti,* in which the direction and timing of colonization as postulated by genetic clock analyses (nuclear and mitochondrial DNA divergence) was shown to be entirely compatible with the independently derived geological data for the timing and sequence of island origins. It should be noted, however, that these clocks may not always run to time. A key problem is exemplified by Clarke *et al.*'s (1996) study of two species of land snails (*Partula taeniata* and *P. suturalis*) on the island of Moorea (French Polynesia). It was found that while the two species exhibit significant morphological and ecological differences, they are genetically close. Clarke *et al.* attributed this closeness to 'molecular leakage' or 'introgression': the convergence of genetic structure through occasional hybridization. Such **introgressive hybridization** is now considered to be quite common in particular island lineages (Clarke and Grant 1996; Chapters 5 and 6). Even low rates (as low as 1 in 100 000) may be enough to upset the phylogenetic trees and molecular clocks upon which many of the emerging scenarios of island evolution discussed in Chapters 5 and 6 now rest (Clarke *et al.* 1996). For further insights into these and other methodological problems of biogeography and systematics, see Myers and Giller (1988), Thorpe *et al.* (1994), and Ridley (1996).

4.3. The geographical contexts of speciation events

4.3.1. Distributional context

In order to appreciate the full gamut of possibilities involved in speciation it is helpful, if slightly artificial, to distinguish the geography of speciation from the mechanism (Table 4.1). The geographical context can be viewed either in distributional or locational terms, i.e. distinguishing the degree of overlap of populations involved in speciation events on the one hand, from the issue of where the evolutionary change is taking place.

The terms sympatry and allopatry denote, respectively, two populations which overlap in their distributions and two populations which have geographically separate distributions. In geographical terms, a new species (or subspecies) may, in theory, arise in one of three circumstances. If the new form arises within the same geographical area as the original, then it may be termed **sympatric speciation**, if it arises in a zone of contact (hybrid zone) between two species it may be termed **parapatric speciation**, and if it arises in a separate geographical

Table 4.1
A framework for speciation patterning

Form	Pattern
Distributional	
Sympatric	Overlapping
Allopatric	Separate
Parapatric	Contiguous
Locational (and historical)	
Neo-endemic	Change on island
Palaeo-endemic	Island form relict
Mechanistic	
Allopatric	Drift and selection
Polyploidy	Changes in chromosome number
Competitive speciation	Other sympatric mechanisms
Tree form (phylogeny)	
Anagenesis	Replacement of original
Anacladogenesis	Alongside original
Cladogenesis	Lines diverge and replace original

area it may be termed **allopatric speciation**. In reality these ideas form a continuum.

It is apparent from much of the debate concerning vicariance versus dispersalist explanations in biogeography (Chapter 3), that allopatric speciation is generally assumed in the literature to be the dominant mode (by some, to the virtual exclusion of other modes). This applies to the treatment of island endemics, where after all speciation is associated with the insularity of the populations involved, i.e. is allopatric in relation to the continental source population. Yet speciation is not of necessity allopatric even in the island context, and the demonstrable occurrence of sympatric forms of speciation exposes the weakness of making such assumptions (Sauer 1990). It is important at the outset to appreciate the scale at which the term is being used. On the regional scale, island and mainland lineages may be allopatric, while at the within-archipelago scale a mix of sympatric and allopatric speciation events occur, as for example, appears to apply to the Hawaiian fruit flies *Drosophila* (Tauber and Tauber 1989) and *Sarona* (a genus of phytophagous insects) (Asquith 1995). Much island speciation has occurred in the condition of island-scale sympatry (i.e. differentiation of two forms occurring on one island), but on the finer within-island scale, this still allows for differing degrees of population overlap (Keast and Miller 1996, p. 111). The problem then arises as to the meaning of the terms allopatry and sympatry applied on a within-island scale, as habitats are commonly locally distinct and spatially segregated. Diehl and Bush (1989) suggest that populations utilizing different habitats should none the less be considered sympatric when all individuals can move readily between habitats within the lifetime of an individual. This must be judged separately for different animal taxa as, for example, a within-island barrier for land snails may not be a barrier to bird populations (Diamond 1977), and would also require modification for use with plant populations.

4.3.2. Locational context—island or mainland change?

In Chapter 3, the distinction between neo-endemic forms (those that have evolved *in situ* on the island) and palaeo-endemic forms (which have become endemic because of the extinction of closely related forms from mainland populations) was introduced. This general idea was recognized by A. Engler as long ago as 1882, but the labelling of particular species or genera as 'relictual' and the relative importance of palaeo- and neo-endemicity on islands are still keenly debated (Bramwell 1979; Cronk 1992; Elisens 1992; Carlquist 1995). Recent developments in phylogenetic techniques promise a resolution of more of these questions in the near future (e.g. Kim *et al.* 1996), but much of the evidence to date rests on more traditional techniques.

Cronk (1987, 1990) discusses the general idea and gives a specific example, of the endemic St Helenan genus *Trochetiopsis*, consisting of two species, Redwood and Ebony. He has compared these species with others in the same family (Sterculiaceae), employing numerical analysis of the morphology of both present-day pollen and ancestral pollen in sediments. He finds that the ancestors of the present species probably arrived by long-distance dispersal at least 9 million years ago, from African or Madagascan stock (Fig. 4.1). The lineage has since diverged into two forms, probably as a result of environmental change at the beginning of the Quaternary (*c.* 2 million years BP), but the degree to which they differ from their mainland relatives is only partly explained by *in situ* change. Evolution and extinction elsewhere among the branch to which the ancestor belonged has meant that the rest of the family has effectively evolved away from the St Helena genus over the 9 million years since colonization. The important point is that in attempting to understand evolution on islands, we cannot assume that over time-scales of millions of years, evolution on the continents that supply the initial island colonists has somehow stood still. We cannot therefore interpret *all* island/mainland differences, and in particular all insular endemicity, as *necessarily* a function of evolutionary change on the island in question. 'There is no doubt that adaptive radiation occurs on islands, but this may be interpreted as the elaboration of a relict ground-plan, without requiring that ground-plan to have arisen by evolution on the island' Cronk (1992, p. 92).

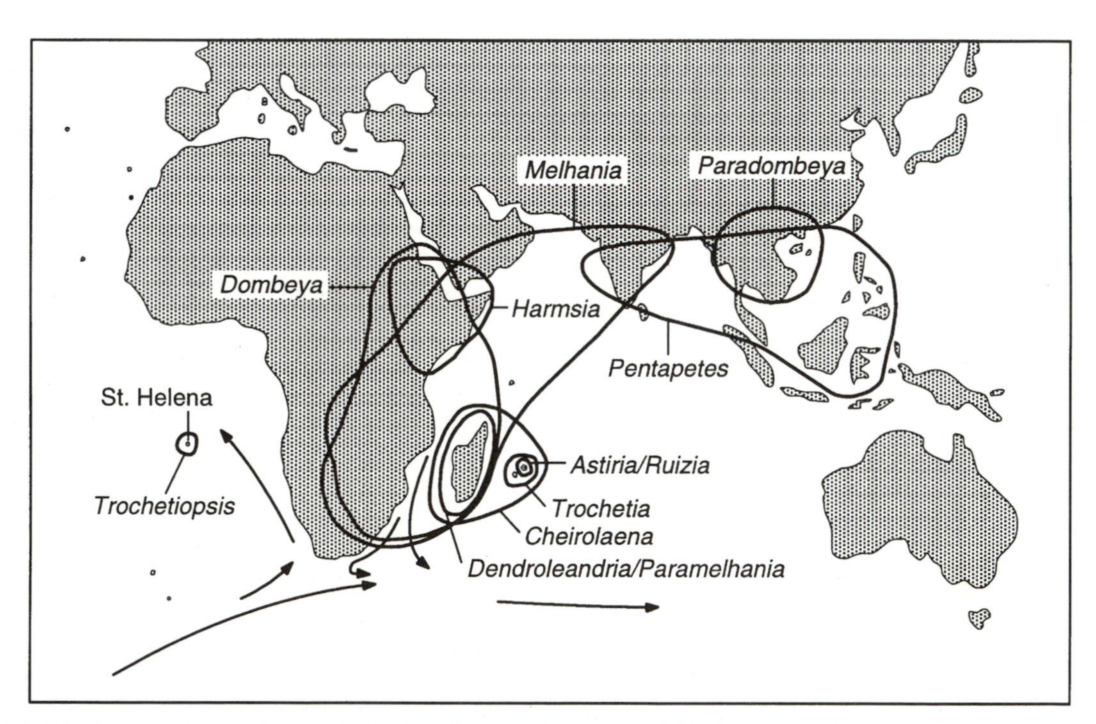

Fig. 4.1. Distribution of genera in the subtribe Dombeyeae (Sterculiaceae). The closest relatives of the *Trochetiopsis* of St Helena are *Trochetia* (Mauritius) and *Dombeya*. The centre of generic diversity is in Madagascar and the Indian Ocean islands. Present day ocean currents (arrows) differ from those predominant at the time the ancestral Dombeyeae stock colonized St Helena at least 9 million years ago, when there was a stronger flow from the Indian Ocean to the South Atlantic. (Redrawn from Cronk 1990, Fig. 3.)

Paulay (1994) comments that relict taxa are well known on large, old, continental islands, the lemurs of Madagascar being classic examples. The primitive weevil family Aglycyderidae of both Pacific and Atlantic islands provides another case of ancient forms restricted to islands. Although short-lived oceanic islands may not be thought to be ideal for long-term survival of relict lineages, if the species involved have good dispersal abilities they may survive by island hopping. Thus three of the six families that comprise the most primitive order of pulmonate land snails are restricted to islands, yet representatives may be found on even the most remote Pacific islands, strongly suggesting their distributional boundaries to be set by biological factors rather than an inability to reach mainland areas.

4.4. Mechanisms of speciation

When examining the mechanisms of speciation rather than focusing on the geographical circumstances, it becomes clear that there are different processes at work. Rosenzweig (1995) summarizes these under the three headings: **geographical** or **allopatric speciation**, **competitive speciation**, and **polyploidy** (involving an increase in the chromosome number). Each form holds relevance to the island context, but as insularity defines the island condition, the allopatric form of speciation will be dealt with first.

4.4.1. Allopatric or geographical speciation

The essence of this model has been summarized by Rosenzweig (1995, p. 87):

- A geographical barrier restricts gene flow within a sexually reproducing population.
- The isolated subpopulations evolve separately for a time.
- They become unlike enough to be called different species.
- Often the barrier breaks down and the isolates overlap but do not interbreed (or they interbreed with reduced success).

The first step of this model suggests a vicariant event (barrier formation), whereas, in the oceanic island context, the starting point will more commonly be dispersal across a pre-existing barrier, i.e. a founding population making chance landfall. But the idea is the same, gene flow is restricted after the founding event and, for the most remote of islands, can effectively be taken to be zero between the mainland source and island populations. The colonizing populations take with them only a subset of the mainland gene pool, so that many alleles are not represented (this is a key part of the so-called founder effect). In this sense founder events represent genetic bottlenecks (discussed further in section 5.1). Genetic drift effects and selective pressures in the novel island context then provide the engine for further differentiation from the mainland forms.

The breakdown of the barrier may not be a frequent feature for oceanic islands, but none the less will happen from time to time, for instance, as a function of sea-level change (Chapter 2). Occasionally, there may also be **double invasions**, where a second invasion of the mainland form occurs notwithstanding the continued existence of an oceanic barrier. This was the hypothesis put forward to explain the occurrence of two species of chaffinches on the Canary Islands (Lack 1947). The endemic blue chaffinch (*Fringilla teydea*) is assumed to have evolved from an early colonization event, to a point where it was sufficiently distinct to survive alongside a later-arriving population of mainland chaffinches (*F. coelebs*).

The barriers that exist between islands *within* an archipelago are undoubtedly important features of remote oceanic island groups. As Rosenzweig (1995, pp. 88–89) puts it 'Suppose propagules occasionally cross those barriers, but usually after enough time for speciation has passed. Then the region and its barriers act like a speciation machine, rapidly cranking out new species.' The radiations that characterize many taxa on the Hawaiian islands reflect this, with interpretations for the relationships among the 700 (or more) species of *Drosophila* involving a number of phases of inter-island movements as a part of this most spectacular of radiations (e.g. Carson *et al.* 1970; Carson 1992). Adaptive radiation is examined in Chapter 6 and so will not be explored in depth here.

An apparent illustration of the contrast between the archipelago context and the single remote island is provided by that most famous of examples of island evolution, Darwin's finches. The Geospizinae are recognized as a distinct subfamily, found in the Galápagos archipelago, which comprises nine islands larger than 50 km^2 (and a larger number of smaller islets), of predominantly desert-like conditions. Here they have radiated into 13 different species. The family is found also on Cocos Island, a lone, forested island of about 47 km^2, which is, just, the nearest land to the Galápagos. Here there has been no possibility of the island hopping which appears to characterize archipelago speciation, and the group is represented by a single endemic species, one of just four species of land bird (Williamson 1981; Werner and Sherry 1987).

Yet, as Sauer (1990) cautions, within an archipelago, in addition to single species being present on different islands, quite commonly several sibling species occur on a single island, where he suggests that they are mainly segregated by habitat rather than being spatially isolated. In many instances, isolation may have been only indirectly relevant as, by screening out competitors, isolation offers opportunities for adaptive radiation to those species that do arrive. Some, at least, of this radiation may take the form of the next class of mechanism, competitive speciation.

4.4.2. Competitive speciation

This is the term that Rosenzweig (1978, 1995) gives to cover a variety of related modes of sympatric speciation. The idea here is that a species expands its niche to occupy an unexploited ecological opportunity, followed by that species' sympatric break-up into two daughters, one in essence occupying the original niche, and the other the newly exploited one. The expansion happens because of a lack of competition with other species, a feature common to remote island faunas and floras. But the break-up into separate species happens because of increased competitive pressure between those portions of the population best able to exploit the two different niches. The operation of these processes must take place over the course of many generations, and observation of the complete process in the field is thus not an option. Rosenzweig illustrates the idea by means of a 'thought experiment' the essence of which, placed in the island context, may be rendered as follows.

A colonizing species of bird will have a particular feeding niche, determined morphologically and behaviourally, and perhaps most clearly expressed by features such as bill size and shape (Lack 1947; Grant and Grant 1989). According to theory, such characters would be expected to have a unimodal distribution of values. The environment of the island may contain so-called empty niches, unexploited ecological opportunities, which for the purpose of our experiment are taken to be distributed bimodally, i.e. there are two resource peaks. The species, unconstrained by competition with the full range of mainland forms, expands from its original modal position somewhere along this resource space so that it occupies each of these niches. Individuals with genotypes that match best to one or the other of these niches become numerically dominant within the population, which has by this stage grown close to its carrying capacity. At this point, disruptive selection kicks in, such that the mid-range phenotypes lose fitness relative to those that are better suited to one or the other of the main feeding niches. Those individuals adapted primarily to a peak in the resource curve may also exploit the valleys either side of that peak. If they do so at all successfully, no opportunity remains for valley phenotypes in-between the two peaks. They will be few in number in the population, and their offspring will have fewer resources to tap. Although they may also reach up to the peaks, they will not be effective competitors in those portions of the resource continuum. Valley genes will thus have little success and will be bred out, as in time members of each 'peak' population that can recognize others of their type (and breed with them alone) will be at a selective advantage. Therefore, isolating behavioural mechanisms develop and at this point the lineage can be viewed as having split. Intentionally, this is a simple rendition of ideas culled from a much larger body of literature, but a fuller review of such work (cf. Otte and Endler 1989) is outside the scope of the present volume. It may be noted, however, that based on laboratory evidence, competitive speciation may occur in as few as 10–100 generations, not as fast as polyploidy (almost instantaneous), but faster than geographical speciation, which seems to require thousands or even hundreds of thousands of years (Rosenzweig 1995).

Grant and Grant (1989) were able to record an early phase in the process of sympatric speciation in a study of a temporary reproductive subdivision within *Geospiza conirostris*, one of the Galápagos (Darwin's) finches occurring on Isla Genovesa. The population was partly subdivided ecologically for a brief period, but this division then collapsed again through random mating. All of the known sympatric species of ground finches differ by at least 15% in at least one bill dimension, and as this has been shown to be a good surrogate for niche differences, there is a minimum niche difference between them. During the temporary subdivision, the two groups of males differed by a maximum of 6% in bill dimensions, and this, occurring as it did over such a brief period, was insufficient to foster any discrimination in a mating context. One of the dry-season food niches sustaining the division declined catastrophically, and this was thought to be a contributory factor to the collapse of the division. None the less, this study shows how sympatric division may be fostered in taxa in which polyploidy does not occur. It should be recognized that Grant and Grant would have had to be extraordinarily lucky to come along just at the right time to record a sympatric split that 'stuck'. They therefore did not regard the failure of this event as evidence against the idea, which they concluded to have some relevance for vertebrates. Rather, they took it to indicate that in order to foster sympatric divergence leading to speciation, the niches or habitats to which different members of a population adapt must be markedly different and display a long-term persistence. This may be comparatively rare, but is more likely to occur on the larger, higher islands.

How important is competitive speciation in the island context? This is difficult to judge as, unlike polyploidy, it is a process that is not obviously distinguishable in the genetic structure of the resulting species from those produced by geographical speciation. It cannot be decided merely by investigating the degree of geographical overlap between sibling species in ecological time, because given the probabilistic nature of dispersal and the magnitude of past environmental change, populations currently isolated from one another may once have been sympatric and vice versa. Indeed, it may commonly be the case that diversification on archipelagos involves a mix of allopatric and sympatric phases of subtly varying combination from one branch of a lineage to the next (cf. Clarke and Grant 1996). The relative roles of allopatric and competitive speciation within the island context cannot therefore be quantified, but one thing remains clear, isolation is crucial, either directly or indirectly (respectively).

4.4.3. Polyploidy

One important class of sympatric speciation is through polyploidy, a condition comparatively common in plants and many invertebrates, but not in higher animals: for instance it is unknown in mammals and birds (Grant and Grant 1989; Rosenzweig 1995). Polyploid species are those that have arisen by an increase in chromosome number. Two main forms of polyploid can be distinguished: if the new species has twice the chromosome complement of a single parent species, it is termed an **autopolyploid**; if it has the chromosomes of both of two parent species (i.e. a crossing of lineages), it is called an **allopolyploid**. According to Rosenzweig (1995), autopolyploids are rare, but at a conservative estimate 25% or more of plant species may be allopolyploids. Barrett (1989) states that current estimates for the angiosperm subset are as high as 70–80% of species being of polyploid origin.

Rosenzweig (1995) presents an intriguing analysis of the proportionate importance of polyploids in different floras, showing that there is a general decrease in the relative contribution of polyploids in floras with decreasing latitude. The best trend was found for continental floras; for islands there was much more scatter in the relationship. Rosenzweig does not distinguish the nature of the islands in his analysis, i.e. whether they were predominantly oceanic or continental in origin, and this may be a confounding factor in his analysis. It is necessary to look a little further into the data to establish whether polyploidy has provided a significant contribution to island evolution and whether it has developed *in situ*. The following studies are cited in order of increasing proportion of polyploidy.

A recent study of the endemic flora of the Juan Fernández Islands (33° S), which are volcanic, oceanic islands off the coast of Chile, involved an examination of about 35% of the 148 endemics, none of which was found to be polyploid (Stuessy *et al.* 1990). If confirmed by further studies, this would be fairly remarkable. Another group of volcanic islands, the Canaries (28–29° N), contain about 580 vascular plant species that are endemic to the islands of Macaronesia (the Azores, Madeira, the Salvage Islands, the Canary Islands, and the Cape Verde Islands), of which 360 are known cytologically (Borgen 1979). Of these, 25.5% were found to be polyploids, a figure that is low in comparison to most known floras and significantly lower than the 36.4% polyploids amongst the 151 non-endemic species for which data were available (Borgen 1979). It appears also that several of the polyploid events were fairly ancient ones, subsequent to which gradual speciation within certain polyploid lineages has occurred (an example being the genus *Isoplexis* in the Scrophulariaceae), but to a lesser degree than within diploid lineages (Humphries 1979). It may be concluded that polyploidy has not been a major evolutionary pathway in Macaronesian lineages. The New Zealand (*c.* 35–47° S) flora consists of approximately 1977 species, 40% of which are known cytologically (Hair 1966). Polyploidy is absent in the gymnosperms, but characterizes 417 (63%) of the 661 angiosperms investigated, a figure approaching the 70–81% polyploidy suggested for oceanic-subarctic and Arctic areas of high northern latitudes. New Zealand is an ancient archipelago, and its flora has generally been assumed to have been primarily of continental origin, relictual of Gondwanan break-up. However, this has been disputed by Pole (1994), who argues against the large-scale continuity of lineages in the New Zealand flora, claiming that palynological data support derivation of the entire forest flora from long-distance dispersal (predominantly from Australia).

Few firm conclusions can be drawn from the data currently available. Polyploidy is more important in some island floras than others. The latitude and age of the islands in question may each be of relevance, although often, as shown, age and origin of the flora are imperfectly known. If the Macaronesian data can be taken as a good guide, it appears that the more spectacular radiations of island plant lineages do not tend to involve polyploidy.

4.5. One more framework: tree form or phylogeny

A number of different models and facets of island evolution have been introduced in this chapter. Too many frameworks may serve to confuse, but there is one more that appears particularly helpful in making sense of island speciation patterns, and while the terms may not be instantly memorable, the ideas seem useful. In some cases, evolutionary change takes the form essentially of the continuation of a single lineage, whereas in other cases the splitting of lineages is involved. Stuessy *et al.* (1990) provide three terms that codify what they see as the chief outcomes (Fig. 4.2). **Anagenesis** is when the progenitor species/form becomes extinct. **Anacladogenesis**

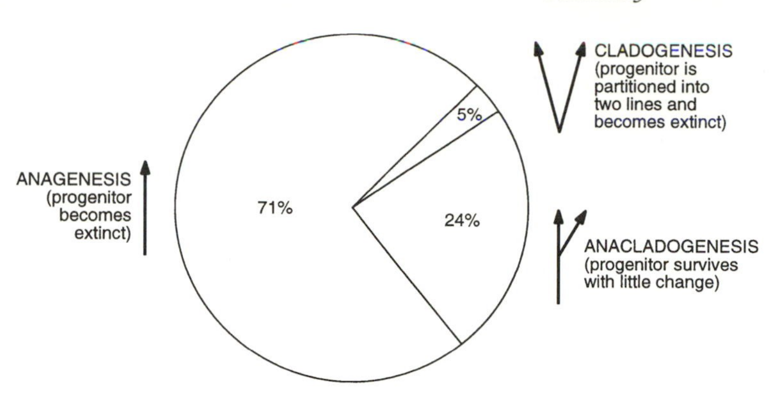

Fig. 4.2. General patterns of phylogeny in the endemic vascular flora of the Juan Fernández Islands. The figures give the percentage of each form of lineage development evident in 123 endemic plant species. (From data in Stuessy *et al.* 1990.)

is when the progenitor survives with little change alongside the derived species. What is meant by alongside is open to different interpretations—it could be within the same island, or less restrictively, within the same archipelago. Finally, **cladogenesis** is where the progenitor is partitioned into two lines and becomes extinct in its original form, a model suggestive of the classic examples of adaptive radiation. This framework sets aside cases where lineages converge or cross, via hybridization. Hybridization is known to occur in both plant and animal taxa on islands (e.g. Sohmer and Gustafson 1993; Grant and Grant 1994; Clarke and Grant 1996), but the extent to which it has led to speciation is difficult to quantify.

4.6. Summary

In order to understand evolutionary change it is necessary to have a basic familiarity with the currency of evolution and knowledge of the problems associated with it. This chapter therefore begins by discussing some of the problems of the species concept, the arguably arbitrary distinction between species and varieties and the higher-order phylogenies of relationships that can be constructed for particular lineages. Recent developments in phylogeny, including the use of cladistic and molecular biological techniques have provided many new insights into long-standing island biogeographical issues, although such techniques should be viewed with proper caution as merely constructing realistic hypotheses of the unknown real genealogies.

A distinction is drawn between distributional, locational (and historical), mechanistic, and phylogenetic frameworks for describing island evolution. It is shown that divergence can take place both through isolation (allopatry) and in conditions of distributional overlap (sympatry), and indeed that these concepts can be applied on different scales: between island and mainland populations, within an archipelago, and within an island. Evolutionary change within a lineage may be accomplished by a variety of genetic mechanisms, including hybridization and polyploidy, and, as will be explored in the next chapter, by more subtle alterations following from founding events, genetic drift, and natural and sexual selection in the novel biotic conditions of oceanic islands. Sympatric speciation may occur through more than one route. Perhaps the classic island examples are those that follow the competitive speciation model of niche expansion and break-up in to daughter populations exploiting different resource peaks. Polyploidy does not seem to have been a major route of speciation on islands, although it undoubtedly occurs. The final framework considered in the chapter is that of phylogenetic trees, which are used to denote whether lineages develop without branching (anagenesis), with branching and extinction of the original form (cladogenesis), or by branching and survival of the original line (anacladogenesis).

5

Arrival and change

Island forms of the same species [of birds] show every gradation from extremely small differences in average size to differences as great as those which separate some of the species. The differences tend to be greater the greater the degree of isolation. Some of them are adaptive and others seem non-adaptive.

(Lack 1947)

In the previous chapter, I examined the nature of the species unit, the varying patterns in which speciation occurs, and in outline form the chief mechanisms of speciation. Now it is time to put some flesh on the bones and to explore in some detail the special insights into evolutionary change that islands have provided. In this chapter I will work through the evidence for evolutionary change on islands, beginning with the circumstances of arrival of a lineage on an island. While some authors stress the importance of chance effects in generating divergence between mainland and island lineages, other studies suggest a variety of directional evolutionary forces within islands. Although much of this literature is concerned with microevolutionary change not necessarily involving speciation (i.e. the recognition of distinct island and mainland species), these processes of arrival and change are the building blocks of the macro-scale models of island speciation that will be considered in Chapter 6.

5.1. Founder effects, genetic drift, and bottlenecks

The founder principle, put forward by Mayr (1954), is that typically an immigrant population to a remote island will establish by means of a small founding population. This contains only a subset of the genetic variation in the source population and it subsequently receives no further infusion. This may immediately provide a bias in the island population, on which other evolutionary processes then operate. Genetic variability can be increased after establishment by mutation and re-sorting. One important form of re-sorting is **genetic drift**, the chance alteration of allele frequencies from one generation to the next, which may be particularly important under sustained conditions of low population size. In addition to such non-selective drift, there may also be distinctive selectional features of the novel island environment. The outcome of these phenomena can, it is thought, be a rapid shift to a new, co-adapted combination of alleles, hence contributing to reproductive isolation from the original parent population.

While it is possible to demonstrate fairly convincingly the occurrence of genetic drift, as Berry *et al.* (1992) have done in their study of a small population of house mice on Faray in the Orkney Islands, the contribution of these various processes to island speciation remains the subject of keen debate (Barton 1989; Clarke and Grant 1996). It is particularly problematic that founding *events* (i.e. colonizations) have been theorized to produce a variety of rather different founding *effects*, and that the latter are difficult to isolate empirically from other microevolutionary processes. In fact, there are at least three distinguishable models of how founder effects may lead to speciation:

1. Mayr's peripatric speciation;
2. Carson's founder-flush speciation; and
3. Templeton's genetic transilience.

These theories place greater degrees of stress, respectively, on:

1. the impact of the general loss of heterozygosity combined with random drift;

2. the role of variations in selection pressure (particularly its relaxation immediately after colonization); and

3. changes in a few major loci within the genome (see discussions in Barton and Charlesworth 1984; Clarke and Grant 1996).

The founding event represents a form of **bottleneck**, a term implying a sharp reduction in population size. Bottlenecks may occur at other points in the life span of a lineage, e.g. following catastrophic disturbance of a habitat, or in the context of species pushed to the brink of extinction by human action. The Hawaiian *Drosophila* (e.g. Carson 1983) have provided much of the data and stimulated much of the argument in relation to founder effects. Studies of *Drosophila* indicate that founder events have occurred and at three levels: in the colonization of the archipelago, of its constituent islands, and of patches of forest habitat ('*kipukas*') within islands (Carson *et al.* 1990; Carson 1992). Significantly, the work of Carson and others indicates that only if the bottleneck is sustained or if it is repeated at short intervals, is there a marked loss of heterozygosity. In contrast, if a bottleneck lasts only one generation and recovery to a large population size is rapid, then increased genetic variance for quantitative traits can be demonstrated in the laboratory.

According to the theories of founder-effect speciation, the narrowing of the gene pool during a bottleneck may serve to break up the genetic basis of some of the older adaptive complexes, liberating additive genetic variance in the new population (Carson 1992). Novel phenotypes might subsequently result from natural selection operating on the recombinational variability thus provided in the generations immediately following the bottleneck. According to a controversial model proposed by Kaneshiro (e.g. 1989, 1995), an important element in this scenario is differential sexual selection. Kaneshiro reported that a minority of male drosophilids perform the majority of matings, and that females varied greatly in their discrimination, and acceptance, of mates. He suggested that in normal, large populations the most likely matings are between males that have a high mating ability (the studs), and females that are less discriminating in mate choice. This provides differential selection for opposing ends of the mating distribution and a stabilizing force maintaining a balanced polymorphism. In a bottleneck situation, however, there would be even stronger selection for reduced discrimination in females, and if this was maintained for several generations, it could result in a destabilization of the previous co-adapted genotypes. In such bottleneck situations, with reduction in mate discrimination by females, the likelihood of a female accepting males of a related species increases, thus allowing hybridization to occur. The term given to the incorporation in this way of genes of one species into the gene pool of another via an interspecific hybrid is introgressive hybridization. With the development of molecular tools, hybridization between animal species (especially insects) has been discovered to occur naturally much more often than reported in the earlier literature (Grant and Grant 1994; Hanley *et al.* 1994; Kaneshiro 1995; Clarke and Grant 1996). Thus, while bottlenecks can serve to reduce the genetic base of a new colony, in some instances they may serve to provide *enhanced* genetic variability within a population, the very stuff on which natural selection may then go to work!

While much of the literature is based on laboratory studies of Hawaiian *Drosophila*, founder effects have been invoked (although not necessarily demonstrated) in other studies (Clarke and Grant 1996). Another Hawaiian example, but from the plant world, is provided by endemic members of the genus *Silene*. Populations appear to be much more polymorphic on the older island of Maui than on the younger island of Hawaii (i.e. the main or Big Island), suggesting that genetic variation had been lost in colonization of the latter (Westerbergh and Saura 1994). The process of restricted gene flow followed by genetic drift also seems to be repeated as younger volcanic terrain is colonized within the island of Hawaii itself.

While some of the advocates of founder effects regard them as a key part of the most likely mode of speciation in most Hawaiian

Drosophilidae (e.g. Kaneshiro 1989, 1995), other workers query their significance. For instance, Barton (1989) observes that enzyme heterozygosity within Hawaiian drosophilids is similar to that of continental species, and queries whether, in practice, founder effects cause much loss in genetic variation. Adopting a rather different—and less direct—approach to the issue, Reyment (1983) considers the relative significance of founder events and random drift versus selection in the size changes evident in Mediterranean island mammals of the Pliocene to sub-Recent period. He points out that while founder events must have been involved, i.e. that the founding populations may be assumed to have been small:

1. the evolution in size is not always as sudden as has been claimed; and

2. that as the directions of the changes are consistent irrespective of the island involved, selective forces (i.e. adaptive changes) are thus indicated to be dominant over drift (cf. Losos *et al.* 1997).

A similar position is adopted by Barton (1989), who regards other elements in the island laboratory as being of greater significance than founder effects, and he mentions specifically what he terms **faunal drift**, the random sampling of species reaching an 'empty' habitat, and thus providing a novel (disharmonic) biotic environment in which the selective forces outlined earlier cause divergence into a variety of new niches (see also Vincek *et al.* 1997).

Sarre and colleagues have examined morphological and genetic variation among seven island and three mainland populations of the sleepy lizard, *Trachydosaurus rugosus,* in South Australia. Their work provides evidence that appears to support the significance of random genetic drift and inbreeding in small isolates (Sarre *et al.* 1990). All but two of their islands were isolated from the mainland between 6000 and 8500 years ago. Contrary to some other recent studies of vertebrates on South Australian islands, their allozyme electrophoresis analysis showed that heterozygosity levels did not vary significantly among the populations, i.e. there has been little erosion of heterozygosity within the insular populations. None the less, the island populations exhibited reduced allelic diversity, with rare mainland alleles tending not to be present on the islands. This may have been due in part to bottlenecks occurring at some point or points since isolation of the populations. There was also a greater degree of genetic divergence shown between island populations than between mainland populations, and this they attributed to genetic drift on the islands.

Allozymes are different forms of an enzyme specified by allelic genes, i.e. by genes that occupy a particular locus. Allozyme electrophoresis has become a fairly conventional method of measuring heterozygosity levels. It is thus of interest that two measures of developmental stability varied significantly among the island populations in a manner distinct from the results of the allozyme study (Sarre and Dearn 1991). The morphological measures used were: fluctuating asymmetry—the degree of random difference between left and right sides in bilateral structures; and percentage gross abnormalities, involving features such as missing ear openings, missing toes, club feet, and deformed head and labial scales. Three of the island populations stood out as having distinctly higher levels of these forms of developmental instability. This was interpreted as a consequence of inbreeding depression resulting from the expression of deleterious genes or the genetic disruption of co-adapted genotypes. One important conclusion drawn from their study is that reduced levels of heterozygosity *per se* are unlikely to be the prime cause of the observed developmental instability. In short, the allozyme technique was failing to reflect the genetic basis for the deleterious morphological features. The relative contribution of environmental and genetic factors on developmental stability in the sleepy lizard cannot at this stage be determined: the allozyme study thus provided just a piece of the island evolutionary jigsaw and not the final solution to it. In contrast to the emphasis given in this study to drift, Thorpe and Malhotra (1996) stress the importance of natural selection for current ecological conditions in their studies of Canarian and Lesser Antillean lizards. It may be significant that Thorpe and Malhotra's islands, unlike Sarre's, are large and contain a considerable range of habitats, i.e. the relative importance of stochastic processes versus directional selection may turn out to be strongly dependent on the environmental context.

In conclusion, what has this brief review of founder effects, genetic drift, and bottlenecks shown? First, the restriction of populations to an extremely small size, either on colonizing an island (the founding event), or subsequently in the history of a lineage (other bottlenecks), may in theory produce a variety of effects. These may, in different circumstances, take the form of:

1. loss of genetic diversity (heterozygosity);
2. addition of novel genetic combinations; and
3. the introgression of genotypic variation from related species (or varieties).

Secondly, it remains difficult to demonstrate these effects in the field, and so while they may each be important, their relative contribution to island evolutionary change remains uncertain. Thirdly, notwithstanding recent developments in molecular biology and genetics, it is hard to be sure how well the specific phenotypic properties upon which the success of a lineage depends are measured by tools such as allozyme electrophoresis (see papers in Clarke and Grant (1996) for further discussion).

The topics covered in this section have a far wider significance than 'merely' the generation of island biodiversity. In particular, the uncertainty surrounding the differing potential outcomes of population bottlenecks are relevant to attempts to predict the species losses likely to result from continuing anthropogenic pressures such as habitat fragmentation (Chapter 9).

5.2. Sex on islands

There are many ways in which plants and animals reproduce, the simplest dichotomy being between sexual and asexual modes. What works best in the island laboratory? How does the possession of a particular breeding system affect the chances of colonization or speciation on islands? Does hybridization create new forms and thus engender diversification, or does it restrict the tendency to radiation by pulling back two diverging forms from the brink of irrevocable parting?

Plants exhibiting sexual reproduction may be classified as monoecious (male and female flowers on the same plant), hermaphrodite (bisexual flowers), or dioecious (plants that, generally, bear either male or female flowers throughout their life span). Bawa (1980) noted that since Hawaii and New Zealand have relatively large proportions of sexually dimorphic species, there may be a correlation between dioecy and the island condition. This is an interesting observation, as it might be assumed to be more difficult for dioecious species to colonize a remote island, given the apparent need to have both sexes present for success (Ehrendorfer 1979; Sohmer and Gustafson 1993). But as Bawa (1980, 1982) has cautioned, too few data were available to control for differences between phytogeographical regions and to establish this high incidence as a reliable general pattern. However, if phytogeographical dissimilarities were ignored, the limited data available indicated that the proportions of dioecious taxa were several times higher in New Zealand and Hawaii than in the flora of the mainland. For instance, data then available indicated that 27.7% of Hawaii's flora is dioecious compared with 9.0% of species on Barro Colorado Island, a recently isolated patch of rain forest in Panama which became an 'island' only when the central section of the Panama canal was constructed to form Lake Gatun: for these purposes it can be considered a piece of mainland. The Hawaiian figure has recently been revised downwards to 14.7% for species which are strictly dioecious, or to 20.7% if a broader grouping of sexually dimorphic species is used—still the highest incidence of dioecy of any well-specified flora (Sakai *et al.* 1995*a*). What might explain a high proportion of dioecious taxa on remote islands? The following possibilities were suggested:

1. Self-compatible hermaphrodites have more often established small populations on islands but selection for outcrossing then favours the evolution of dioecy *in situ*. This is the **autochthonous development hypothesis** (Thomson and Barrett 1981).

2. Dioecious taxa may have been disproportionately more successful in colonizing the islands, which might, in turn, be a function of the established correlation between dioecism and seed dispersal by avian frugivores (Sakai *et al.* 1995*b*). Birds are important agents for the establishment of island interior floras. Thus, the high proportion of dioecious species may reflect little on the nature of the breeding system but instead be an indirect effect of better dispersal. This might be termed the **dispersal syndrome hypothesis**.

3. Once dioecious taxa reach an island they may have a better chance of establishment, not withstanding the general need for two sexes to be present, because of a further feature linked evolutionarily with dioecy, that is the tendency for dioecious plants to be pollinated by small generalist insects. This might be termed the **pollination syndrome hypothesis**.

4. As dioecism results in outbreeding, with its assumed gains in fitness, dioecious lineages may survive better than other colonists. This might be termed the **enhanced survival hypothesis**. A comparison of the average number of species per genus for the Hawaiian flora reveals that dioecious genera are almost twice as speciose as hermaphroditic counterparts. Apparently, dioecious forms have arisen successfully more frequently than those of hermaphrodite ancestry (cf. Carlquist 1974).

5. As was known to Darwin, dioecism is more prevalent among trees and shrubs than among herbs, and within the island floras the proportion of perennial woody species is generally larger than for the world flora as a whole, thus leading to another indirect effect, which might be termed the **growth form hypothesis**.

Sakai *et al.* (1995*a*) have advanced the analysis using the recently completed re-appraisal of the Hawaiian flora. They found that dioecy (strictly dimorphy) is frequent in part because many colonists were dioecious, but that autochthonous evolution of dioecy has occurred in at least 12 lineages, including several species-rich lineages. Indeed one-third of currently dioecious species derive from monomorphic colonists. The high incidence of dioecy is thus not because dioecious colonists evolved more species per colonist than monomorphic colonists. The generality of the pattern emerging from Hawaiian studies has yet to be established. The issue of dioecy in island floras is included here not because it is a clearly established pattern, but almost for the reverse reason: it is largely unknown, but potentially fascinating. The fascination lies in the potential for understanding the evolutionary assemblage of island floras via the hierarchical links between functional characteristics such as plant growth form, dispersal, colonization, and survival, which may ultimately help us to understand why some lineages have radiated on islands while others are absent or are present but have failed to radiate (Sakai *et al.* 1995*b*; Barrett 1996). Similar ideas are relevant to understanding the structure in the assemblage of island floras on ecological time-scales (Whittaker *et al.* 1997; Chapter 8).

Reproductive mode may also be tied up with colonization probabilities in certain animal taxa. Hanley *et al.* (1994) note that **parthenogenetic** (asexually reproducing) lizard species are relatively common on islands. This may be because of: an enhanced colonization ability; the opportunity to flourish in a less biotically diverse environment (in which the relative genetic inflexibility that might be associated with parthenogenesis is not too disadvantageous); or the opportunity to escape from hybridization or competition with their sexual relatives. Their study examined the coexistence of parthenogenetic and sexual forms of the gecko *Lepidodactylus* on Takapoto atoll in French Polynesia. The two forms of *Lepidodactylus,* although closely related and hybridizing to a limited extent (through the activities of males of the sexual form), are considered separate species. Their coexistence appears to be stable in the short term at least, as the proportions of their two populations were similar in 1993 to a previous survey in 1986. Hanley *et al.* note that the asexuals are actually extremely heterozygous relative to the sexuals. This may seem odd at first glance, but apparently all parthenogenetic vertebrates with a known origin have arisen from a hybridization event. This in large part accounts for their considerably higher levels of heterozygosity than closely related sexuals, but other mechanisms may well be involved. In this particular case, far from the asexual form suffering from competition with the sexual, it is the parthogenetic form that has the wider distribution within the atoll (perhaps indicative of hybrid vigour), with different clones having overlapping but different habitat preferences, and with the sexual form more or less restricted to the lagoon beach. In this case the coexistence of the two species appears to be facilitated by the specialization of the *less* successful sexual taxon to beach habitats. This particular parthenogenetic lizard is clearly not competitively inferior at the present time.

According to figures cited by Grant and Grant (1994), speciation through hybridization may account for as many as 40% of plant species (all allopolyploids arise in this way), but is considered to be much rarer in animals. However, cross-breeding in animals is not so rare, and there have been numerous studies of hybrid zones, wherein hybridization takes place between two recognized species. When successful, it provides a form of gene flow between them (cf. Kaneshiro 1995, above). Grant and Grant (1994) studied the morphological consequences of hybridization in a group of three interbreeding species of Darwin's finches on Daphne Major (Galápagos) between 1976 and 1992. *Geospiza fortis* bred with *G. scandens* and *G. fuliginosa*. The hybrids in turn backcrossed with one of the recognized species. Interbreeding was always rare, occurring at an incidence of less than 5% of the populations, but none the less was sufficiently frequent to provide new additive genetic variance into the finch populations two to three orders of magnitude greater than that introduced by mutation. During the course of their studies B. R. Grant and P. R. Grant (1996) noted a higher survival rate among hybrids following the exceptionally severe El Niño event of 1982–83. This event led to an enduring change in the habitat and plant composition of the island, which appears to have opened up the food niches exploited by the hybrids, and provided for most of the backcrossing observed during the study. Their work thus demonstrates that significant fluctuations in climate, albeit modest on a Quaternary scale, can have important ecological and evolutionary consequences.

Depending on the nature of the species involved in hybridization, this process may either render evolution in a new direction less likely or, where the species involved differ allometrically (i.e. in form), may provide the starting point of a new evolutionary trajectory. Hybridization is also considered to be common within the Hawaiian flora. Indeed, even in the species-rich genus *Cyrtandra*, which was once thought not to hybridize much (if at all), recent studies suggest that it might in fact be quite common (Sohmer and Gustafson 1993).

5.3. Niche shifts

The **ecological niche** has been variously defined, leading to difficulties in using the term without ambiguities creeping in. When first introduced into ecology it reflected the figurative usage of a place in the community, but developed (through Grinnell and Elton's ideas) into specifying the types of opportunities that occur for species to use resources and avoid predators (Schoener 1989*a*). Hutchinson's (1957) usage was of a different form: the *n*-dimensional hypervolume of environmental axes within which a species occurs. A plant or animal has a certain range of an environmental resource, such as temperature, at which growth is optimal. Outside this optimal range it may survive but not thrive, until a point is reached, the limits of its species tolerance, beyond which the species cannot persist. Each resource gradient forms a part or dimension of the hypervolume.

In the island evolutionary context, the term **empty niche** is often used where forms that might be met on a mainland are absent from the island, providing evolutionary opportunities for forms that do reach the island: the idea being that a fundamental niche exists and it is just a matter of filling it. This harks back to the debate about disharmony (Chapter 3) and whether there are really architectural blueprints for the way communities are constructed. While such concepts as the empty niche are of intuitive value, they can prove extremely difficult to apply.

The literature discussed in the following sections mostly refers to a form of niche theory that derives from the resource gradient idea. Typically, it does not encompass a multidimensional characterization of the niche, but is limited to a single or very limited set of characters, such as mobility, body size, or jaw size, each of which provides a surrogate measure for one dimension of the *n*-dimensional hypervolume of the organism. The ambiguities of the niche concept are avoided in many such studies by describing the distributions of species populations along resource spectra without the label 'niche' being applied. As Schoener (1989*a*) points out, much of this literature is, in essence, concerned with competition, chasing the elusive question (Law and Watkinson 1989) of the extent to which competing species can coexist. The answer remains elusive in the island evolutionary context because it is easier to detect functional changes (particularly morphological changes) in island forms, than it is to determine the role of interspecific competition in relation to other potential causes of such change. The best chance here is to examine contemporary processes, which may then be inferred to hold a wider historical relevance. The greatest methodological difficulties come when it is the 'ghost of competition past' that is being sought, i.e. when researchers are setting out to deduce past events from current pattern (Law and Watkinson 1989).

5.3.1. The loss of dispersability

For a distant (oceanic) island there will clearly be a strong bias towards groups that are highly dispersive, as these will, by definition, have a greater chance of colonizing. This was reflected in the biogeographical patterns discussed in Chapter 3, of filter effects, the gradual loss of taxa with increasing distance from continental source areas, and of disharmony. The relevance of these effects are broad, and wide ranging, from the evolutionary opportunities on islands like Hawaii, to the ecological structuring in less remote islands such as the Krakatau group, discussed in Chapter 8. Island biogeography thus bears the imprint of dispersal structuring of faunas and floras almost as its first principle.

Once on an isolated island, however, there may be a strong selective force against dispersal ability in certain taxa. The loss of dispersiveness has long been considered a common feature of evolution on remote islands. Although most birds and insects are capable of flight (i.e. are volant), some are only weak or reluctant fliers, and others are flightless. Examples given by Williamson (1981) of flightless forms include the 20 endemic species of beetle on Tristan de Cuhna, all but two of which have reduced wings. A number of ideas have emerged in explaining such tendencies. One put forward by Darwin (letter to J. D. Hooker, 7 March 1855; Roff 1991) is that highly dispersive propagules are more liable to be lost from the gene pool, for instance by a highly dispersive plant or insect being blown off to sea, so selecting for less-dispersive forms among the island population. A second idea, applicable to flightless birds such as the famously extinct dodos, is that the lack of predators means there is no longer a selective advantage in flight in ground-feeding species. Indeed, flight might be disadvantageous in that the energy used in maintaining and using flight muscles is effectively wasted. However, as Williamson (1981) cautioned in citing these explanations, flightless insects and birds (e.g. several ratites, such as the ostrich) can also be found on continents. The condition is thus not unique to islands. Before attempting to test ideas concerning loss of dispersal, it might be prudent to confirm the relationship statistically, i.e. to demonstrate that the incidence of flightlessness on islands is higher than expected by chance alone.

Roff (1991, 1994) has undertaken just such analyses for insects and for birds. Darwin's original observations on insects were based, of course, on relatively few data, and in his analysis of the much larger data sets now available Roff (1991) found that statistically there is no evidence of an association between oceanic islands and insect flightlessness. He argues that the large size of most oceanic islands makes flightlessness of little positive selective value: only a small fraction of the population is likely to be lost from such sizeable land masses (see also Ashmole and Ashmole 1988). That his statistical analyses of the incidence of flightlessness on islands has overturned the 'conventional wisdom' he therefore regards as unsurprising. It should be stressed, however, that Roff's analyses are based on a dichotomy between forms capable of flight and those that are incapable of flight. He does not take into account island forms that have evolved a reduced flight capability or behaviour without actually losing the ability altogether. None the less, his analysis provides another example of the importance of actually attempting to test the supposed peculiarities of island forms.

In his second article, on birds, Roff (1994) points out two important factors that complicate the analysis. First, in the case of most bird groups, the small sample size limits the potential for formal statistical analysis. Secondly, the phylogenetic constraints on flightlessness must be considered. The condition may have evolved once, followed by subsequent speciation within a flightless lineage, or it may have evolved several times, in extreme cases once for each flightless species. For example, among important flightless bird groups, the 18 known penguin species are considered to be monophyletic, having speciated from a common flightless ancestor, while the 41 ratite species (ostriches, etc.) represent between one and five evolutionary transitions to flightless forms, depending on whose phylogeny one accepts. In contrast, flightlessness in rails has most probably evolved separately for each of the 17 (or more) flightless species, among the 122 (or more) recognized rail species. Truly flightless rails (Fig. 5.1) are only found on islands and they represent about half of the island rail species (Roff 1994). More ratites occur on islands than on continents, but the taxonomic distribution does not indicate that flightlessness is more likely to evolve on islands in this group. Examples of flightlessness from other groups of birds include the following island forms: the kagu, the two dodos, the kakapo (the only flightless parrot), the Auckland Islands teal, the Galápagos cormorant, the three Hawaiian ibises, and four Hawaiian ducks; these 13 species representing at least eight evolutionary transitions. Taking all these factors into account, Roff did conclude that a tendency to flightlessness is, in general, more likely to occur on islands (and also, incidentally in aquatic species), although only within the rail group could the association with insularity be established by a formal statistical test.

Fig. 5.1. A classic island endemic: the Aldabran flightless white-throated rail, *Dryolimnas cuvieri* ssp. *aldabranus*. Flightless rails have evolved on numerous oceanic islands. Most are now extinct. (Photo: Clive Hambler 1983.)

On the topic of flightless rails, which seems to be the group best exemplifying the trend in island birds, Diamond (1991*a*) reported a new species, *Gallirallus rovianae*, which is believed to be either flightless or weak-flying. It was collected in 1977 from New Georgia, Solomon Islands. *Gallirallus rovianae* possibly occurs on four nearby islands, which were joined at times of low sea levels in the Pleistocene. The ancestral form is volant and more boldly patterned than the new species. Diamond regards it and its relatives on other oceanic islands as cases of convergent evolution, whereby rails have independently evolved weak-flying or flightless forms and also, in several cases, reduction in boldly patterned plumage, on widely scattered oceanic islands. In addition to the 11 groups of rails showing independent evolution of multiple flightlessness or weak-flying derivatives (subspecies or species) which Diamond recognizes, he points out that subfossil remains of others, now extinct, have been found on most of those Pacific islands that have been explored by palaeontologists (see also Steadman 1997*a*). These weak- or non-flying rails do not fare well in contact with people, as half of those extant at the time of European discovery have subsequently become extinct. Diamond concluded that this form of evolution in rails is probably even more common than currently demonstrated, with many more extinct (and perhaps even some more extant) forms awaiting discovery (Chapter 10). In the case of rails, then, there is clearly a strong selection pressure resulting in reduced flying ability, and this appears to be related to the energetic and weight burden of unnecessary flight muscle on islands lacking terrestrial mammalian predators.

From the plant world, the genus *Fitchia* (in the Asteraceae), which is endemic to the Polynesian islands of the south Pacific, illustrates a reduction in dispersability (Carlquist 1974; Williamson 1981). Its closest continental relatives are herbs with spiked fruits transported exozoically (i.e. by external attachment to an animal), but in *Fitchia* the spikes are relict, the fruits are much larger, the plants are trees, and their fruits drop passively to the forest floor. Another example is provided by endemic Pacific island species of *Bidens* (also in the Asteraceae) which illustrate various steps in the loss of barbs and hairs and other features favouring dispersiveness, these changes being accompanied by ecological shifts from coast to interior and, in some cases, wet upland forest habitats (Carlquist 1974; Ehrendorfer 1979). Of course, it could be argued that these examples suffer from a certain circularity of reasoning. To allay such concerns, a demonstration is needed of such changes actually occurring. Once more, the Asteraceae provide the evidence.

Cody and Overton (1996) monitored populations of *Hypochaeris radicata* and *Lactuca muralis*, both wind-dispersed members of the Asteraceae, on 200 near-shore islands in Barkley Sound (Canada). The islands ranged from a few square metres in area up to about 1 km^2. Significant shifts in diaspore morphology occurred within 8–10 years of population establishment, the equivalent in these largely biennial plants of no more than five generations. The diaspore in these species consists of two parts, a tiny seed with a covering (the achene), which is surrounded by or connected to a much larger ball of fluff (the pappus). The clearest findings were for *Lactuca muralis*, the species with the largest sample size. Founding individuals had significantly smaller seeds (by about 15%) than mainland populations, illustrative of a form of founder effect (see above). Seed sizes then returned to mainland values after about 8 years, while pappus volumes decreased below mainland values within about 6 years. If the pappus is thought of as a parachute and the seed the payload, then the ratio between the two indicates the dispersability of the diaspore (Diamond 1996). Figure 5.2 provides a schematic representation of the trends in dispersability. Cody and Overton were able to quantify these phenomena because they were working on a sort of microscale version of the oceanic island examples usually cited, i.e. many of their populations were very small, occupying tiny areas, and their species were short-lived—ideal for monitoring short-term evolutionary change (Diamond 1996).

5.3.2. Gigantism and nanism

It presents something of a puzzle at first glance that two conflicting traits are viewed as common in island forms: **gigantism**, examples of which include the Galápagos and Indian Ocean tortoises (Fig. 5.3) (Darwin 1845; Arnold 1979), and **nanism** (dwarf forms), illustrated by island subspecies of ducks, which are commonly smaller than mainland species (Lack 1970). Of course, size variations within taxa are not restricted to islands, and so there is once

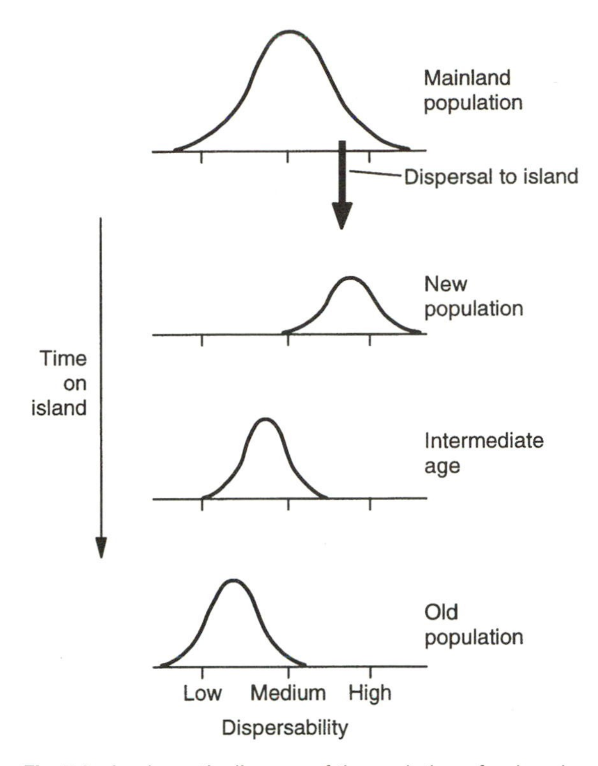

Fig. 5.2. A schematic diagram of the evolution of reduced dispersability of wind-dispersed plants on islands, as suggested by the changes recorded in *Lactuca muralis* (Asteraceae) on near-shore islands in British Columbia, Canada. Maximal dispersability characterizes the youngest island populations, as the founding event(s) will be from the upper end of the dispersal range of the mainland population. Thereafter selection acts to decrease dispersability of propagules on islands as more dispersive diaspores are lost to sea. (Simplified from Cody and Overton 1996, Fig. 1.)

Fig. 5.3. The giant tortoise *Geochelone gigantea*—a classic island giant; Aldabra Atoll, Indian Ocean. (Photo: Clive Hambler 1983.)

again a danger of inferring a general rule from a selection of a few striking cases. For instance, all giant tortoises are considered as belonging to the genus *Geochelone*, the modern island forms reaching up to 130 cm long on Aldabra and the Galápagos. However, giant tortoise populations are readily eliminated by humans, and other species of *Geochelone* occur, or occurred, in South America, Madagascar, the Seychelles, the Mascarene Islands, and on the African mainland. It is not certain whether large size evolved after reaching islands or whether it merely bestowed the capabilities of dispersing to islands as remote as the Galápagos (Arnold 1979). However, there is now a reasonable basis for stating that size changes do commonly occur in island evolution and for an intuitively satisfactory, if mostly unproven, understanding of why the changes occur in the direction observed.

Foster (1964) and Lomolino (1985) have examined size data for insular species or subspecies of mammals. Compared with mainland forms, small mammals tend to become larger, and vice versa. For instance, Foster's data suggested that around 85% of island rodents are larger, possibly because of the absence of predators, with a resulting increase in intraspecific competition, and selection supposedly favouring larger individuals (Adler and Levins 1994). It is worth cautioning that in other circumstances, competition has been invoked as a force producing size *reductions* in small vertebrates (specifically in lizards; section 6.2.3). In relation to the size increase displayed by small mammals, it may have been the case that while terrestrial carnivores were lacking from Mediterranean islands, birds of prey would have been present (Reyment 1983). Birds of prey take small mammals, and an increase in size among rodents may have served to reduce predation in an environment lacking land carnivores (see also Schüle 1993). At the other end of the scale, carnivores and artiodactyls (ruminants of the even-toed type) become smaller in about the same proportion of cases, an outcome of the advantage in small size in the restricted insular environment with its limited food resources (as Lomolino 1985). Damuth (1993) has analysed data on continental mammals, concluding that there is an optimum body size for energy acquisition at about 1 kg. He observes that in the absence of a full array of competitors and predators, it should be expected that island populations will evolve towards a body size most advantageous for exploiting resources in their diet category. This he regards as part of the reason for the island trends, although he recognizes that other factors are important.

Reyment (1983) reviewed these ideas with respect to the so-called HEDR fauna of the Mediterranean islands—the dwarf hippopotami, elephants and deer, and the giant rodents of the Pliocene to sub-Recent period. If these atypically sized animals are unfamiliar to the reader, it is because none of these endemic forms survived to the present era (Schüle 1993). Reyment (1983) regards the explanation for the size changes as being in strong selection pressure, following similar lines of reasoning as above. For small forms, the absence of predators is assumed to allow intrapopulation competition to select larger forms, which will dominate reproductive effort in the population. In the animals which start large, like elephants, it may be that islands provide insufficient resources to sustain such large body weights. Even infrequent scarcity could provide strong selection pressure towards smaller forms. Some of the best evidence for nanism seems to be for fossils of the Cretan form of dwarf elephant, which show a wide size range, interpreted as representing stages in the development of the dwarf form. 'The flatland, lumbering ancestral form, with good swimming capabilities, had been irrevocably transformed into a clever climber of rocky slopes' (Reyment 1983, p. 302). The prize for the greatest degree of dwarfing in an island elephantid goes to the Maltan *Elephas falconeri*, which was barely more than a metre in height (Lister 1993) and only a quarter the size of the putative ancestor, *E. namadicus*. Schüle (1993) has suggested that low reproduction rates, dwarfing, and the development of more efficient dentition accompany each other in the endemization of island ungulates. The lack of carnivores allows dense populations of the few large herbivores to exert strong pressure on the island vegetation, resulting in semi-desertification. This, in turn, adds to the selective pressures towards nanism in a system of positive feedback. Plausible, but unproven.

The general view is that the remarkable decreases in size occurred over a relatively short period of time. A recent revelatory find from Wrangel Island, some 200 km off the coast of north-eastern Siberia, is of particular interest in terms of the time-scale of change. Vartanyan *et al.* (1993) reported remains of the woolly mammoth, which appear to have remained on the island when it became separated from the mainland *c.* 12 000 BP. Survival of the mammoth was permitted here by the climatic regime which maintained its favoured steppe vegetation. Whereas elsewhere, in its continental range, the mammoth became extinct by around 9500 BP (either because of climate change, or humans, or a combination of both), radiocarbon dates of teeth indicate its presence on the island between 7000 and 4000 BP. Not only did the population thus attain relictual status, but during the period between 12 000 and 7000 BP it appeared (on the basis of fossil teeth) to shrink in body size by at least 30%. However, size changes need not be so rapid, nor follow so soon after colonization. Sicilian elephants probably took between 250 000 and 600 000 years to shrink to about half the dimensions of the ancestral *Elephas antiquus* and about 1 million years to reach one quarter of the ancestral size.

It is important in interpreting such size changes to build up a picture of the dynamics of evolutionary change on islands. If elephants can colonize once, why not subsequently? According to Reyment (1983) the available evidence for the Mediterranean islands indicates that in each case investigated the colonization event was fairly short, and there were no further successful arrivals of ancestral forms after the new adaptations had become established; however, the partial nature of the fossil record limits the confidence one can have in such statements. The disharmonic composition of the Pleistocene island faunas indicates that large mammals with good swimming capabilities (the deer, elephants, and hippopotami) had a clear advantage in colonizing islands. Given the distances involved, a combination of active swimming and drifting must have been involved—such biotas have been labelled '**waif biotas**'. The presence of certain small mammals can probably be explained by passive transport on floating flotsam, i.e. by **rafting**. (Where islands are spread out with large patches of sea separating them, the concept of a **sweepstake route** has been invoked, suggesting a large degree of chance as to which species of a particular taxon make it to distant islands.) To explain the apparent patterns of single phases of colonization in the larger fauna, Reyment (1983) suggested that some kind of catastrophic event, competition, or overpopulation might have driven animals to undertake such journeys. The late Miocene saw the Messinian salinity episode, indicating a marked eustatic fall in sea level, with the Mediterranean becoming isolated from the world oceanic system, and large parts drying to form salt beds. The event peaked about 5 million years ago, well before the dramatic sea-level fluctuations that accompanied Pleistocene glacial–interglacial cycles. Some of the large Pleistocene mammals, like *Myotragus* (dwarf antelope or island goat) on the Balearic Islands and *Prolagus* (a giant rodent) on Sardinia, could have been relicts of the desiccation on the Miocene/Pliocene border. But it appears that the hippopotami, elephants, and deer reached the islands during the Lower Pleistocene. As most of the suggested Quaternary land bridges to Mediterranean islands are insupportable on recent geological evidence, it follows that these animals must have swum (Schüle 1993).

It is a generally held view that the arrival of continental species typically displaces island forms, and according to Schüle (1993) there is some evidence for this in the Mediterranean mammal data. For instance, *Naemorhedus,* a form of dwarf antelope, disappeared from Corso-Sardinia following elephant and hippopotamus colonization during the Lower Pleistocene. Because of their lack of patterns of self-defence (unnecessary traits in the absence of predators), all Pleistocene ungulates, other mammals of any considerable size, flightless birds, and giant tortoises on the Mediterranean islands were almost certainly wiped out by the first humans to arrive. This is, however, only recognizable in the archaeological record on Cyprus and Sardinia. It appears that the extant wild ungulates on the Mediterranean islands were all introduced by man.

While the tendencies to nanism and gigantism appear to be fairly clearly established, many species do not exhibit these traits and some contradict the general trends. Against the apparent tendency to size increases in small vertebrates, there is fossil evidence for island lizards from the Caribbean and elsewhere, of size reductions during the Holocene (Pregill 1986). Explanations for these trends have been varied. In particular cases, the evidence is strongly supportive of some form of coevolutionary competitive pressure brought about by formerly allopatric taxa coming into sympatry (cf. Roughgarden and Pacala 1989). In other cases, the activities of humans, such as habitat alteration and the introduction of terrestrial predators, may be responsible not only for the extinction of larger lizards from within island assemblages but also for the reduction in size within species evidenced by the Holocene fossil record (Pregill 1986). Some support can be found for this idea from historical and contemporary data. For instance, in the Virgin Islands, populations of the St Croix ground lizard (*Amereiva polyps*) have declined steeply over the past century, and body size of mature adults has also decreased. The ecology of these islands has been greatly affected by humans, and in particular, the introduction of the mongoose has been credited with an important role in the extinction of the lizard from the main island (Pregill 1986). Associations between human colonization and the extinctions of larger species from within insular lizard assemblages are also known from the fossil records of the Canary Islands, Mascarenes, and Galápagos archipelagos. Indeed, Pregill (1986) concludes that insular lizards that draw our attention because of their exceptional size may have been rather ordinary in prehistory. Fossils of extinct large frogs and snakes are also known from the Caribbean.

Food chain links may be important in selecting for both smaller and larger forms. An interesting example is Forsman's (1991) study of variation in head size in adder *Vipera berus* populations on a group of islands in the Baltic Sea compared to the Swedish mainland. In accord with the general trend for smaller vertebrates noted above, it was found that relative head length of adders was smallest in the mainland population. An interesting further feature of this study was that, within the island set, relative head length increased on the islands with increasing body size of the main prey, the field vole, *Microtus agrestis*. This was interpreted as the outcome of stabilizing selection for head size within each population, that is to say an evolutionary response to the geographic variation in body size of the main prey species and the small number of alternative prey species available on islands.

As noted above, while plausible explanations may be put forward for size changes in island forms, many remain untested. Schwaner and Sarre (1988), in their study of Australian tiger snakes *Notechis ater serventyi*, set out to distinguish which of three hypotheses (adapted from an early study by T. J. Case of iguanas) best accounted for large body size in populations on Chappell Island in the Bass Strait, Southern Australia. The three hypotheses were:

1. **Predation hypothesis**: either there is selective release if no predation occurs, or there is a selective advantage to escape the 'window' of vulnerability for smaller prey if top predators are absent.

2. **Social-sexual hypothesis**: due to the usually high densities that occur among island populations, intraspecific competition among males and females selects for large body size.

3. **Food availability hypothesis**: increases in the mean and variance in food supply/demand ratio select for gigantism.

The tiger snake is the only snake on Chappell Island, and the only other predators are a small number of feral cats and two pairs of raptors. Humans may also have an occasional role in killing snakes. In general, the size of tiger snakes on offshore islands varies in relation to the body size of available prey, a finding similar to that reported for the study of the Baltic islands. However, the Chappell Island snakes are exceptionally large whereas their potential prey are no larger than those of other islands in the region which support large snakes. A systematic mark–recapture study was undertaken on Chappell Island in November 1985, measuring a variety of features of size, weight, and muscle strength. Although predation was not measured directly, indirect evidence, such as frequencies of injuries, suggested that snakes were attacked infrequently and that predation had not contributed to the evolution of gigantism in this case.

The social-sexual hypothesis argues for selection for larger males as a consequence of male–male combat for mates. However, in this and other island studies of tiger snakes, there was no evidence for such combat, and the relative degree to which male size exceeded females was less than for other islands supporting large snakes, and no different from mainland populations. The hypothesis was not supported.

Schwaner and Sarre's (1988) third hypothesis related to food availability. Survival of adult snakes is fairly heavily dependent on young mutton birds *Puffinus tenuirostris*, which are available as prey for only a few weeks each year. It was calculated that during the brief season, the mutton birds provided a potential larder of some 891 chicks/hectare, or 35 chicks/snake. Mutton bird chicks are thus a saturating, albeit highly seasonal resource. The selective advantages of larger body size in the tiger snake might be in larger clutch size, more frequent reproduction, and, critically, greater fat storage enabling survival through long periods of fasting. Although other prey are available on the island, their densities are low in comparison to the mutton birds. Adult snakes thus eat an abundant, high-quality resource, which requires little searching time to locate, but is highly seasonal. Schwaner and Sarre concluded that *seasonality* of food resources, rather than quantity *per se*, best explains the extreme size of the tiger snakes of Chappell Island. They do not claim to have finally refuted the other two hypotheses, merely to have identified which of the three seems most reasonable. Notwithstanding the tentative nature of the conclusion, this is the sort of study that will produce evidence to distinguish between competing hypotheses and thus advance our understanding of island evolutionary change.

5.3.3. Character displacement

Character displacement can, and frequently does, refer to size changes among island forms. What distinguishes these niche shifts from those discussed above is the causal interpretation: that it is competition between two fairly similar varieties or species that brings about selection, in one or both, away from the region of resource overlap (Brown and Wilson 1956). Diamond *et al.* (1989, p. 675) define ecological character displacement as 'the effect of competition in causing two initially allopatric species to diverge from each other in some character upon attaining sympatry'. Darwin (1859) termed this 'divergence of character', and it has also been termed 'character coevolution' or 'coevolutionary divergence', although there is actually not one, but a suite of related theoretical ideas centred around this notion (Otte 1989).

Caribbean *Anolis* lizards provide examples of character displacement, Schoener (1975) finding shifts in perch height and diameter consistent with competitive effects. Species of similar size were found to affect one another more than dissimilarly sized species, while larger species affected smaller ones more than the reverse (cf. Roughgarden and Pacala 1989). Furthermore, according to a distributional simulation using Monte Carlo analysis, the distribution of lizard species over microhabitat types on satellite islands of the Greater Antilles was also found to be consistent with competition (Schoener 1988). However, the results of this analysis depended on the assumptions built into the null model—a general problem with null models in ecology (Chapter 8).

There is also some evidence for competitive displacement in Hawaiian crickets. Otte (1989) presents an array of observations, including data on song variation in *Laupala*, a genus of swordtail crickets. The songs were found to be hyperdispersed, the divergence in signals being greater between coexisting populations than between allopatric populations, suggesting that competitive interactions are important. Otte concluded that character displacement is probably extremely common in the Hawaiian crickets and, by extension, elsewhere, although it is difficult to demonstrate unequivocally. While the most convincing demonstrations of character displacement tend to be from systems involving few interacting species (cf. the *Anolis* studies), such effects may also be detectable, and certainly are relevant, where a somewhat larger network of interacting species is involved. Interactions among species, even if leading to niche shifts in one or both, do not necessarily have to lead to coevolutionary divergence and radiation, indeed it should be noted that this model has been rejected and a different interpretation put forward for the *Anolis* of the Northern Lesser Antilles (Chapter 6).

Given the difficulty in validating ideas of character displacement (Connell 1980), it would be ideal if an experiment were to be devised that controlled for other complicating factors. The principal problem is that character changes with time are generally not observed, but are merely inferred by comparison of the characteristics of populations in allopatry with those in sympatry. Such studies provide corroboration but not validation. Diamond *et al.* (1989) have reported a natural experiment which comes close to solving this problem, in demonstrating divergence within populations in sympatry for less than three centuries. In the mid-seventeenth century, Long Island, off New Guinea, erupted violently. The eruption may have exceeded that of the 1883 Krakatau eruption in scale; more than 30 km^2 of ejecta was released, ash covering New Guinea to a distance of 500 km. Long Island itself suffered caldera collapse, and the resulting doughnut-shaped island (328 km^2 and 1280 m high) was covered in over 30 m of volcanic ejecta. The nearby islands of Tolokiwa and Crown were also affected by the eruptions and it may be concluded from geological, biological, and other evidence that the islands were effectively stripped of their former inhabitants (whether sterilization was complete or not is unimportant). By analogy with the recolonization of Krakatau (cf. Whittaker and Jones 1994*a*,*b*; Thornton 1996) it is reasonable to assume a delay of a few decades for revegetation to proceed to the point at which a forest bird assemblage would have developed. Thus the bird populations of the three islands can be taken to have been founded less than 300 years ago. The avifauna contains two different-sized species of honeyeaters of the genus *Myzomela*—*M. pammelaena* and *M. sclateri*. Fortuitously, for present purposes, they are the only islands on which the two coexist. Allopatric populations on other islands can be identified as the probable sources for the founding populations of the Long group. Thus, the serendipitous 'experiment' has a maximum duration of about three centuries, for which two otherwise allopatric species have co-occurred. Furthermore, the two species do occur together within the Long group, being abundant together in all habitats and at all altitudes, often being found in the same flowering tree. They are thus sympatric even at the smallest within-island scale.

Diamond *et al.* (1989) compared morphological measurements of all available specimens from the two species and found that the Long (sympatric) populations are significantly more divergent from each other than are the allopatric source populations. The form this takes is that the larger species, *M. pammelaena,* is bigger still on Long than in the source populations, and the smaller *M. sclateri* is even smaller. Comparing samples of the two species from allopatric populations, the weight ratio was found to lie between 1.24 and 1.43, but for the sympatric Long populations the value was 1.52, i.e. the difference was greater. Diamond *et al.* (1989) considered alternative explanations for these patterns but concluded that character displacement is the most parsimonious answer. If this is accepted, character displacement is thus shown to be capable of developing in relatively short periods of time. Indeed, the 300-year time span may be unremarkable by analogy with similar niche shifts in other circumstances. Conant (1988) found that Laysan finches (*Telespyza cantans*) introduced to Pearl and Hermes Reef underwent measurable shifts in bill shape within 18 years, in adjusting to their local food supply. Similarly, significant change in diaspore size in *Lactuca muralis* in Barkley Sound took place within just five plant generations (see above; Cody and Overton 1996). Diamond *et al.* (1989) went on to consider 13 other island groups where two or more myzomelid honeyeaters occur together on the same island. In some cases they were found to differ in habitat (such as altitude), but where co-occurring in the same habitat (**syntopy**), their weight ratio was always 1.5 or larger. Size may affect coexistence through effects on dominance, productivity requirements, prey size, perch position, and the economics of hovering. The data thus provide, in the specific case, good evidence for character displacement without apparently plausible alternative explanations, and in other cases, corroborative support for similar effects where myzomelid honeyeaters co-occur.

5.3.4. Ecological release

Ecological release is almost a converse of competitive displacement. The response to release from competitors or from other interacting organisms, such as predators, can take perhaps two main forms: the loss of 'unnecessary' features (e.g. defensive traits, bold patterning); or the rather different form of an increase in variation within the species. Sadly, the lack of defensive traits commonly involved a lack of fear of humans, undoubtedly a contributing factor to the demise of island species such as the dodos (Chapter 10). The Solomon Islands rail, *Gallirallus rovianae*, discussed earlier as an example of flight-loss, also exemplifies the loss of bold patterning (Diamond 1991*a*). Similar trends have been noted in earlier studies of island birds: with the loss of elaborate or distinctive morphology, and a return to simpler song types (Otte 1989). The general interpretation offered for these changes is that organisms released from competition with closely related species no longer require such accurate mating barriers, and so these features are gradually lost. The second form of release in the absence of close competitors on an island is to free a colonist from constraining selective pressure, thereby allowing it to occupy not only different niches but also a wider array than the continental ancestral form (Lack 1969; Cox and Ricklefs 1977).

This form of ecological release, the increase in niche breadth, forms a key part in the reasoning of the competitive speciation model (section 4.4.2) and is an important part of many scenarios for island evolution (e.g. adaptive radiation). This is therefore an important concept. As with many similar ideas, most studies invoke it without being able to offer clear proof. There is some evidence for ecological release of insular species as a result of the absence of predators, indeed Lomolino (1984*a*) argues that it may be more widespread than generally recognized and, conversely, that competitive release may be overestimated. He cites examples of Fijian fruit bats of the genus *Pteropus* being more diurnal in the absence of predatory eagles than those on less isolated Pacific islands, and of the meadow vole *Microtus pennsylvanicus* being essentially indiscriminant of habitat type on islands without one of its major predators, *Blarina brevicauda*.

Island finches provide classic examples of evolutionary increases in niche breadth within lineages. In illustration, granivorous finch species on remote islands such as Hawaii and the Galápagos are highly divergent in terms of beak sizes (and seed sizes utilized) compared with finches on continents and, plausibly, this can be related to an absence of competitor taxa on the islands (Schluter 1988). Increases in niche breadth within an island lineage are not always morphologically apparent. The ancestor of the Darwin's finch of Cocos Island, *Pinaroloxias inornata*, colonized an island where 'empty' niche space and the lack of interspecific competition provided opportunity for ecological release. This endemic finch has diversified behaviourally and, while showing little morphological variation, exhibits a stunning array of stable individual feeding behaviours spanning the range normally occupied by several families of birds (Werner and Sherry 1987). This intraspecific variability does not appear to be attributable to difference in age, sex, gross morphology, or to the opportunistic exploitation of patchy resources, but appears instead to originate and be maintained year-round behaviourally, possibly via observational learning.

5.3.5. Other niche shifts and syndromes

Although the preceding sections cover some of the main categories of niche shift, there are other features that deserve mention. Hawaii provides an illustration of the trend to woodiness in elements of its flora. It is generally considered that most colonists were weedy, herbaceous, or shrubby, with only a minority of tree-like plants: subsequently, woodiness has developed in the insular lineages. Recent genetic and cladistic studies have supported this view (Carlquist 1995). The Amaranthaceae and Chenopodiaceae provide classic cases, in which the endemic species are only woody (or in cases suffrutescent—woody only at the base of the stem), whereas in continental source areas they are principally herbaceous (Carlquist 1974; Sohmer and Gustafson 1993; Wagner and Funk 1995). For the Atlantic islands, in contrast, Sunding (1979) has interpreted woodiness as a primitive feature, and the Macaronesian vascular flora as thus containing many species of palaeo-endemic character (as defined in section 3.5.1). However, analyses of the nuclear rDNA have since shown that the woody Macaronesian *Sonchus* and five related genera were derived ultimately from continental herbaceous ancestors, most probably from a single founding species (i.e. the alliance is monophyletic). During the adaptive radiation of the lineage, it has undertaken a limited number of inter-island transfers, has diversified in to differing habitats, and has developed considerable morphological diversification, including the development of the woody lineage (Kim *et al.* 1996). The molecular data thus seem to be swinging the argument convincingly against the relictual interpretation of woodiness.

In birds, changes in feeding niche, loss of flight, and loss of defensive instincts on remote islands have already been alluded to, as have size reductions in ducks. An example of size increase is provided by island races of wrens around the British isles, which are typically slightly larger in terms of wing length than their mainland counterparts (Williamson 1981). Another feature, which has been observed by Lack in the birds of the Orkney isles, is the selection of a wider range of nesting sites than is normal on the British mainland (Table 5.1).

Although the various evolutionary changes in ecological niche in island populations have been

Table 5.1

Alterations in nesting sites in some Orkney birds first noted by David Lack (after Williamson 1981, Table 6.4). In all cases the normal mode is also found in Orkney

Species	Normal mode	Orkney mode
Fulmarus glacialis (fulmar)	On cliffs	On flat ground and sand dunes
Columba palumbus (woodpigeon)	In trees	In heather
Turdus philomelos (song thrush)	In bushes and trees	In walls and ditches
Turdus merula (blackbird)	Woods and bushy places	Rocky and wet moorland
Anthus spinoletta (rock pipit)	Sea cliffs	Out of sight of sea
Carduelis cannabina (linnet)	Bushes and scrub	Cultivated land without bushes; reedy marshes

considered separately so far, it will be apparent to the reader that many of the changes evolve together. A proper understanding of individual changes requires models that incorporate several such phenomena simultaneously. Adler and Levins (1994) set out to construct such a model for island rodents, synthesizing results from a range of empirical studies, principally of mice and voles. They noted that island populations of rodents tend to evolve higher and more stable densities, better survival, increased body mass and reduced aggressiveness, reproductive output, and dispersal. It is striking that while there are exceptions, island rodent populations of different species and from disparate geographic areas often demonstrate similar sets of patterns. Adler and Levins termed these collective differences the **island syndrome**. They summarize the main traits and some of the more likely explanations for them in tabular form (Table 5.2).

The higher densities of rodent populations indicated by the crowding effect are an important part of this analysis. Williamson (1981) has pointed out that because for their area islands tend to have fewer species on them than continental areas, population densities will differ between islands and mainlands, in effect as a statistical artefact. He distinguishes the notions of **density compensation** and **density stasis**. Density compensation is where island communities have the same total population density but distributed over fewer species, i.e. with large population sizes per species. Density statis is where the overall population of the community on the island is less than that of the reference mainland system, such that population sizes per species are the same as the mainland. Distinguishing between the two conditions is tricky and necessitates an appropriate experimental design. According to Adler and Levins (1994) a number of studies of island rodents have avoided the pitfalls identified by Williamson, demonstrating that higher average densities of rodent populations do indeed occur in some cases on islands. They cite several examples of experimental studies using fenced enclosures, which have shown that rodent populations in such circumstances can reach abnormally high densities and often destroy the food supply (the 'fence effect' of Table 5.2), thereby indicating dispersal to be an important regulator of population densities under normal circumstances in which there is no (complete) fence. As rodent populations on small islands lack an adequate dispersal sink, the fence effect may be invoked as a force providing selective influence on the population.

One of the mainstays of island *ecological* analysis in recent decades has been the study of the relative significance of island area and isolation (Chapter 7). Adler and Levins' (1994) scheme extends to dealing with how these factors may work in an island *evolutionary* context (see also Adler *et al.* 1986). They hypothesize that average population density might, in general, increase with increasing isolation and hence with reduced dispersal, but that this fence

Table 5.2
Short-term and long-term changes in island rodents and proposed explanations (from Adler and Levins 1994, Table 2)

Island trait	Proposed explanation
Reduced dispersal	Immediate constraint (short-term response) and natural selection against dispersers (long-term response)
Reduced aggression	Initially, reduced population turnover, greater familiarity with neighbours, and kin recognition. Long-term directional selection for reduced aggression
Crowding effect	Isolation ('fence effect' resulting from reduced dispersal) and reduced number of mortality agents such as predation, both of which result in crowding of individuals and consequently higher population densities
Greater individual body size	Initially, a norm of reaction as a response to higher density. Long-term directional selection for increased body size in response to increased interspecific competition
Lower reproductive output per individual	Initially, a reaction norm as a response to increased density. Long-term directional selection in response to decreased mortality
Greater life expectancy (higher survival probabilities for individuals)	Reduced number of mortality agents such as predation

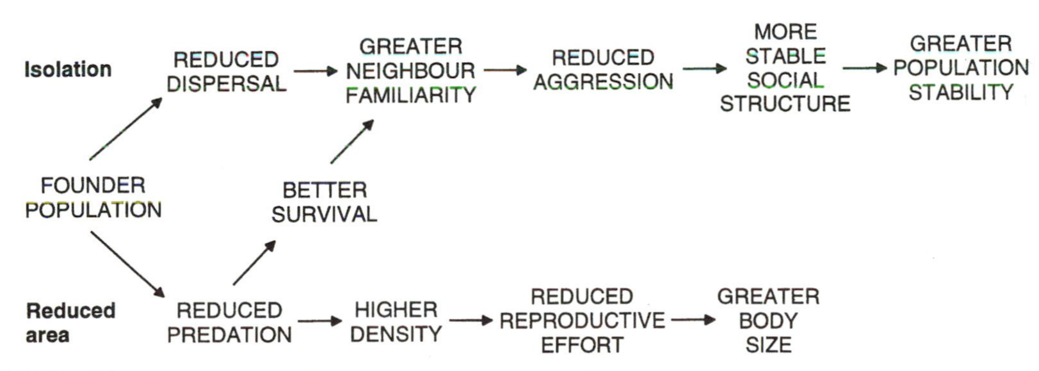

Fig. 5.4. Schematic diagram showing the initial short-term effects of island isolation and area on rodent populations. Long-term effects of insularity are directional selection for increased body size, reduced reproductive output, and reduced aggression. (Redrawn from Adler and Levins 1994, Fig. 2.)

effect might decline with an increase of island area to the point where it disappears altogether as islands more closely resemble a mainland. Thus the island syndrome might be expected to occur only in islands which are: (1) sufficiently isolated, (2) not too large so as to resemble mainlands, and (3) not so small that they cannot support persistent populations. Figure 5.4 illustrates the differing effects which area and isolation are envisaged as having. Effects which may be expressed largely as reaction norms in the short term, i.e. within the range of phenotypes of the founding population, may, under sustained selection pressure, produce rapid evolutionary change and locally adapted island populations that differ from mainland populations. The effect of isolation is a direct one, on dispersal, while that of area is less direct. Larger islands and mainlands typically have more predators, competitors, and habitat types, and because of this (and especially because of predation effects), densities are depressed compared with those of small islands. While Adler and Levins' (1994) model is based on studies of rodents, there is no particular reason why it could not be extended to and tested on other systems.

5.4. Summary

The founder principle is that small bridgehead populations bring with them a biased subset of the genetic variation of the source population. Such bottlenecks appear to be important in the rapid evolutionary divergence of mainland and island lineages, although the scale of founder effects in island evolution remains a contested issue. Reproductive behaviour and the sexual systems of colonists provide further insights into evolutionary change on islands; for instance, it is now known that dioecy has evolved autochthonously in at least 12 lineages of Hawaiian plants, but relatively few comparative data are available and few generalizations are possible at this stage. Hybridization appears to have been more important in island evolution than once realized.

Island forms have undergone a wide variety of forms of niche alteration, including shifts and expansions in feeding niche, alterations in nesting sites in birds, loss of defensive attributes, loss of dispersability, tendencies to size increase in small mammals, and to size reduction in large mammals. Each of these trends requires careful evaluation prior to acceptance of a significant island effect. In some cases niche shifts can be due to character displacement caused by competitive interactions, or to ecological release in the absence of the usual array of competing species. Particular syndromes of traits can be identified which can be understood in relation to the geographical context of the island systems under examination. Each of these changes may take place without speciation being recognized as having occurred, each may also be part of the macro-scale island evolutionary patterns and models discussed in Chapter 6.

6

Emergent models of island evolution

Studies of island biotas are important because the relationships among distribution, speciation, and adaptation are easier to see and comprehend.

(Brown and Gibson 1983, p. 11, citing one of A. R. Wallace's general biogeographic principles.)

Having considered the evidence for microevolutionary change on islands in the previous chapter, we now move on to consider how these building blocks fit together to produce neo-endemics on islands. Although other categorizations are possible, I identify three emergent patterns or models of evolutionary change on islands. The first involves speciation without much radiation of lineages, and may be most applicable to isolated islands. The other two models, the taxon cycle and adaptive radiation, find their most dramatic illustrations within archipelagos. This chapter will explore the properties of these models and evaluate the evidence for their operation. The final section attempts to place the principal island evolutionary concepts and models into a simple framework of land area versus island or archipelago isolation.

6.1. Speciation with little or no radiation: anagenesis

The Juan Fernández archipelago contains two principal islands, Masatierra and Masafuera. These are volcanic islands of age 4 and 1–2 million years, respectively, located some 600 km west of mainland Chile. Masatierra is closest to the mainland, Masafuera is located some 150 km further west. Of the native vascular plants, 18% of the genera and 67% of the 148 species are endemic to the archipelago. There is even one endemic family, the Lactoridaceae, represented by its only species *Lactoris fernandeziana*. Using modern techniques of genetics (chromosomal inventories, studies of flavonoid evolution, etc.) Stuessy *et al.* (1990) estimate that the presence of the 69 genera with endemic taxa on these islands may be explained by 73 original colonization events: in most cases only a single introduction per genus. The relatively simple configuration of the two islands has provided a neat 'experiment' in the evolutionary development of lineages. As introduced in Chapter 4, Stuessy *et al.* (1990) have combined an analysis of phylogeny with the distributional patterns within the archipelago (Fig. 4.2). Of the endemic flora, 71% correspond to the **anagenesis** model, in which the progenitor form becomes extinct; 24% to the **anacladogenesis** model, where the progenitor survives little changed while a peripheral or isolated population diverges rapidly, perhaps while entering new habitats; and 5% to the **cladogenesis** model, in which the lineage divides into two lines and the original form fails to survive. A number of assumptions have to be made in order to derive such figures, yet the results are open to testing, for instance, by further studies of DNA.

The Juan Fernández Islands thus provide some intriguing patterns of endemicity. The matter of how these patterns have arisen is equally interesting. Earlier studies of isozymes among six species of *Dendroseris* (Asteraceae) by the same research group had revealed relatively slight genetic differences between these related taxa, despite their being morphologically quite distinct. Similar results have also been reported for studies of four Hawaiian genera (with exceptions in one line of the genus *Dubautia*). That fairly rapid morphological evolution in oceanic islands may be accompanied by small genetic change (cf. Sarre and Dearn 1991), Stuessy and colleagues regard as compatible with the results of their phylogenetic analyses, in which cladogenesis was of relatively little importance and anagenesis was overwhelmingly the dominant form of change. A general scenario might be that of a founder population 'capitalizing' on available ecological regimes (niches), and undergoing rapid initial morphological adaptation. Change would continue as the overall biotic environment evolves with the arrival of further plant and animal colonists. The occurrence of vicariant events within the islands appears to have been generally low, and the motor for speciation is thus the further arrival of forms and the rapidly changing environments experienced by the previous colonists as a consequence. Presumably, changes in the abiotic environment throughout the Quaternary have also contributed to the evolutionary dynamics.

The notion that later immigrating species alter the biotic environment, and thus have an important role in the further evolutionary change of earlier colonists, has been an important one in island evolutionary theory. It is most fully developed in the next model to be considered, the **taxon cycle**.

6.2. The taxon cycle

One of the striking patterns in island biogeography is that islands appear to represent a form of evolutionary 'blind alley'. Populations that establish and adopt the insular ways seem doomed to eventual extinction, resulting in a general unidirectional movement of species from continents to islands over evolutionary time-scales. The taxon cycle helps explain why this may be so.

The first thing to say about the taxon cycle is that it is a model that has evolved as it has colonized different islands through different taxa and authors. It is thus not a discrete hypothesis, allowing easy refutation. The term was coined by Wilson (1961) in his studies of Pacific ants, although similar models had been proposed earlier (Ricklefs 1989). The shared features of different invocations appear to be: that the evolution takes place within an island archipelago, in which immigrant species undergo niche shifts, which are in part driven by competitive interactions with later arrivals, such that the later arrivals ultimately may drive the earlier colonizing inhabitants towards extinction. The methodology employed in all the early studies was that patterns of geographic distribution and taxonomic differentiation provided the empirical basis from which were inferred cycles of expansion and contraction in the geographical distribution, habitat distribution, and population density of species in island groups (Ricklefs and Cox 1978).

6.2.1. Melanesian ants

As developed by Wilson (1959, 1961), the taxon cycle described 'the inferred cyclical evolution of species [of Melanesian ants], from the ability to live in marginal habitats and disperse widely, to preference for more central, species-rich habitats with an associated loss of dispersal ability, and back again' (MacArthur and Wilson 1967, glossary).

Wilson (1959, 1961) recognized differences in the ranges of the ponerine ants as a function of their habitat affinities (Fig. 6.1). Marginal habitats (littoral and savanna habitats) were found to contain both small absolute numbers of species and higher percentages of widespread species. These data were interpreted as a function of changes in ecology and dispersal as ants moved from the Oriental region, and particularly its rain forests, through the continental islands of Indonesia, through New Guinea, then out across the Bismarck and Solomon Islands, on to Vanuatu, Fiji, and Samoa. Wilson bundled the species into three stages. Stage 1 species are those that dominate in the marginal habitats: open lowland forest, grassland, and littoral habitats (Fig. 6.2). They have a greater ecological amplitude than the other species, occurring also in other habitats. They tend to be trail-making ants, nesting in the soil. Stage 1 species typically show a continuous distribution

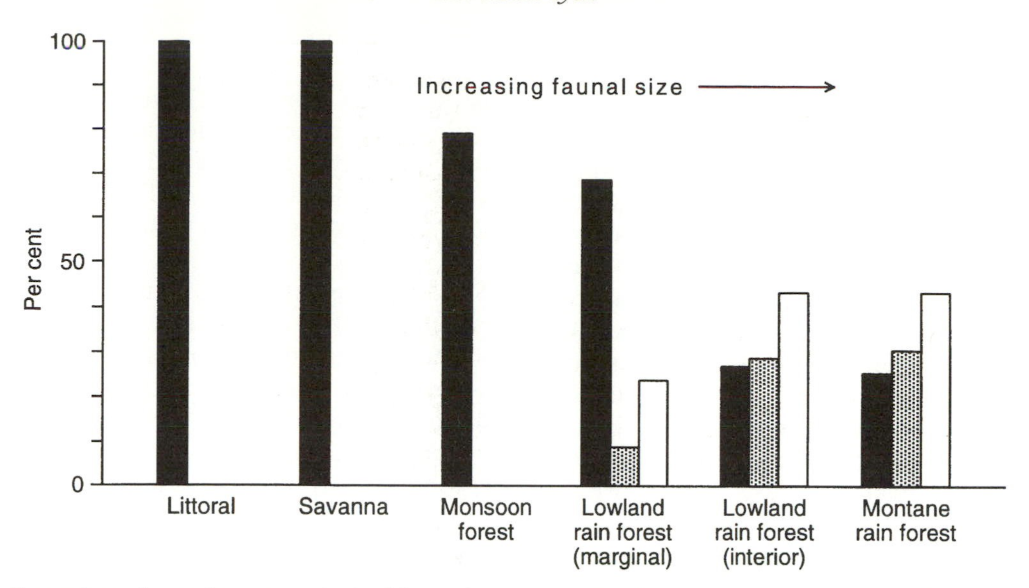

Fig. 6.1. Proportions of ponerine ant species in different habitats, as a function of their geographical distribution. The marginal habitats, to the left, contain both smaller absolute numbers of species and higher percentages of widespread species. Dark columns, species widespread in Melanesia; stippled columns, species restricted to single archipelagos in Melanesia but belonging to groups centred in Asia or Australia; blank columns, species restricted to single archipelagos and belonging to Melanesia-centred species groups. (Redrawn from Wilson 1959.)

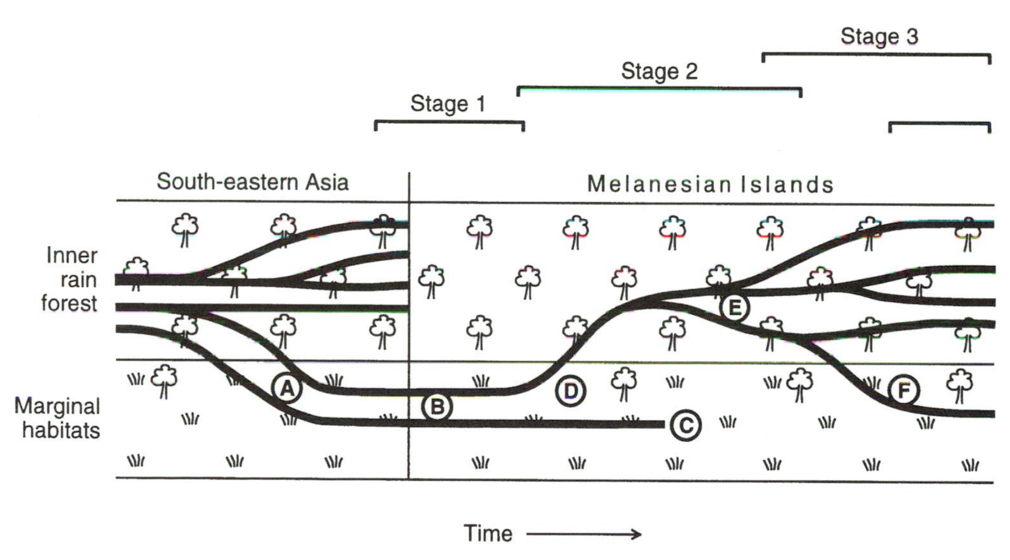

Fig. 6.2. The inferred taxon cycle of ant species groups in Melanesia, in which is traced the hypothesized histories of groups derived originally from Asia. The following sequence was postulated: stage 1—species adapt to marginal habitats (A) in the mainland source region (SE Asia) and then cross the ocean to colonize these habitats (B) in New Guinea. These colonizing populations may become extinct in time (C). Alternatively, the cycle enters stage 2—the ants invade the inner rain forests of New Guinea and/or surrounding islands (D). If successful in re-adapting to inner rain forest habitats they, in due course, diverge (E) to species level. Stage 3—diversification progresses within Melanesia, while the group remaining in Asia may, in cases, be contracting its range, such that the lineage becomes centred in Melanesia. A few members of these lineages, especially those on New Guinea, may re-adapt to the marginal habitats (F), and expand secondarily. (Redrawn from Wilson 1959.)

with no tendency to break into locally distinct races. Stage 2 of the process sees ants 'returning' to the dominant vegetation types, of interior and montane forest, within which they are more likely to be found nesting in logs or similar habitats. If they succeed in adapting to the inner rain forests they eventually differentiate to species level within Melanesia, forming superspecies or species groups. As they differentiate, they are liable to exhibit reduced gene flow across their populations. Stage 3 species occupy similar habitats to those of stage 2, but evolution has now proceeded to the point where the species group is centred on Melanesia and lacks close relatives in Asia. This may be, in part, a function of change in Melanesia, and in part the contraction in the group remaining in Asia. Progress within the cycle is thus marked by a general pattern of restriction of species to narrower ranges of environments within the island interiors. Meanwhile, these lineages have been replaced by a new wave of colonists occupying the beach and disturbed habitats.

Pivotal to this inferred evolutionary system is that the process is driven by the continuing (albeit infrequent) arrival of new colonist species. Their colonization initiates interactions that push earlier colonists from the open habitats. This is because the earlier colonists are likely to have become more generalized in the time since they themselves colonized, as they spread into additional habitats, thus losing competitive ability in their original habitats (Brown and Gibson 1983). This model is persuasive in that it provides answers to some seemingly puzzling phenomena. For example, it helps explain why species inhabiting interior forests of the oceanic islands are more closely related to species of disturbed habitats on New Guinea, than to those species that have similar niches and which occur in mature forest. Although largely a unidirectional model, it was suggested that some lineages did find a way out of their alleyway, perhaps temporarily. Occasionally, a stage 3 species, particularly those on New Guinea, may readapt to the marginal habitats, becoming a secondary stage 1 species, and expanding in its distribution (Fig. 6.2).

6.2.2. Caribbean birds

Ricklefs and Cox (1972) put forward a taxon cycle model to account for the biogeography of the West Indies avifauna. Their version distinguishes an additional stage (Table 6.1, Fig. 6.3), but the basics are shared with the ant study.

Stage I The first stage is the invasion of an island or archipelago by a mainland species. In the Caribbean, this means a species spreading across many islands. It is thus taken to be a dispersive form with interchange occurring between mainland and islands and among islands. The species is likely to be one of disturbed or coastal habitats, as such species typically have good dispersive ability. There is therefore little differentiation initially between mainland and island forms.

Stage II The colonist may then expand its niche, invading other habitats, and becoming more generalized in the use of resources. Species at this stage have more spotty distributions as selection against mobility reduces gene flow. They gradually evolve local forms and become restricted to a subset of the islands. Species then become vulnerable to being outcompeted in their original colonizing niche by further specialized colonists and may become restricted to interior forest habitats so that their niche breadth narrows again.

Stage III As they proceed through the stages of the cycle, species become highly differentiated endemics that ultimately become extinct and are replaced by new colonists from the mainland. Stage III species thus evidence a longer history of evolution in isolation, being found as scattered endemic forms.

Stage IV The final distributional pattern in the cycle is when a highly differentiated endemic species persists as a relict on a single island. The final step is when the form has no distribution other than in the fossil record: it is extinct.

One reason for the success of colonists is that a recent arrival may have left behind predators, parasites, and competitors on colonizing an

Table 6.1
The taxon cycle as applied to the avifauna of the Caribbean by Ricklefs and Cox (1972, 1978) (from Ricklefs 1989)

(a) *Characteristics of distribution of birds in the four stages of the taxon cycle*

Stage of cycle	Distribution among islands	Differentiation between island populations
I	Expanding or widespread	Island populations similar to each other
II	Widespread over many neighbouring islands	Widespread differentiation of populations on different islands
III	Range fragmented due to extinction	Widespread differentiation
IV	Endemic to one island	N/A

(b) *Number of species of passerine bird on each of three islands*

	Jamaica	St Lucia	St Kitts
Stage I	5	8	6
Stage II	10	7	6
Stage III	8	9	2
Stage IV	12	2	0
Total	35	26	14
Area (km^2)	11 526	603	168
Elevation (m)	2257	950	1315

island, enabling it to flourish despite the existence of local forms with a longer period of evolutionary adjustment to the conditions on that island. An important part of the mechanism for the cycle as put forward by Ricklefs and Cox (1972) was the evolutionary reaction of the pre-existing island biota to new immigrants. This idea is termed **counter-adaptation**. Over time the longer-established members of the biota begin to exploit or compete with the immigrant more effectively, thus lowering the competitive ability of the latest arrival. The subsequent arrival of additional, competitively superior immigrants, will then tend to push the earlier colonists into progressively fewer habitats and reduce their population densities. By this reasoning, range of habitat use and population density should each be diminished in species at an advanced stage of the taxon cycle.

Ricklefs and Cox (1978) examined these propositions from their earlier paper on the basis of standardized frequency counts taken in nine major habitats on three Caribbean islands: Jamaica, St Lucia, and St Kitts. Each species was first assigned to one of their four taxon cycle categories (Table 6.1a). Non-passerines did not show distinct trends in relation to stage of cycle, but the data for passerines was more interesting. The most common stage I species on Caribbean islands demonstrated habitat breadths and abundances rarely attained by mainland species, indicating that colonization of islands must involve some degree of ecological release (cf. Cox and Ricklefs 1977), although the phenomena appeared to be confounded by variation in the proportion of species in each stage of the taxon cycle. Passerine species at 'later' stages of the cycle tend to have more

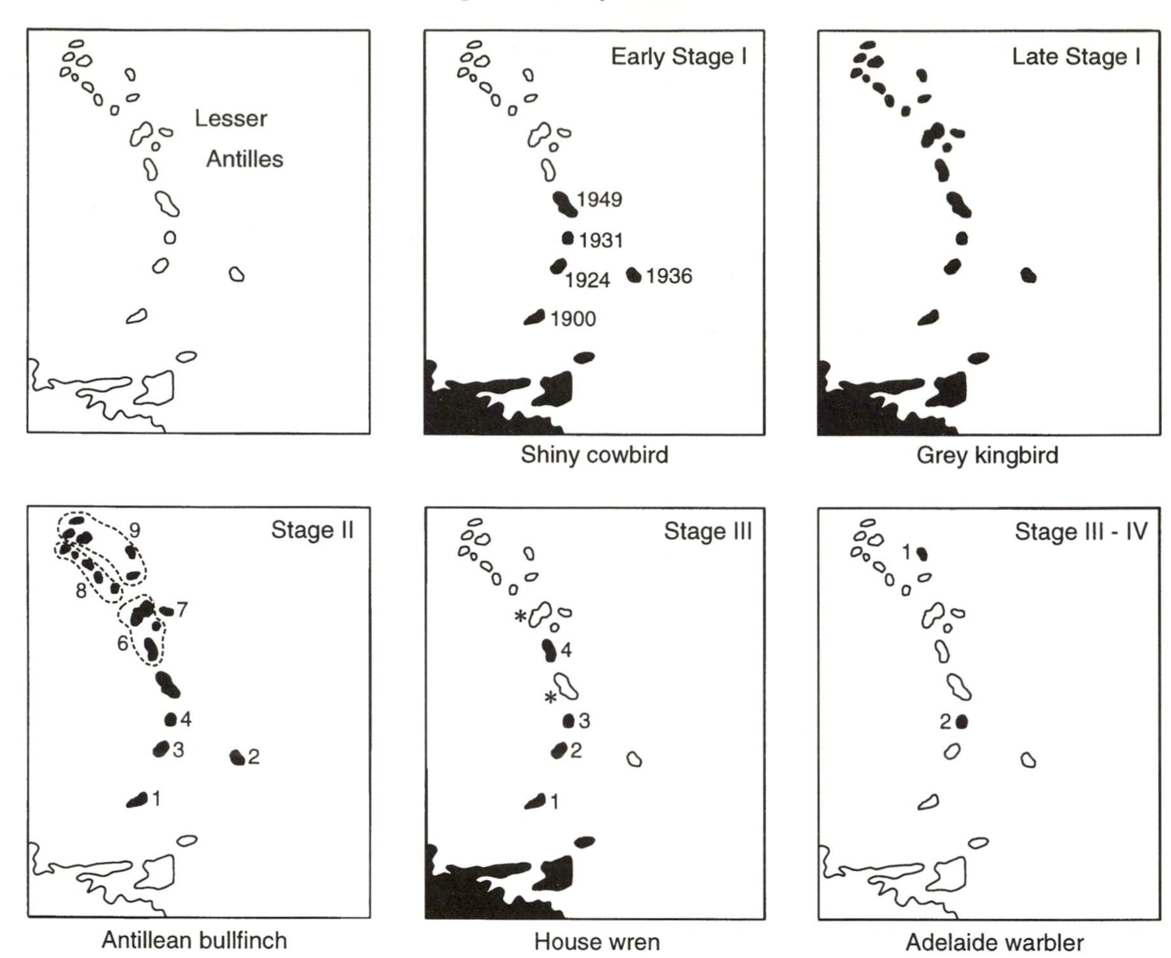

Fig. 6.3. The distribution of members of the Lesser Antillean avifauna, illustrative of the proposed stages of the taxon cycle. Species are as follows: shiny cowbird, *Molothrus bonairiensis*; grey kingbird, *Tyrannus dominicensus*; Antillean bullfinch, *Loxigilla noctis*; house wren, *Troglodytes musculu*; and Adelaide warbler, *Dendroica adelaidae*. The small figures indicate differentiated populations (subspecies). The house wren has become extinct on the islands marked* within the present century. Dates refer to first colonization. (Redrawn from Ricklefs and Cox 1972, Fig. 1.)

restricted habitat distributions and reduced population densities. Early-stage species tend to occupy open, lowland habitats, while late-stage species tend to be restricted to montane or mature forest habitats. Similar findings were reported for birds on the Solomon Islands by Greenslade (1968). Ricklefs and Cox (1978) found these trends to be clear for Jamaica, with 35 passerines, but they are difficult to detect on St Kitts, with 13 species, none of which is endemic.

Ricklefs and Cox's analyses are fascinating, in part because they encompass a dynamic vision of evolution at work. However, their approach did not provide critical tests of the model in a framework in which the ideas compete with alternative explanations, and it is not entirely clear which line of evidence, from those considered, would be capable of leading to model rejection. Similar criticisms could be levelled at Greenslade's (1968) taxon cycle model for the Solomon Islands. A particular problem with the Caribbean model is that many of the endemics of Jamaica, particularly those endemic at the generic level, are widespread and abundant, contrary to the original idea expressed above.

Ricklefs and Cox (1978) took this as illustrating that species on Jamaica have been able to diverge sufficiently in ecological terms for these generic-level endemics to avoid competition from more recent colonists to the island; which seems to be a case of special pleading.

The Rickleffs and Cox model was subsequently criticized by Pregill and Olson (1981), who contested that the four stages are simply criteria that define a set of distributional patterns and that almost any species in any archipelago would fall into one of these categories. The distribution patterns shown in Fig. 6.3 and Table 6.1b do not, of themselves, indicate that each species will exhibit each of these patterns in turn during its history. To examine this they turned to fossil data. It appears from this line of evidence that very similar distributional patterns can arise from quite different evolutionary histories. In illustration, the smooth-billed ani (*Crotophaga ani*) is absent from fossil deposits in the West Indies, whereas the white-crowned pigeon (*Columba leucocephala*), a species that is practically endemic to the West Indies but which, none the less, has a stage I distribution like that of the ani, occurs commonly in Pleistocene deposits. Another problem is that not all endemic species belong to groups that were once widespread: a number appear to have colonized an island from the mainland and differentiated at the species level without having dispersed to other islands, an example being the Jamaican becard (*Platypsaris niger*). Pregill and Olson go on to criticize the idea of counter-adaptation, stating that 'Ecological doctrine and good sense revolt at the idea that a species with a long history of adaptation to a particular environment would be at a competitive disadvantage with newly arriving colonists' (p. 91). It is not clear to me why they hold this view so strongly; adaptation is not a process of optimization in the best of all possible worlds. Moreover, the history of human introductions to oceanic islands appears to indicate that newly arrived colonists can do rather well at the expense of indigenous forms, although the circumstances involved (habitat alterations, etc.; Chapter 10) may be sufficiently different to weaken the force of this argument.

Generally, analyses of faunal turnover in island studies are based on the unstated premise that environments are effectively stable, i.e. that biotic forcing factors, and not environmental factors, are responsible for turnover. In contrast, Pregill and Olson's (1981) analyses of the fossil record and of relict distributions, particularly amongst the xerophilic vertebrates of the Caribbean, indicate that conditions of climate and habitat were appreciably different during the last glaciation and indeed earlier in the Pleistocene (cf. Buskirk 1985). Lowered sea level combined with climatic changes to produce connections between islands presently separated and also produced a general increase in the extent of xeric habitats, such as arid savanna, grassland, and xeric scrub forest. They list several examples from the fossil record of species characteristic of dry, open country, which once had a much wider distribution than they have had in late Holocene times; these include the burrowing owl (*Athene cunucularia*), the Bahaman mockingbird (*Mimus gundlachii*), thick knees (*Burhinus*), falcons and caracaras (*Polyborus* and *Milvago*), curly-tailed lizards (*Leicophalus*), and rock iguanas (*Cyclura*). The significance of such losses of species on particular islands is exemplified by the figures for New Providence Island, where in total 50% of the late Pleistocene avifauna and 20% of the fossil herpetofauna no longer occur, many of these species being xerophilic.

Pregill and Olson (1981) provide a number of other examples of species distributions which they regard as most parsimoniously explained by reference to long-term environmental changes. For instance, they list a number of pairs of Hispaniolan birds which they regard as most probably having differentiated at times during the Quaternary when Hispaniola was divided into north and south islands. This list includes the todies (*Todus subulatus* and *T. angustirostris*) and the palm-tanagers (*Phaenicophilus palmarum* and *P. poliocephalus*). Thus they appear to have identified some serious problems with the taxon cycle model as applied to West Indian birds. They favour a different model, that the distributional patterns identified might be primarily an evolutionary outcome of climates becoming wetter (and warmer) since the end of the Pleistocene rather than being interpretable simply in terms of interspecific interactions.

Within the taxon cycle model it is assumed that a widespread species moves out in a single colonization phase from a mainland source, and spreads out according to a stepping-stone process of colonization through the archipelago, producing a monophyletic group within which birds on islands close to one another should be more closely related than those distant from one another. Klein and Brown (1994) point out that the ornithologist James Bond (after whom the fictional character was named) proposed multiple colonizations from mainland sources for West Indian island populations of widespread species. Multiple colonization events would mean that the representatives of a species on different Caribbean islands would not form a monophyletic group relative to samples from the various mainland source pools. Modern techniques allow Bond's idea to be tested.

Klein and Brown (1994) studied mitochondrial DNA (mtDNA) from specimens of yellow warbler (*Dendroica petechia*) collected from North, South, and Central American sites, as well as from the West Indies. The most parsimonious tree constructed from their phylogenetic analyses indicated colonization of some islands in the Lesser Antilles by Venezuelan birds, and colonization of the Greater Antilles by Central American birds. Furthermore, they found evidence of multiple colonizations not only of the West Indies as a whole, but also of individual islands. Such events can involve the introgression of characters from one lineage to another. Moreover, the phylogenetic data indicated that birds of adjacent islands are not always each other's closest relatives, against the assumptions of a stepping-stone model for colonization. The multiple colonizations of the West Indies by yellow warbler, which are indicated by the phylogenetic data, thus provide several points of inconsistency with Ricklefs and Cox's (1972, 1978) taxon cycle model, at least for this particular species of bird. At least some of the differences between populations are a consequence of what could be thought of as a series of founder effects, rather than exclusively because of *in situ* selective pressures and drift. Klein and Brown (1994) note that a number of recent studies of bats in the Caribbean provide similar lines of evidence for multiple colonizations of individual islands, and of one group of related haplotypes being widespread while another was confined to the Lesser Antilles.

Klein and Brown's (1994) data, as they emphasize, apply only to a single bird species, which leaves a good number still in dispute. How can the various opposing views of the biogeographical patterns described in relation to the taxon cycle framework be reconciled? In large part the disputes relate to the actual historical patterns, and in part to the causal forces at work: are they biotic (dispersal, colonization, competition, adaptation, etc.) or abiotic (sea-level change, climate change, changing habitat availability, etc.)? The answer must lie in more such tests, using differing sources of data—geological, phylogenetic, palaeo-ecological, distributional, ecological, and experimental. The next case study represents an example of just such an integrative approach.

6.2.3. Caribbean anoles

Roughgarden and Pacala (1989) have revised the interpretation of some of the data for the *Anolis* genus of lizards in the Caribbean. The anoles comprise about 300 species, or between 5 and 10% of the world's lizards. On Caribbean islands they are extremely abundant, replacing the ground-feeding insectivorous birds of continental habitats. The distribution of species/subspecies relates best not to the present-day islands but to the banks on which the islands stand, separated from other banks by deep water. The anoles have long provided one of the classic illustrations of **character displacement**, whereby island banks have either one intermediate-sized species or two species of smaller and larger size respectively. This has been taken to indicate that where two populations have established, interspecific competition between them has resulted in divergence of size of each away from the intermediate size range so that they occupy distinct niche space (Brown and Wilson 1956). This progresses to the point at which the benefit of further reduction in competition is balanced by the disadvantage of shifting further from the centre of the resource distribution (Schoener and Gorman 1968; Williams 1972). However, Roughgarden and Pacala (1989) contend that within the islands of the Antigua, St Kits, and Anguilla banks in the northern Lesser Antilles a series of independently derived facts contradict this model, and instead allow for an interpretation of a different sort of dynamic model; in short, a form of taxon cycle (see also Rummel and Roughgarden 1985).

The *Anolis* taxon cycle of the northern Lesser Antilles begins with an island occupied by a medium-sized species, which is joined by a larger invader from the Guadeloupe archipelago. Rather than the two species moving away from the middle ground, as in the character displacement scenario, both species evolve a smaller size. The medium-sized species becomes smaller as it is displaced from the middle ground by its larger competitor. This opens up niche space in the centre of the resource axis, selecting for smaller size in the invader, which thus approaches the medium (presumably optimal) size originally occupied by the first species. The range of the original resident species contracts, culminating in its extinction. At this point, the invader has taken up occupancy of the niche space of the first species at the outset of the cycle. The islands remain in the solitary state until (hypothetically) another large invader arrives.

The key points in the Roughgarden and Pacala (1989) analysis are as follows. Intensive scrutiny of the systematics of *Anolis* over a period of four decades has resulted in a phylogenetic tree in which five main groups have been identified. The northern Lesser Antilles are populated by the *bimaculatus* group, which is subdivided into two *series* at the next level of the hierarchy (Fig. 6.4). The *wattsi* series comprise small brown lizards that perch within a few feet of the ground. The *bimaculatus* series comprise large or medium-sized green or grey-green lizards, that are relatively arboreal. Islands (for which read *island banks*) have either one species or two living in the natural habitat, i.e. excluding small enclaves of introduced anoles living near houses. Anguilla, Antigua, and St Kitts are the island banks having two species, in each case a larger *bimaculatus* and a smaller *wattsi*.

Earlier research by Williams (1972, cited by Roughgarden and Pacala 1989) had produced two empirical 'rules' concerning the sizes and ratios of the anoles:

- Rule 1: species from islands in which only one anole is present are intermediate in size (snout–vent length) between the body sizes seen on islands in which two anole species coexist. This rule is correct for 11 of 12 island banks having a solitary species (they range in length from 65 to 80 mm) and it applies throughout the eastern Caribbean.

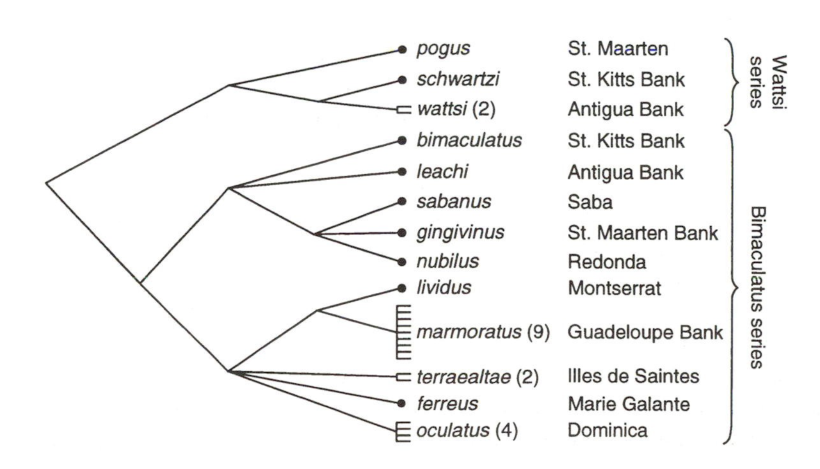

Fig. 6.4. Phylogenetic tree for *Anolis* of the Bimaculatus group (containing the Wattsi series and the Bimaculatus series) in the northern Lesser Antilles. Those end-points represented by dots are monotypic species, the others are geographic races or subspecies according to current nomenclature. This tree is merely one trunk of a larger phylogenetic tree for eastern Caribbean *Anolis*, constructed on the basis of detailed systematic/phylogenetic data by Roughgarden and Pacala (1989). (From Roughgarden and Pacala 1989, Fig. 1.)

- Rule 2: species on a two-species island differ in body length by a factor of 1.5–2.0, such that the larger species exceeds 100 mm and the smaller is less than or equal to 65 mm in length, the size of the smallest solitary species. This rule is correct for four of the five two-species island banks, and again applies throughout the eastern Caribbean.

These patterns appeared to be consistent with the character displacement model. It was also thought that the original residents were members of the *bimaculatus* series, and that the invading species were members of the *wattsi* series, with both invading from Puerto Rico. Although these ideas appeared to explain much of the variation in *Anolis* through the Lesser Antilles, there were some exceptions and further research produced more problems in respect to the northern islands (Roughgarden and Pacala 1989):

1. On St Maarten there are two anoles of nearly the same body size. By the character displacement hypothesis, one or both should be a new arrival. However, both are differentiated (visibly and biochemically) from other anoles in the region.
2. Medium-sized lizards should become larger but Pleistocene cave deposits demonstrated unequivocally that larger anoles have become smaller, and no fossil data show a medium-sized species becoming larger.
3. The two-species community supposedly has its origins in two colonists of similar size, which diverge in size; but evidence from experimental studies of competition, and accidental introductions of anoles, demonstrate that an invader very similar in size to an established resident does not easily establish itself. A second species arriving on an occupied island needs to be larger than the occupant in order to become established.

Having dismissed the original model, Roughgarden and Pacala (1989) put forward 12 strands of evidence in further support of their taxon cycle model, from which the following points are taken. The only known extinction of an anole in the Lesser Antilles is of the smaller species from a two-species island. North of Guadeloupe, only the *wattsi* series has sufficient geographic variation within an island bank to have led to a subspecific nomenclature, the Antigua bank having *A. wattsi wattsi* on Antigua proper and *A. wattsi forresti* on Barbuda. This may be interpreted as indicating a longer presence of the smaller-sized lineage, the *wattsi* series, than the larger *bimaculatus* series on the northern Lesser Antilles. This is contrary to the original model but is required to be the case by the taxon cycle model. The source for the *bimaculatus* series appears to be Guadeloupe and this is geographically sensible given that the prevailing currents run towards the northern islands. In short, data on the ecology of colonization, the phylogenetic relationships among anoles, their biogeographical distribution, and the fossil record, refute the character displacement explanation for the northern Lesser Antilles, but are consistent with the taxon cycle model. Yet, in advocating this model, Roughgarden and Pacala (1989) caution that such a cycle does not appear to happen often, and there is no evidence for its operation for anoles in the Lesser Antilles other than in these northern islands.

Phylogenetic analyses shed further light on this, by showing that the Lesser Antilles are occupied by two distinct lineages of *Anolis*. Dominica and islands to the north contain species related to those on Puerto Rico, whereas islands to the south harbour species with South American affinities (Losos 1990, 1996). The patterns found of character displacement in conditions of sympatry in the northern Lesser Antilles contrast with the evidence for change in body size in the southern Lesser Antilles. In the latter case change appears to be unrelated to whether a species occurred in sympatry with congeners, and appears instead to result from a process of ecological sorting, such that only dissimilar-sized species can colonize and coexist (Losos 1996).

6.2.4. Evaluation

As presaged in the introduction to this section, the taxon cycle has proved to be difficult to evaluate, in part because of the fluidity of the theoretical framework. The possible influence of environmental changes, and the role humans may have had in 'messing up' the biogeographical signal, should also be considered for each particular case. It has, for instance, been postulated that size reductions and extinctions of some larger lizards in the Caribbean were linked not to intrinsic biological processes so much as to human colonization of the islands (Pregill 1986)—an entirely believable proposition. Thus the extent to which species inhabiting island archipelagos go through a series of distributional and evolutionary changes consistent with taxon cycle models—and the factors that drive these changes—remain unquantifiable.

6.3. Adaptive radiation

The word 'adaptation' gives an erroneous impression of prediction, forethought or, at the very least, design. Organisms are not designed for, or adapted to, the present or the future—they are consequences of, and therefore abapted by, their past.

(Begon *et al.* 1986, p. 6)

Island evolutionary forces reach their most spectacular embodiment in the patterns termed 'adaptive radiations'. The Hawaiian honeycreepers (sometimes 'honeycreeper-finches') and drosophilids and the Galápagos finches are the most famous illustrations from the animal kingdom. Hawaii and the Canaries provide some of the better-known plant examples. But we have seen that selection and adaptation on islands does not necessarily lead to radiation. In common usage, adaptive radiation is not so much an evolutionary model, more a biogeographical pattern. It is therefore important at this point to establish what it embodies.

The availability of vacant niche space is one important feature in radiations, as it allows a form of ecological release, allowing the diversification which sometimes, but not always, leads to speciation within a lineage. As suggested by its name, *Metrosideros polymorpha* is very diverse and multiform. It occupies habitats ranging from bare lowlands to high bogs, and occurs as a small shrub on young lava flows and as a good sized tree in the canopy of mature forest. It is the chief tree of the Hawaiian wet forests and, as an aside, at least some individuals can be found in flower at any time of the year, a pattern supported by enough of the nectar-producing flora to have allowed the evolution of the nectar-feeding lineages among the honeycreepers. Despite its radiation of forms, all members of this complex were originally allocated to *M. polymorpha*. Some highly distinct populations have now been recognized as segregate species, but active hybrid swarms also exist (Carlquist 1995).

Having noted that not all adaptations are radiative, and that not all the products of radiations are recognized as full species, it will come as no surprise to the reader to find that the application of the term 'adaptive' has also been questioned. One element of this is set out in the quote heading this section, and while in part it is a matter of semantics, the point made is a serious one. A second line of criticism has been developed by Gittenberger (1991). He has argued that there should be no automatic labelling of radiations as adaptive, but that a degree of specialization of species in to different niches should be involved before the term is used. Should the radiation occur without a clear degree of niche differentiation, such as may be the case when radiations have occurred allopatrically in fragmented habitats, it should be considered whether they are better termed **'non-adaptive radiation.'** His interpretation of observations of the land snail genus *Albinaria* within the island of Crete is that they have diversified into a speciose genus *without much* niche differentiation, to occupy more or less the same or only a narrow range of habitats, yet only rarely do more than two *Albinaria* species live in the same place. He thus posits non-adaptive radiation as a logical alternative to adaptive radiation, while recognizing that the process of radiation may involve a blend of the two over time (cf. Lack 1947). Cameron *et al.* (1996), in their study of land snails from Porto Santo

Island (Madeira), also concluded that the majority of the 47 endemic species recorded had arisen from radiation of an essentially non-adaptive character. Barrett (1996) suggests that the term non-adaptive radiation may also be an appropriate way of describing diversification in particular plant lineages, citing specifically Aegean island populations of *Erysimum*.

Once these points are registered, there is actually little to the core theory of adaptive radiation that has not already been touched upon in earlier sections. Adaptive radiation invokes not so much a particular process of evolutionary change as the emergent pattern of the most spectacular cases, the crowning glories of island evolution. For this very reason the best cases are liable to involve a wide range of the evolutionary processes introduced as separate topics earlier.

The data in Table 6.2 provide recent estimates for the degree of endemism and an idea of the degree of radiation involved in the biota of Hawaii. These figures are not cast in stone, as they depend on the taxonomic resolution and assumptions involved in their calculation. For example, Paulay (1994), cites an estimate that there may be as many as 10 000 Hawaiian insect species, and that they may have evolved from 350–400 successful colonists. Notwithstanding such uncertainties, Hawaii clearly provides spectacular examples of radiations in taxa as diverse as flowering plants, insects, molluscs, and birds. Within the flowering plants, Fosberg (1948) hypothesized 272 original colonists (more recent estimates vary from 270 to 280 (Wagner and Funk 1995)) for the *c.* 980–1000 flowering plant species, the pattern of radiation being indicated by the following statistics: there are 88 families and 211 genera, the 16 largest genera account for nearly 50% of the native species, about 91% of which are endemic (Sohmer and Gustafson 1993).

The biogeographical circumstances in which radiations take place are reasonably distinctive. Radiations are especially prevalent on large, high, and remote islands lying close to the edge of a group's dispersal range (Paulay 1994). MacArthur and Wilson (1963, 1967) termed such peripheral areas the **radiation zone**. Here the low diversity of colonists, and the disharmony evident in the lack of a normal range of interacting taxa, facilitate *in situ* diversification (Diamond 1977). In illustration of this, it has been noted that ants dominate arthropod communities across most of the world, but in Hawaii and South-East Polynesia they are absent (or were, prior to human interference). In their place, there have been great radiations of carabid beetles and spiders, and even caterpillars have in a few cases evolved to occupy predatory niches (Paulay 1994).

Table 6.2

Number of presumed original colonists, derived native species, and endemic species for a selection of the Hawaiian biota (Sohmer and Gustafson 1993)

Animal or plant group	Estimated number of colonists	Estimated number of native species	% Endemic species
Marine algae	?	420	13
Pteridophytes	114	145	70
Mosses	225	233	46
Angiosperms	272	*c.* 1000	91
Terrestrial molluscs	24–34	*c.* 1000	99
Marine molluscs	?	*c.* 1000	30–45
Insects	230–255	5000	99
Mammals	2	2	100
Birds	*c.* 25	*c.* 135	81

Of course, for less dispersive taxa, their radiation zones may coincide with less remote archipelagos, which have a greater degree of representation of interacting taxa than the most remote archipelagos. Hence, the circumstances for radiation reach their synergistic peak on the most remote islands, the epitome being Hawaii (below). Examples that fit this idea of maximal radiation near the dispersal limit include birds on Hawaii and the Galápagos, frogs on the Seychelles, gekkonid lizards on New Caledonia, and ants on Fiji; while exceptions—taxa that have not radiated much at remote outposts—include terrestrial mammals on the Solomons and snakes and lizards on Fiji (MacArthur and Wilson 1967). Equally, it is clear from even the most remote island archipelagos that not all lineages within a single taxon have radiated to the same degree. Nearly 50% of the *c.* 1000 native flowering plant species of Hawaii are derived from fewer than 12% of the *c.* 280 successful original colonists (Davis *et al.* 1995). Most of the rest of the colonists are represented in the present flora by single species. While to some extent such differences may reflect the length of time over which a lineage has been present and evolving within an archipelago, this does not provide a general explanation. In short, while these geographical circumstances may be conducive to radiations, they are not the only factors of significance and they do not inevitably lead to radiation within a lineage.

6.3.1. Darwin's finches and the Hawaiian honeycreeper-finches

Although the Galápagos are renowned for other endemic groups, notably the tortoises (Arnold 1979) and plants (Porter 1979), the most famous group of endemics must be Darwin's finches (Emberizinae; *Geospiza* spp.). The context within which these birds and the other creatures have evolved is as follows. The Galápagos are in the east Pacific, 800–1100 km west of South America (Fig. 6.5). There are some 45 islands, islets, and rocks, of which nine are islands of area greater than 50 km^2. Isabela, at 4700 km^2, represents over half of the land area and is four times the size of the next largest island, Santa Cruz. Isabela and Fernandina have 1500 m high volcanoes, but most of the islands are relatively low. Volcanic in origin, they are true oceanic islands, never having been connected to mainland (Perry 1984). They remain volcanically active, many lava flows being recent and still unvegetated. Geological evidence suggests that true islands have existed in the Galápagos for at least 3.3 million and not

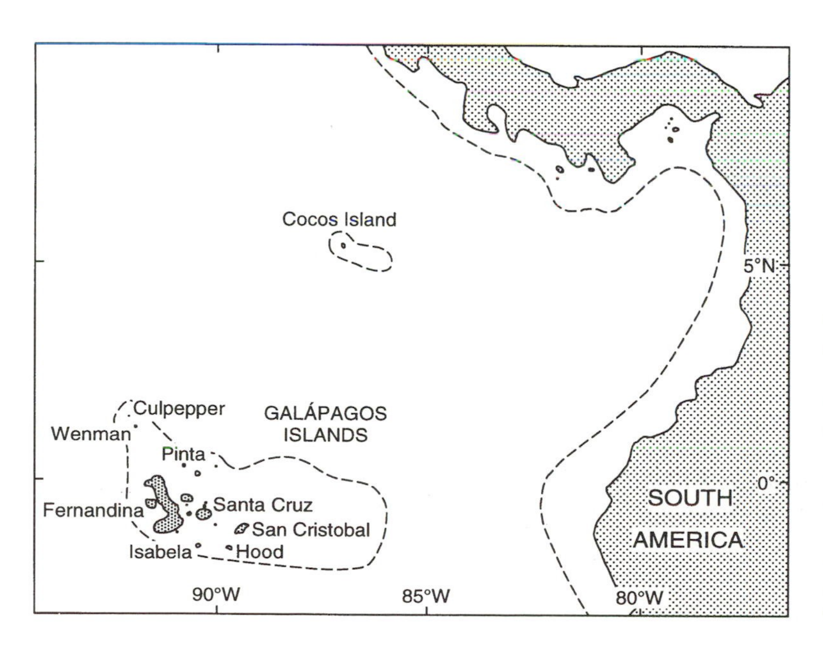

Fig. 6.5. The locations of Cocos Island and the Galápagos islands relative to South and Central America. Cocos Island is the only place outside the Galápagos where a member of Darwin's finches, the Emberizinae, can be found. The dashed line approximates the 1800 m depth contour. Five degrees on the equator represent approximately 560 km. (Redrawn from Williamson 1981, Fig. 9.1.)

more than about 5 million years (Simkin 1984). The four westernmost islands have geological ages ranging from 0.7 to 1.5 million years. Although equatorial, they are comparatively cool and average rainfall in the lowlands is less than 75 cm/year (Porter 1979).

The finches are not alone. Twenty-eight species of land birds breed on the Galápagos, of which 21 are endemic. There are four mockingbirds, which are recognized as separate species on the basis of morphological differences, although as they have allopatric distributions, it is not known if they could interbreed or not (Grant 1984). It is also unclear whether their genus (*Nesomimus*) is sufficiently distinct morphologically to warrant separation from the mainland genus (*Mimus*). Similar problems exist with the finches. As noted by Darwin (1845) in the quote at the start of Chapter 4, such uncertainties do nothing to diminish their interest to the 'philosophical naturalist'. Darwin's finches will be the focus of attention here as they have radiated to the greatest degree, thirteen Galápagos species being generally recognized. In fact it is extremely difficult to identify all the finches, as the largest members of some species are almost indistinguishable from the smallest members of others (Grant 1984). Collectively, they feed on a remarkable diversity of foods: insects, spiders, seeds, fruits, nectar, pollen, cambium, leaves, buds, the pulp of cactus pads, the blood of seabirds and of sea-lion placenta. It is principally through changes in beak structure and associated changes in feeding skills and feeding niches that the differentiation between the finches has come about (Lack 1947; Grant 1994). In illustration, the woodpecker finch (*Camarhynchus pallidus*) uses a twig, cactus spine or leaf petiole as a tool, to pry insect larvae out of cavities. Small, medium and sharp-billed ground finches (*Geospiza fuliginosa, G. fortis,* and *G. difficilis*) remove ticks from tortoises and iguanas, and perhaps most bizarre of all, sharp-billed ground finches on the northern islands of Wolf and Darwin perch on boobies, peck around the base of the tail and drink the blood from the wound they inflict.

A comprehensive theory for this radiation was developed by Lack (1947) and has been updated and summarized by Grant (1981, 1984; and see Vincek *et al.* 1997). The key points are as follows. One of the islands was colonized from the mainland, or possibly from Cocos Island (the only other place where a member of the group is found). Genetic analyses suggest that the effective population size of the founding flock was at least 30 individuals (Vincek *et al.* 1997). The founding population expanded quite rapidly, undergoing selective changes and/or drift. After some time individuals of this population colonized another island in the archipelago, where conditions would have been slightly different. Further changes would have taken place through random genetic changes, drift, and selection. The significance of adaptive changes is evident in the differences in feeding niches which have already been referred to. A degree of differentiation between allopatric populations would then have been evident. At some point, individuals of one of the derived populations flew to an island already occupied by a slightly differentiated population. The result of this might sometimes have been an infusion of the newcomers into the established population, but when sufficient behavioural differences had already developed hybridization was limited, and hybrids, when occurring, largely unsuccessful. Selection would have favoured members of the two groups which fed in different ways from each other and so did not compete too severely for the same resources. Recent research has shown that female finches appear able to distinguish between acceptable and unacceptable mates on the basis of beak size and shape and also on the different patterns of songs (Grant 1984). This may have been the means by which females were able to select the 'right' mates, thus breeding true, and producing progeny that corresponded with the peaks rather than the valleys of the resource curves (section 4.4.2). Such characteristics would have a selective advantage in the populations over time, thus allowing two or more species to exist in sympatry. To put these points together: the radiation of the lineage has taken place in the context of a remote archipelago, presenting extensive 'empty niche' space, in which the considerable distances between the component islands has led to phases of inter-island exchange only occasionally. Differing environments have apparently selected for different feeding niches both between and within islands. Thereafter, behavioural differences between forms maintain sufficient genetic distance between sympatric populations to enable their persistence as distinctive lineages.

However, as Grant (1984) cautioned, this is not the only way in which speciation can occur. Theoretically a lineage may have split into two non-interbreeding populations on a single island, as suggested by the model of competitive (sympatric) speciation (section 4.4.2), provided the environment was sufficiently heterogeneous. Further progress has been made on the interactions and exchange that may occur between sympatric populations of Galápagos finches, highlighting the importance of sympatric episodes in lineage development (P. R. Grant and B. R. Grant 1996), but before discussing such work, it is necessary to outline the case of the Hawaiian equivalents to Darwin's finches.

The Hawaiian islands have formed as a narrow chain from a hot-spot, which appears to have been operational for over 70 million years, although the oldest high island of the present group, Kauai, is only about 5.1 million years old (section 2.2). It is believed that the hot-spot has never been closer to North America than it is today and its position relative to Asia has probably also been stable (but see Myers 1991). Molecular clock data suggest that few lineages exceed 10 Ma, i.e. the pre-Kauai signal in the present biota of the main islands is actually rather limited (Wagner and Funk 1995; Keast and Miller 1996).

At least 20 natural avian colonizations have been suggested (Tarr and Fleischer 1995), and endemics comprise approximately 81% of the native avifauna. Other radiations less well known than the Drepanidinae include the following. The elepaio (*Chasiempis sandwichnesis*) is a small active flycatcher endemic to Hawaii. Distinctive subspecies occur on Kauai and Oahu, while a further three occur within the youngest and largest island, Hawaii itself (Pratt *et al.* 1987). The native thrushes of Hawaii are placed in the same genus, *Myadestes*, as the solitaires of North and South America. Five species are recognized by Pratt *et al.* (1987), each occurring on its own island or island group. One was last seen in 1820, and three of the remaining species are severely endangered, with populations—if they survive—numbering fewer than 50 individuals (Ralph and Fancy 1994).

The Hawaiian honeycreepers (or honeycreeper-finches), the endemic subfamily Drepanidinae, have shown an even greater radiation than Darwin's finches. For a long time it was believed that this radiation had resulted in 23 species in 11 genera. From a single type of ancestral seed-eating finch, the group had radiated to fill niches of seed-, insect-, and nectar-feeding species, with a great variety of specialized beaks and tongues; thus providing one of the most popular illustrations of evolutionary radiation (Fig. 6.6; Carlquist 1974; Williamson 1981). It is now known that there were actually many more species in the recent past. Estimates of the number of species known historically range from 29 to 33, with another 14 having recently been described from subfossil remains (more are likely to follow): most extinctions occurring between the colonization of the islands by the Polynesians and the arrival of Europeans on the scene (Olson and James 1982, 1991; James and Olson 1991; Tarr and Fleischer 1995). As noted, different genera are generally recognized within this subfamily, including *Psittirostra* (the Hawaiian name of which is Ou) which have short, conical beaks, for seed eating (essentially they are still finches), and *Pseudonestor* (Maui parrotbill), which uses its powerful beak to tear apart twigs in order to reach wood-boring beetles. The main evolutionary line, however, is quite different. Its members have longer, narrower bills, and they feed on insects (in the case of *Paroreomyza*, the Molokai creeper) and on nectar (*Loxops* (Akepa) and *Hemignathus* (e.g. Akialoa)).

Tarr and Fleischer (1995) observe that genetic differentiation between species of drepanidines is less than would be expected on the basis of their morphological divergence, indicating that they are the product of a relatively recent arrival (in the order of 3.5–8 million years), followed by rapid adaptive radiation.

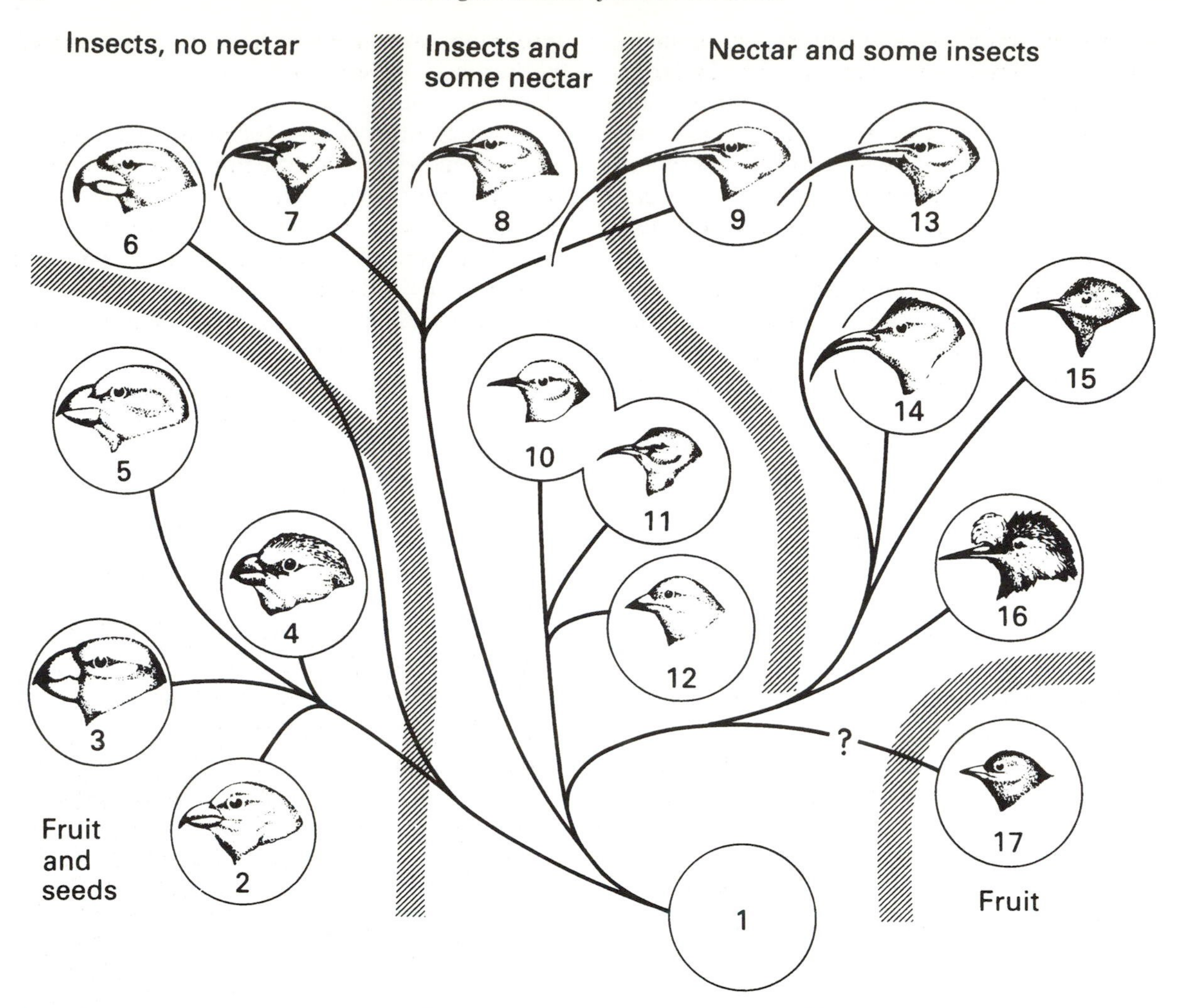

Fig. 6.6. The inferred pattern of evolution of dietary adaptations as represented by 16 of the Hawaiian honeycreepers (family Fringillidae, Subfamily Drepanidinae). 1, Unknown finch-like colonist from Asia; 2, *Psittirostra psittacea*; 3, *Chloridops kona (Psittacirostra kona)* (extinct); 4, *Loxioides bailleui* (*Psittacirostra bailleui*); 5, *Telespyza cantans* (*Psittacirostra cantans*); 6, *Pseudonestor xanthophyrs*; 7, *Hemignathus munroi* (*H. wilsoni*); 8, *H. lucidus*; 9, *H. obscurus* (*H. procerus*); 10, *H. parvus* (*Loxops parva*); 11, *H. virens* (*Loxops virens*); 12, *Loxops coccineus*; 13, *Drepanis pacifica* (extinct); 14, *Vestiaria coccinea*; 15, *Himatione sanguinea*; 16, *Palmeria dolei*; 17, *Ciridops anna* (extinct). Source: Cox and Moore (1993, Fig. 6.11). The taxonomy preferred here follows Pratt *et al.* (1987), with that given in Cox and Moore (1993) in brackets where different.

They concluded that evidence for an allopatric model of speciation was fairly convincing for some of the honeycreepers, but they were not able to rule out sympatric speciation (particularly in the light of recent fossil finds). *Himatione sanguinea sanguinea* (the Apapane) and *Vestiaria coccinea* (the I'iwi), two of the more widespread taxa, show an intriguing lack of differentiation, and it has been suggested that this may reflect relatively recent range expansion. This may have occurred in response to the arrival and spread of the tree *Metrosideros polymorpha*, or possibly because of the loss of competitors due to the extinction of other avian taxa. If these interpretations are correct, these two species may represent the early stage of a new cycle of dispersal and differentiation (Tarr and Fleischer 1995).

Having set out the basics of the Hawaiian picture, we can pick up the thread of the discussion of the interactions and exchanges between sympatric populations by reference to comparative studies between the Hawaiian and Galápagos lineages. Morphological variation within populations of Darwin's finches has been the subject of detailed studies over the past 20 years. From these it has emerged that there is considerable variation in body size and beak traits within populations of ground finches (*Geospiza*), and that the amount of variation itself differs among populations (Grant 1994). In studies of six of these species, Grant and Grant (1994) found evidence for the limited occurrence of so-called introgressive hybridization, allowing gene flow between populations of different species. This was suggested to be an important factor contributing to intraspecific variation. This is of considerable interest in island evolution for the reasons outlined above, and in terms of the relative importance of allopatry and sympatry as contexts for speciation.

Grant (1994) set out to establish from the study of 524 museum specimens of the seven species of Hawaiian honeycreepers (Carduelinae; Fringillidae) with finch-like bills (which might be termed honeycreeper-finches) whether the same hybridization process was at work in a lineage that has diversified to a much greater extent. The measurements demonstrated that variation within populations of Hawaiian honeycreeper-finches was *less* than among the *Geospiza*. The single Hawaiian species with both sympatric and allopatric populations did not show greater variation in the sympatric population, as would have been expected if hybridization was effective. It was concluded that there was no evidence of hybridization occurring within the past 100 years. Grant reported that G. C. Munro had suggested in 1944 that *Rhodacanthis flaviceps* might be a hybrid produced from *R. palmeri* and *R. psittacea*, and indeed the only morphological hint of hybridization among the Hawaiian species came from the close similarity found between *R. flaviceps* and *R. palmeri*. If *R. flaviceps* is indeed of hybrid origin, then hybridization must have been going on for a long time, as the species is recognizable in fossil material.

Grant (1994) put forward two hypotheses which might explain why the hybridization evident in the Galápagos finches was not apparent in the Hawaiian equivalents.

1. Geological and electrophoretic studies indicate that the honeycreeper-finches have been present in the Hawaiian archipelago for a much longer period, perhaps three times longer than have the finches of the Galápagos, and over this time have diversified further and have evolved prezygotic (behavioural) and/or postzygotic isolating mechanisms.

2. The Hawaiian honeycreeper-finches have evolved greater dietary specializations in the generally less seasonal and floristically richer Hawaiian islands (Fig. 6.7). In an environment

Fig. 6.7. The typical dissected terrain of an old volcanic island, St Louise Heights, above Honolulu, Oahu Island, Hawaii. Although this forest is disturbed and contains both exotic plants and animals, it is still possible to see representatives of the Hawaiian honeycreepers in this area. However, about half of the lineage is now known to be extinct (see text). (Photo: RJW 1991.)

with more distinctive resource peaks, and following the line of reasoning outlined in the section on competitive speciation, stabilizing selection for specialist feeding may have provided strong selective pressure against an ecological niche intermediate between the two parental species.

Grant put forward a model to capture the important distinctions in a hypothetical phylogeny for the radiation of sympatric taxa (Fig. 6.8). An early phase of divergence is characterized by occasional genetic contact through interbreeding, followed by a genetically independent phase. The duration of the phase of introgression will depend on the ecological isolation attained, which will be related to environmental heterogeneity. From genetic distance measures of finch species known to hybridize from the Galápagos and from North America, Grant (1994) estimates the apparent duration of the phase of introgression, i.e. of hybridization, as up to 5 million years. He speculates that it may have been occurring over the whole of the estimated 2.8 million years of the radiation of Darwin's finches and quite possibly for much of the 7.5 million years estimated by some (but see Tarr and Fleischer 1995) for the diversification of the Hawaiian honeycreepers. All such figures involve error margins and in the case of the Hawaiian species the evidence for such a model is very largely indirect. Hybridization has also been observed between the Galápagos land and marine iguanas which, although of monophyletic origin, are actually placed in separate genera; however, analysis of the DNA of populations on Plaza Sur Island suggested that if gene exchange occurs it must be at a very much lower rate than in the finches (Rassmann *et al.* 1997). Further lineages will have to be studied if the applicability of Grant's model is to be established.

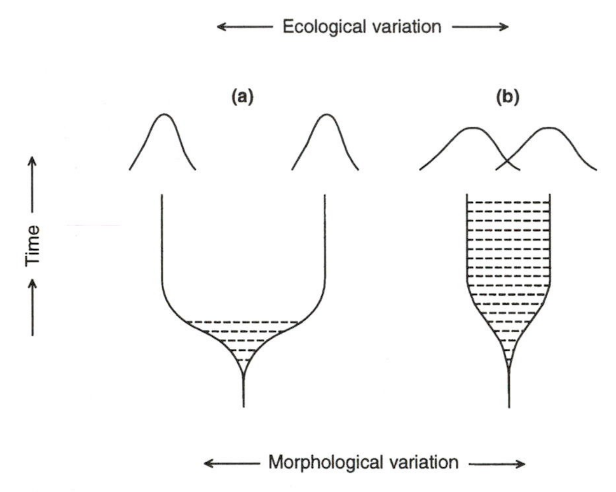

Fig. 6.8. A model illustrating the relative significance of introgressive hybridization during divergence, as a function of resource space. In the early introgressive phase of divergence, newly formed species remain in genetic contact through occasional interbreeding, but this is followed by a genetically independent phase. The length of time over which introgression occurs may vary because of ecological factors and the heterogeneity of the resource base. Resource-use curves are shown at the top of the diagram. Species that have become resource specialists (model a) are less likely to hybridize than are generalists (model b). These hypothetical representations were offered by Grant (1994) for the Hawaiian honeycreepers (model a) and Galápagos finches (model b), respectively. (Redrawn from Grant 1994, Fig. 5.)

The studies reviewed in this section allow us to conclude that both allopatric and sympatric episodes can be involved in radiations of island archipelago birds. While events must vary from one lineage to another, P. R. Grant and B. R. Grant (1996) suggest the following general scenario for sympatric congeners, based largely on their work in the Galápagos. First, following colonization of different islands, there is an initial phase of differentiation of allopatric populations. Secondly, inter-island movements re-establish sympatry, following which further differentiation takes place between the two populations. Over a period of several million years, occasional introgressive hybridization occurs and, at least in some cases, the recombinational variation imparted plays a creative role, facilitating further divergence. Eventually, when the lineages have diverged far enough, this form of exchange effectively ceases to play a role.

6.3.2. Hawaiian crickets and drosophilids

The Hawaiian endemic crickets are considered to be derived from as few as four original colonizing species, each being flightless species arriving in the form of eggs carried by floating vegetation. The ancestral forms were a tree cricket and a sword-tail cricket from the Americas, and two ground crickets from the western Pacific region (Otte 1989). Three of the successful colonists have radiated extensively and Hawaii now has at least twice as many cricket species as the continental United States. Much later, a further eight species have colonized, but these are considered to have been introduced by humans and will not be discussed further.

The tree crickets (Oecanthinae) have been calculated on phylogenetic grounds to have colonized Hawaii about 2.5 million years ago, radiating into three genera and 54 species (43% of the world's known species), with the greatest diversification seen within the older islands, which were occupied earliest (Otte 1989). They have radiated into habitats not occupied by their mainland relatives. Distributional and cladistic data suggest that virtually all speciation takes place within islands, much of it probably in locally isolated habitats on volcanoes. This emphasis on within-island speciation concurs with data for the plant genus *Cyrtandra* and the land snail genus *Achatinella*, each of which is represented by over 100 species on Oahu alone. As discussed earlier, Otte considers competitive displacement to have had a role in the within-island evolution of the Hawaiian crickets. Other factors have been important, and the eruptive characteristics of Hawaiian shield volcanoes are significant for crickets, as well as for the drosophilids (Carson *et al.* 1990; Carson 1992). The lava islands, lava tubes, ridges, and valleys characteristic of these high islands promote within-island isolation.

Drosophila and several closely related genera in its subfamily include about 2000 known species, of which Hawaiian drosophilids (the closely related genera *Drosophila* and *Scaptomyza*) account for some 600–700 species (Brown and Gibson 1983, after Carson *et al.* 1970)—although, as noted earlier, the eventual figure could be as great as 1000 species (Wagner and Funk 1995). The ancestors of the Hawaiian drosophilids (a single, or at most two species) probably arrived on one of the older, now submerged islands and have radiated within and between islands. Carson and colleagues have studied the chromosome structure of members of the picture-wing group, using the karyotypic phylogeny as a means of reconstructing the radiation. The sequence, as might be expected, appears to stem from the oldest island, Kauai (age 5.1 million years), and, in general, the older islands to the west contain species ancestral to those of the younger islands to the east, i.e. most, but not quite all, of the inter-island colonizations have been from older to younger islands. In undertaking this analysis, the islands of the Maui complex (Maui, Molokai, and Lanai) may be taken as a single unit, as they have fused and separated at least twice in their short history, due to sea-level changes, adding another element of complexity to the picture. It was deduced by Carson that, starting from Kauai, a minimum of 22 inter-island colonizations are required to account for the phylogeny of the picture-wing species found on the islands of Oahu, the Maui complex, and the big island of Hawaii itself. The precise details of this sequence of events may vary according to the means used to construct the phylogeny and as the biogeographical picture is updated, but the broad pattern is clearly established (Fig. 6.9). Furthermore, the trend of colonization from old to young islands found for these drosophilids is the most frequent found amongst the lineages investigated to date (Wagner and Funk 1995). However, the majority of speciation events among the drosophilids have occurred within islands and, for example, Carson (1983) notes that of 103 picture-wing *Drosophila* species, all but three are endemic to single islands or island complexes. Brown and Gibson (1983, citing Carson *et al.* 1970) report that nearly 50 species of *Drosophila*, having leaf-mining larvae, have been reared from the leaves of *Cheirodendron*. If species can currently coexist in a *Cheirodendron* forest, and even be found in the same leaf, it is difficult to rule out the possibility that in some cases populations may have diverged while in sympatry. However, as already indicated, these islands have provided tremendous opportunities for within-island isolation, and it is now generally considered that isolation of this form has been crucial to lineage development.

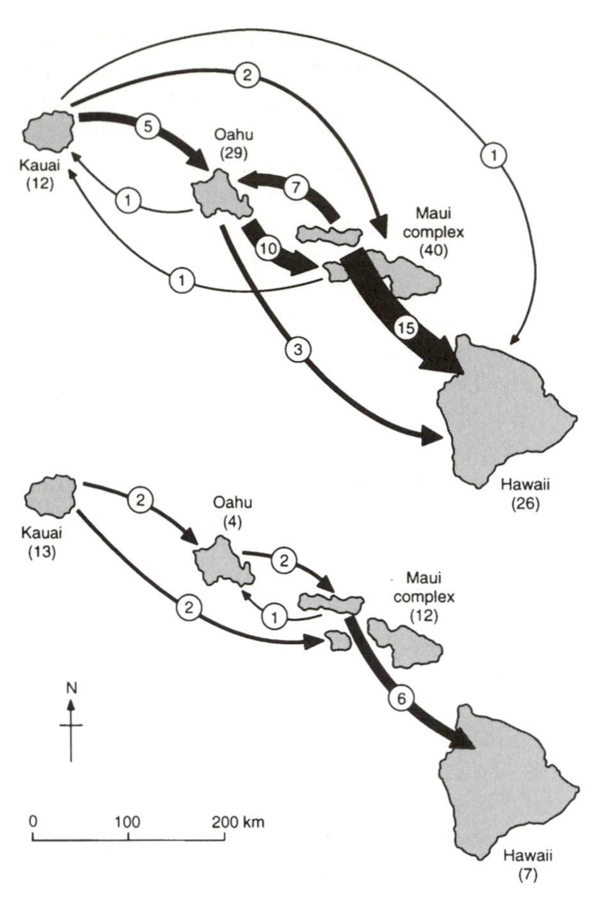

Fig. 6.9. The dispersal events among the Hawaiian islands that are suggested by the inter-relationships of the species of picture-winged *Drosophila* flies (above) and of silverswords (tarweeds) (below). Arrow widths are proportional to the number of dispersal events, and the number of species in each island is shown in parentheses. (Redrawn from Cox and Moore 1993, Fig. 6.9, after an original in Carr, G. D. *et al.* (1989). Adaptive radiation of the Hawaiian silversword alliance (Compositae—Madiinae): a comparison with Hawaiian picture-winged *Drosophila*. In *Genetics, speciation and the founder principle*, (ed. L. Y. Giddings, K. Y. Kaneshiro, and W. W. Anderson), pp. 79–95. © Oxford University Press, New York).

One form of intra-island isolation prevalent on Hawaii is when lava flowing down the flanks leaves undamaged 'islands' of forest, termed *kipukas*, that can vary from a few square metres to many hectares in area. Whether *kipukas* maintain their isolation long enough to allow for speciation may vary between taxa, but in one particularly well-studied picture-wing, *Drosophila silvestris*, it has been found that different isolates show a remarkable degree of genetic differentiation. Carson *et al.* (1990) and Carson (1992) have suggested that the many phases of development of these shield volcanoes provide conditions of continuing division and flux, acting as a general crucible for parallel active evolutionary change in the associated organisms. 'The forces that created the Hawaiian islands and their long-ago foundered predecessors not only formed the firmaments upon which life could diversify but also may have played a heretofore unappreciated direct role in the acceleration of evolutionary processes as they operate in local populations' (Carson *et al.* 1990, p. 7057). Geological evidence suggests that the surface of Mauna Loa (the main volcano of the island of Hawaii), has been replaced at an average rate of about 40% per 1000 years during the Holocene, and accordingly the populations of *Drosophila* and other species of the volcano flanks must have been repeatedly 'on the move' (see also Wagner and Funk 1995). It is thus entirely reasonable to posit that species which are sympatric today were at least locally allopatric at the time their lineages diverged. At this sort of scale, however, the terms allopatry and sympatry are probably inadequate, being hypothetical endpoints, potentially involving little or no interchange on the one hand, and free interchange on the other. Given the dynamism of these environments in evolutionary time, and the fluxes in population sizes and connectivity which are implied, the reality may lie somewhere between these absolutes, with each species existing in the form of metapopulations (Chapter 9), i.e. units experiencing a degree of interchange and of supply and re-supply as local isolates wax and wane (Carson 1992).

6.3.3. Adaptive radiation in plants

Figure 6.9 shows that the general trend of colonization from old to young Hawaiian islands in *Drosophila*, is matched by similar trends in the radiation of the silversword alliance or tarweeds, which are a group of 28 species in three genera

(*Dubautia*, *Argyroxiphium*, and *Wilkesia*) in the Asteraceae. Illustrations of adaptive radiation come most famously from the Galápagos and Hawaii (Carlquist 1974; Wagner and Funk 1995), yet to restrict the treatment of the topic to these islands would be negligent. Excellent plant examples can be found in Macaronesia, the collective term for the Azores, Salvage Islands, Madeira, Canary Islands, and the Cape Verde Islands. These islands have some 3200 species of native plants, of which about 680, or 20% of the total, are endemic (Humphries 1979). The Canary Islands are the largest in area and the richest, with 460 endemics, about 45% of the native flora. Eighteen Macaronesian lineages are endemic at the generic level also, with 17 of these being restricted to the Canaries. Some have radiated to a significant degree, with *Argyranthemum* (22 species; Asteraceae or daisy family), *Monanthes* (16 species; Crassulaceae or houseleek family), and *Aichryson* (14 species; Crassulaceae) providing the most species-rich examples. Other lineages, which are not endemic at the generic level, have radiated to even greater extents, notably *Aeonium* (36 species; Crassulaceae), *Sonchus* (29 species; Asteraceae), and *Echium* (28 species; Boraginaceae or viper's bugloss family). Within the Macaronesian flora as a whole, most genera are represented by only one or two species, and most of those with over four occur in the Canary Islands (Fig. 6.10). In general, radiation of lineages has been greatest on the larger islands with the greatest diversity of habitats. The island with the greatest number of endemic species, 320, is Tenerife, and it also has the largest numbers of *Aeonium* (12 species), *Sonchus* (11 species), and *Echium* (9 species). Most endemic species have a restricted distribution, 48% occurring on a single island, a further 15% on two. From this and other considerations, Humphries (1979) concluded that allopatric speciation has been crucial in the flowering plants. Furthermore, from studies of a sample of 350 of the endemic species, he concluded that 90% had evolved by gradual divergence following the break-up of a population (including migration from one area to another), with the remainder accounted for by abrupt speciation involving hybridization, polyploidy, and other forms of sudden chromosomal change.

Support for the use of the term adaptive radiation in several of the above Macaronesian genera comes from studies of features such as habit, leaf morphology, and habitat affinities. Humphries (1979) argues that evidence of similar morphology in species of the same or different genera (termed parallel or convergent morphology) occurring in similar habitats, is supportive of a model of selection having favoured particular adaptive outcomes. One of the best-documented examples of adaptive radiation within a Macaronesian lineage is provided by Humphries' work on the genus

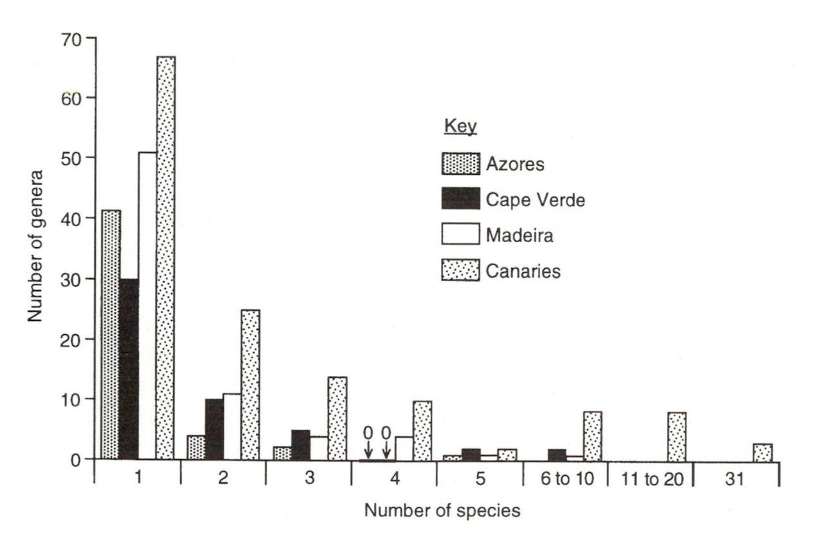

Fig. 6.10. The distribution of numbers of species per genus for the Macaronesian phanerogamic flora, wherein each bar represents the number of genera represented by the appropriate number of species, for the Azores, Cape Verde Islands, Madeira, and Canary Islands. Twenty per cent of the 3200 native species of the Macaronesian flora are endemic to just one of these archipelagos. (Data from Humphries 1979.)

Table 6.3
Adaptive radiation in the genus Argyranthemum *(22 spp.; Asteraceae) in the Macaronesian islands (data from Humphries 1979)*

Environment	Adaptive features	Islands	Species
Broadleaved forests	**Type A**: shrubby, unreduced leaves	Tenerife and Gomera Madeira	*A. broussonetii* *A. adauctum* ssp. *jacobaeifolium* and *A. pinnatifidium*
South-facing, lowland arid	**Type B/C**: reduced lignification, habit, capitulum size, and leaf area	Tenerife Gran Canaria	*A. frutescens* ssp. *gracilescnes* and *A. gracile* *A. filifolium*
Sub-alpine and high montane	**Type D**: dies back after flowering each year, leaves very dissected, hairy	Tenerife	*A. tenerifae* and *A. adauctum* ssp. *dugourii*
Exposed northern coastal areas	**Type E**: reduced habit, large capitula, increased succulence	Canaries	*A. coronopifolium*, *A. frutescens* ssp. *canariae*, ssp. *succulentum* and *A. maderense*

Argyranthemum (note: once again a genus in the Asteraceae). Table 6.3 outlines, for a subset of the 22 species of the genus, some of the adaptive features that characterize species restricted to particular habitats within the Macaronesian islands.

In terms of the genetic changes involved, a substantial amount of work has been undertaken on particular lineages, with the consistent finding that adaptively radiated groups all show the same chromosome number (Humphries 1979). It appears that those notably few groups which have adaptively radiated into a wide range of different conditions have done so without major chromosomal change, i.e. from generic changes rather than major chromosomal reorganization. For example, experimental evidence reveals that *Argyranthemum* species are in many cases capable of producing hybrids, although only two natural examples are known (Humphries 1979). In 1965 a tunnel was constructed connecting the formerly isolated Teno peninsular to the Buenavista region. Shortly afterwards, hybrids between *A. coronopifolium* and *A. frutescens* were found on scree produced during tunnelling. Introgression of *frutescens* genes into *A. coronopifolium* occurred in an area spreading out from the original hybridization event, to such an extent that pure forms of the latter had almost disappeared from the vicinity by the end of the 1970s.

In general it appears that the 'species swarms' represented by the species-rich Macaronesian genera are essentially interfertile, i.e. they readily hybridize, particularly in disturbed areas. However, stabilized hybrids are rare and so it seems that the ecological differentiation between species is sufficient to maintain species integrity in most groups: it follows that while gene flow may occur between species, few speciation 'events' have been involved (Humphries 1979). Thus while details differ, it appears that the essential features of evolution in the Macaronesian flora, in cases involving spectacular radiations, are common to other studies encountered in this section (see also Kim *et al.* 1996). In terms of Rosenzweig's (1995) three modes of speciation, the data support a role for each of the geographical, competitive, and polyploid models in declining order of significance.

6.4. Observations on the forcing factors of island evolution

Darwin was right in 1845 to point to oceanic islands as of great interest to the 'philosophical naturalist'. Evolutionary change on islands has been shown in this and Chapter 5 to come in many forms, from the microevolutionary alterations of morphology in near-shore mice populations to the macro-scale radiation of lineages. Many geographical, ecological, and genetic mechanisms have been invoked in explanation of the evolutionary changes recorded. At first sight this diversity is bewildering, but on reflection a number of common patterns emerge. Much can be explained by reference to the degree of isolation, on the one hand, and the area and diversity of habitats, on the other. But, there is a third axis which ought to be considered as part of an evolutionary analysis, time itself.

6.4.1. Environmental change as a driving force

The importance of environmental change has been stressed in Chapter 2 as crucial to an understanding of island biogeography. Island evolution does not take place in controlled laboratory conditions, but on platforms which themselves evolve and which have been subjected to changing degrees of isolation, varying area, altered sea levels, and pronounced climatic oscillations. One simple illustration of the significance of this is provided by McDonald and Smith's (1990) study of the speciation of the Hispaniolan palm-tanagers, *Phaenicophilus palmarum* and *P. poliocephalus*, in which the genetic evidence suggests a divergence into the two species during a period of raised sea levels within the Pleistocene when Hispaniola was separated into separate northern and southern islands, i.e. the divergence in these birds occurred during a period of allopatry. Island evolutionary models that postulate sequences of events on time-scales of thousands or millions of years, and which focus on biotic factors to the exclusion of the abiotic, are likely ultimately to be found wanting. The debate on the taxon cycle (above) is illustrative of this problem.

Interesting insights into the potential importance of environmental change have emerged from Porter's (1979) analyses of endemism in the flora of the Galápagos archipelago. He recognized seven principal vegetation zones, related to coastal proximity, rainfall, and altitude. Most low islands have only two or three, with only the larger, higher islands having high rainfall and all seven zones. The vascular flora consists of 543 indigenous taxa (species, subspecies, and varieties) and 192 introduced species. Several lineages have radiated to an impressive degree, most notably *Scalesia*—once again the Asteraceae providing a showpiece of island radiation! There are seven endemic plant genera and approximately 43% of the indigenous flora is endemic (Fig. 3.9).

If endemism is examined by vegetation zone (Porter 1979, 1984), a very interesting pattern emerges:

1. the littoral zone has 10 endemic taxa, just 4% of the endemics;

2. the arid zone has 154 endemic taxa, 67% of the endemics; and

3. the mesic zone has 67 endemic taxa, 29% of the endemics.

A low degree of endemism in the littoral zone is explicable because of high rates of immigration by sea-dispersed plants, which may provide sufficient gene flow to slow the pace of divergence and to saturate the available strand-line habitats (Sauer 1969). Not only are the majority of endemics from the arid region, but nearly all of the evolutionary radiation within lineages has taken place in the arid region. Why should this be so? Arid zone habitats today are relatively extensive and are found on all islands, furthermore, palaeo-environmental evidence indicates a continuity in their availability. Studies of the past 25 000 years from lake sediments on Isla San Cristobel suggest a dry period of 15 000 years when mesic habitats were greatly reduced in area, followed by 10 000 years when more mesic habitats have been present (Chapter 2; Colinvaux 1972). This seems to have been mirrored in the distribution and radiation of the endemic vascular flora. The mesic regions appear to have been populated mainly by dispersal from South America, rather than by plants evolving from related species of the arid lowlands of the Galápagos themselves (Porter 1984). Endemism may therefore be low in the mesic zone because these habitats have only recently become fairly extensive within the archipelago and have had to be colonized principally by long-distance dispersal. The vegetational history, set in the context of the apparent infrequency of successful colonization events (below), thus provides a plausible reason for the relatively low proportion of mesic-zone endemics. The continued addition of introduced species supports the notion that the indigenous flora was impoverished (i.e. non-equilibrial) (Fig 3.9) and, given sufficient time, might have increased further in diversity by natural means.

6.4.2. Rates of speciation vary through time and between taxa

How long does it take? Generalizations are difficult, although particular case studies provide at least some indications of what is possible. For instance, Losos *et al.* (1997) were able to demonstrate significant adaptive differentiation over a 10–14 year period in populations of *Anolis sagrei* introduced experimentally into small islands in response to varying vegetation of the recipient islands. Diamond's study of character displacement in the myzomelid honeyeaters is another example of the sort of context in which a fairly precise time frame, a maximum of *c.* 300 years, could be put to a given evolutionary change (Diamond *et al.* 1989). The limitation of such studies is that they are concerned with microevolutionary changes rather than full speciation: for the latter we depend upon geological and biological dating techniques. Repeated speciation is demonstrated by the 70 species of *Mecyclothorax* beetles endemic to 1 million-year-old Tahiti, and the picture-wing *Drosophila*, in which 25 of the 26 species on the island of Hawaii are restricted to that approximately 0.6–0.7 million-year-old island (references in Paulay 1994). Assuming that potassium–argon dates do indeed provide reliable guides to actual island ages, such statistics provide an illustration of the degree of evolutionary change that has taken place on islands during the Quaternary period. However, it should not be assumed that speciation rates on any particular island are constant through time. Indeed, it is reasonable to suppose that rates might decline through time as an island accumulates species through colonization and through speciation of earlier arrivals. This was suggested for the study of the Juan Fernández Islands discussed earlier. Such studies benefit from use of the 'genetic clock' which, on the assumption of a more or less constant rate of accumulation of mutations and thus of genetic differences between isolates, allows estimates of the dates of lineage divergence (but note the possible failings of such methods, discussed in section 4.2). Additional studies from the Juan Fernández Islands by the same research group add further support to the idea of high initial rates of radiation. In illustration, electrophoretic data for the endemic genus *Robinsonia* (Asteraceae) suggest that the founding population arrived early in the 4.0 million year history of Masatierra Island, radiating and speciating rapidly after colonization (Crawford *et al.* 1992).

Kaneshiro *et al.* (1995) come to the same conclusion in their most recent analyses of species groups within the picture-wing *Drosophila* of Hawaii:

> Most of these species, like many other extant terrestrial endemic fauna, show a very strong but by no means exclusive tendency to single-island endemism. Most species thus appear to evolve on an island early in its history and thereafter remain confined to that island. Colonists arriving at newer emerging islands tend to form new species, a finding that has led to the serious consideration that speciation may be somehow related to founder events.
>
> (Kaneshiro *et al.* 1995, p. 71).

The Hawaiian islands have been likened to a conveyor belt, in which the islands themselves undergo a pattern of birth, growth, maturity, and decline, eventually to form atolls or disappear altogether beneath the waves. The more recent lava flows (past 100 years or so) within the youngest island are thus, today, the sites in which novel species and adaptations are most apparent.

While such large islands as Hawaii were once assumed to be fairly well buffered from climatic variations, the large-scale global climatic fluctuations of the Pleistocene are now believed to have affected them (Chapter 2; Nunn 1994) and must have some bearing on the biogeography of such island systems (as suggested for the Galápagos flora, above). On ecological timescales, present-day weather conditions can, after all, be highly significant determinants of breeding success in bird species, including, it has been found, in one of the Hawaiian honeycreepers, the Laysan finch (Morin 1992). Some of the most interesting insights in this field come from the remarkable studies of hybridization in Darwin's finches by B. R. Grant and P. R. Grant (1996), showing how influential the climatic anomaly of the 1982–83 El Niño was in affecting gene flow through hybridization. Thus while speciation rates may indeed be higher at early stages in the life of an oceanic island, it will be interesting, as more studies of different ages of island become available, to see whether there is evidence for alteration in evolutionary rates as a function of Quaternary climatic changes (cf. Kim *et al.* 1996).

Many island lineages, probably the majority, have not radiated spectacularly at the species level (Fig. 6.10). It is intriguing that certain types of organisms appear to have done so repeatedly, in widely separated archipelagos. Two examples are:

1. finch-like birds on Hawaii and Galápagos; and

2. genera within the Asteraceae (Compositae) on Hawaii, Galápagos, and Macaronesia.

Does this indicate that they are particularly prone to rapid genetic change? Perhaps. But, if so, it is not universal. Two observations provide what may be an important clue. First, it has been noted by Helenurm and Ganders (1985) that the 19 species of Hawaiian *Bidens* exhibit more morphological and ecological diversity than this diverse genus does in the Americas. Yet the genetic diversity of this entire insular radiation is comparable to that found among populations within particular American species. The second observation is that the Cocos finch appears to have radiated ecologically in comparison to mainland relatives, but not to have diversified into separate breeding populations (Werner and Sherry 1987). The suggestion offered was that the key to its 'radiation' was that it possesses the genetic basis for a degree of plasticity in behaviour. Both these rather different responses to the condition of insularity are forms of plasticity. Plasticity may be crucial to success in the circumstances of large, remote islands, presenting as they do spatially varied, temporally variable, and confined environments. Taxa that happen to have the genetic template to vary will do so in the oceanic island setting. Others form only a limited number of endemics.

6.4.3. Dispersal ability and endemism

Dispersal ability has been discussed at many points within this chapter. Evidence from archipelagos has shown the importance of variations in dispersal ability between and within taxa. This is fundamental to understanding which forms colonize (and just as importantly which do not), and the relative importance of within-island and between-island speciation. Some of these points can be illustrated by comparison of ferns and seed plants. Changes in dispersal ability after colonization also form a consequential part of the island evolutionary theatre (above), but will be passed over here.

The two main islands of the Juan Fernández archipelago are about 150 km apart. Within the angiosperm genera containing endemic species, 31% have at least one species on both islands, the corresponding figure for pteridophytes (ferns) is 71% (Stuessy *et al.* 1990). This is one of a number of features of the flora that can be interpreted in terms of the relative ease of dispersal of ferns and seed plants. The superior dispersability of ferns is evident also in data from Galápagos (below), Hawaii, and from the

Caribbean (Bramwell 1979), as well as from near-shore continental islands, such as Krakatau (Whittaker *et al.* 1997).

Porter's (1979, 1984) analyses of the Galápagos flora are again illuminating in respect of the dispersability of different elements. The following proportions of the indigenous flora are endemic:

1. 8 of the 107 pteridophytes;
2. 18 of the 85 monocotyledonous plants; and
3. 205 of the 351 dicotyledonous plants.

The low proportion of endemic ferns may be explained as a function of their better dispersability. This is indicated by the data in Fig. 6.11, which illustrate not only the tiny size of fern spores, but that seed sizes of grass and sedge (monocotyledonous plants) are intermediate in size compared to the classic island endemics of the Asteraceae. The figures for endemicity on the Galápagos are comparatively low compared with those for the Hawaiian flora, in which the difference between pteridophytes and angiosperms is again evident—70% and 91%, respectively, are endemic according to current estimates (Table 6.2). Furthermore, if only species, as opposed to subspecies and varieties, are considered, the level of endemism on the Galápagos falls to 34% of the indigenous

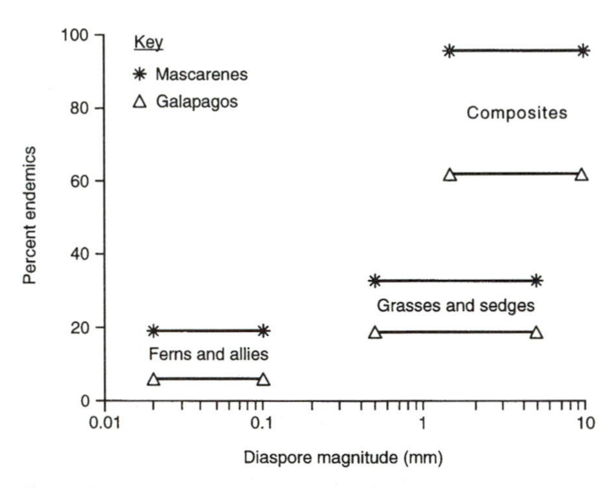

Fig. 6.11. Percentage of endemism for selected plant taxa of the Galápagos and the Mascarenes, against diaspore magnitude (the rough length of the dispersal element). (Adapted from Adsersen 1995, Fig. 2.2.)

flora (Porter 1984). Reasons for this may include the more recent origin of the Galápagos, the greater isolation of Hawaii, and that the Galápagos are generally drier. Porter (1984) has calculated that the Galápagos flora could be accounted for by 413 natural colonists. On this basis and assuming the age of the archipelago to lie between 3.3 and 5 million years (see above; Simkin 1984), the present native vascular plant flora of the islands could be accounted for by one successful introduction every 7990–12 107 years. The error margins on such estimates may be large, but they serve to indicate that successful dispersal events to such a remote location must naturally be very infrequent, judged on ecological time-scales. This turns out to be important to an understanding of the distribution of endemics within the archipelago (above).

Modern genetic studies are being used increasingly to test and refine such models. For instance, Elisens (1992) has employed allozyme electrophoresis in a study of genetic divergence between the Galápagos and mainland species of *Galvezia* (Scrophulariaceae). Allozyme variation was minimal within the endemic *G. leucantha*, suggestive of a bottleneck—possibly a single colonization event—less than 1.5 million years ago. The phylogeny suggested that the Galápagos species is of a relictual genotype from an extinct South American progenitor.

The concentration here on comparisons of ferns and seed plants is intended to illustrate that lumping together taxa such as 'all plants', or even all flowering plants, or all insects, is of limited power in evaluating the dispersal structuring of island biotas. This was illustrated for the flowering plants of Hawaii by Carlquist (1974), who attributed Fosberg's hypothesized set of original colonists of the archipelago to their most likely means of long-distance dispersal (Table 6.4) (see also Keast and Miller 1996). As some plant taxa in the source regions lack the adaptations that enable such vast distances to be crossed by these means, they typically do not colonize even moderately remote islands. The ecological correlates that may accompany such dispersal structuring were discussed in the section on dioecy, and will be given further attention in Chapter 8.

Table 6.4
Most probable means by which the original flowering plant colonists dispersed to Hawaii (data from Carlquist 1974)

Dispersal mode	Percentage of colonists
Birds	
Mechanically attached	12.8
Eaten and carried internally	38.9
Embedded in mud on feet	12.8
Attached to feathers by viscid substance	10.3
Oceanic drift	
Frequent (able to float for prolonged periods)	14.3
Rare (most likely to have arrived by rafting)	8.5
Air flotation	1.4

6.4.4. Biogeographical hierarchies and island evolutionary models

Williamson (1981) has suggested redefining island types as oceanic where evolution is faster than immigration, and as continental if immigration is faster than evolution. He also offered the observation that it is useful for any particular island context to distinguish between those groups for which the island is continental, readily reached by dispersal, and those for which the island is oceanic, leading to little dispersal. Thus the Azores could be regarded as oceanic for beetles, intermediate for birds, and continental for ferns. The gradual diminution in the range and numbers of taxa that have been able to cross ever-increasing breadths of ocean means that, in general, the more remote the island the more disharmonic the assemblage, and the greater the amount of vacant niche space. Thus, the greatest opportunities for adaptive radiation are towards the edge of the range of particular taxa, i.e. in the radiation zone of MacArthur and Wilson (1967), reaching their synergistic peak in the most remote archipelagos. Thus, in the absence of ants (amongst other things), there have been great radiations of carabid beetles and spiders in Hawaii and South-East Polynesia (Paulay 1994). It appears likely that while some diversification in certain lineages has been enabled by such disharmony, the absence of particular interacting organisms, pollinators for plant species for instance, may have restricted the colonization, spread, and evolution of other lineages. That is to say, hierarchical relationships within the island biota, partly predictable and partly the chance effects of landfall (e.g. the highly improbable arrival of the first drosophilids on Hawaii at a particular point in time), must have a strong influence on the patterns that have unfolded.

Once on an archipelago, the relative importance of inter-island and intra-island speciation may vary (Paulay 1994). Only a handful of isolated archipelagos show large endemic radiations of birds or plants. The occurrence of a number of islands in proximity to one another appears to be particularly important to birds, suggesting that inter-island effects must be important (cf. P. R. Grant and B. R. Grant 1996). Intra-island speciation is restricted to islands large enough to allow effective segregation of populations within the island. This is taxon dependent, so that for land snails and flightless insects an island of a few square kilometres may be sufficiently large.

The historical impact of humans may well have been underestimated in the evolutionary and ecological literature, and provides an alternative hypothesis to set alongside the more traditional biotic hypotheses such as competitive displacement and the taxon cycle (Pregill and Olson 1981; Pregill 1986). As several authors emphasize, different island groups have their own special circumstances and histories, and few theories can span them all. A corollary of this is that each theory may have its own constituency. The solution, as elsewhere, must be to set up multiple working hypotheses and seek evidence to distinguish between them. Yet, frequently, a particular lineage reflects not the operation of a single process, but of several. Therefore it may not be possible to explain the overall biogeography of an island lineage by means of a single model, such as character displacement, a double invasion, or the taxon cycle.

Having established the relevance of factors such as the regional biogeographic setting, environmental change, dispersal differences between taxa, island area, island habitat diversity, and island isolation and configuration, it would seem to be valuable to attempt to place at least some of these into a common framework. Adler and Levins' (1994) model of the respective roles of area and isolation in the insular syndrome in rodents (Fig. 5.4) provides a starting point. Figure 6.12 is offered as an attempt to place

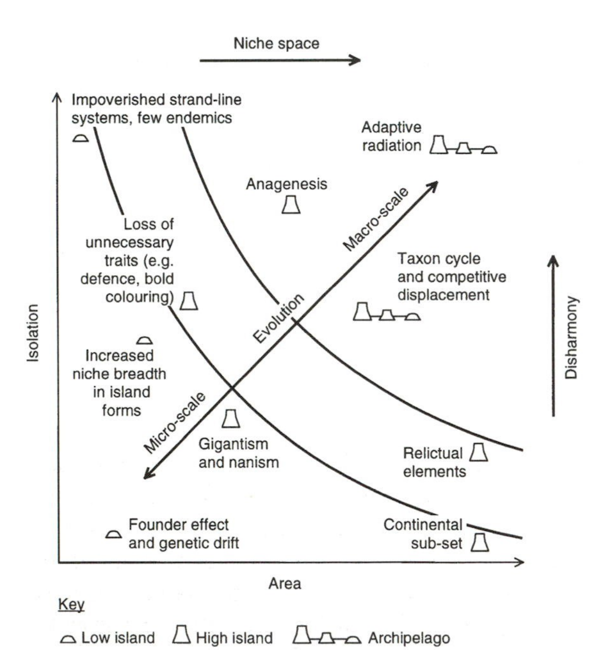

Fig. 6.12. Conceptual model of island evolution. This diagram attempts to highlight the geographical circumstances in which particular evolutionary phenomena are most prominent. The scalars will vary between taxa.

island evolutionary ideas and models into a simple island area–isolation context. The aim is to highlight the geographical circumstances in which particular evolutionary phenomena are most prominent. This is not to imply that, for instance, founder events and drift are insignificant in large and isolated oceanic islands, as they do form elements of the macroevolutionary models. Rather, it is that they emerge as the main evolutionary storyline for small, near-mainland islands which, as in the rodent model, are not so small that they cannot support persistent populations. Working up the isolation axis, small islands of limited topographic range and degree of dissection are liable to be poorly buffered from pronounced environmental fluctuations, such as are associated with El Niño events or cyclonic storms. Those species or varieties that become endemic on such islands are likely to have relatively broad niches, at least in so far as tolerating disruptions of normal food supplies is concerned, as illustrated by the Samoan fruit bat, *Pteropus samoensis* (see section 8.6; Pierson *et al.* 1996).

Large, near-continent islands are likely to have very low levels of endemicity; mainland–island founder effects and drift are unlikely to be prominent in their biota. Islands such as (mainland) Britain fit this category, having essentially a subset of the adjacent continental biota, most species having colonized prior to the sea-level rise in the early Holocene which returned the area to its island condition. Large islands with a somewhat greater (but not excessive) degree of isolation and considerable antiquity are the sites where relictual elements are most frequently claimed in the biota (but see Pole 1994; Kim *et al.* 1996). As frequently noted, the greatest degree of evolutionary change occurs on remote, high islands. Where these islands are found singly, or in very widely spread archipelagos, speciation is frequent, but often without the greatest radiation of lineages, fitting the model of anagenesis.

Non-adaptive radiation features principally in the literature on land snails, for high and remote islands. The adaptive radiation of lineages is best seen on large, diverse, remote archipelagos, in which inter-island movements

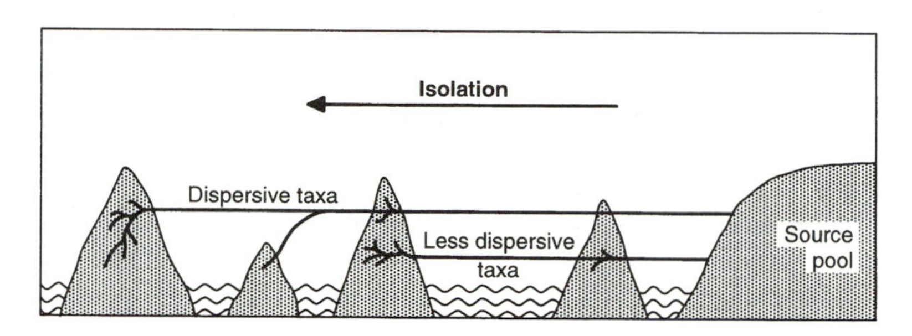

Fig. 6.13. Dispersive taxa radiate best at or near to their effective range limits (the radiation zone), but only moderately, or not at all, on islands near to their mainland source pools. Less dispersive taxa show a similar pattern, but their range limits are reached on much less isolated islands. The increased disharmony of the most distant islands further enhances the likelihood of radiation for those taxa whose radiation zone happens to coincide with availability of high island archipelagos.

within the archipelago allow a mix of allopatric and sympatric speciation processes. Somewhere in the middle of this factor space comes the taxon cycle, or at least, the patterns associated with it. Most cases proposed for the taxon cycle are from archipelagos which are strung out in stepping-stone fashion from a continental land mass (the Antilles) or equivalent (the smaller series of islands off the large island of New Guinea). Here the degree of disharmony is less than for systems such as Hawaii and the Galápagos, and, crucially, there are likely to be repeat colonization events by organisms quite closely related, taxonomically or ecologically, to the original colonists.

Of course, Fig. 6.12 is a simplification. Further dimensions might be added for other factors, such as environmental change. Williamson's redefinitions of oceanic and continental islands on a per taxa basis provide another useful ingredient, as underlying the differences in domain identified in the figure are the between-taxa differences in dispersability, and thus in degree of disharmony. Thus the scalars for area and isolation for Fig. 6.12 should vary between taxa. This may again be illustrated in simple diagrammatic form. Figure 6.13 makes the simple point, long-recognized in island biogeography, and encapsulated in MacArthur and Wilson's (1967) notion of the radiation zone, that taxa are most likely to radiate towards the limits of their dispersal reach. Such radiation is, however, best accomplished on higher islands, and archipelagos of such islands, than it is on low islands or single high islands. A single island system may thus provide for an island ecological focus in one taxon, a taxon cycle in another, and radiation in a third.

6.5. Summary

Three distinctive emergent patterns are recognized by which island endemism may be understood. First, it is noted that some island speciation occurs with little or no radiation (anagenesis), but instead by change in the island form away from the colonizing phenotype along a singular pathway (change may also occur on mainlands, such that the island forms are in part relictual). A second important model is that termed the taxon cycle, wherein evolution takes place within the context of an island archipelago, driven by a series of colonization events of taxonomically and/or ecologically related forms. Immigrant species are believed to undergo niche shifts, which are in part driven by competitive interactions with later arrivals. Through time, the earlier colonists lose mobility, decrease in distributional range, and ultimately may be driven to the brink of extinction by later colonists. Evidence for the taxon cycle is largely, but not exclusively, distributional, and at least one of the prime examples, that of West Indian avifauna, has been the subject of criticism for its apparent failure to agree with historical evidence for environmental and distributional changes. The most spectacular evolutionary patterns, the radiation of lineages, are to be found on the most isolated of large, oceanic islands. While some radiations have been postulated to be essentially non-adaptive (featuring 'drift' in within-island isolates rather than clear niche changes), most radiations involve clear alterations of niche and may be termed adaptive. Examples of the former appear to flow mostly from studies of land snails, whereas the latter include the best-known plant, bird, and insect examples from Hawaii, Galápagos, and Macaronesia.

Recently, phylogenetic evidence has allowed some of the ideas and models outlined above to be exposed to testing and has greatly improved the detail of many evolutionary scenarios. It appears, on theoretical and empirical grounds, that the fastest rates of radiation occur among early colonists at early stages of the existence of an island. The best data are from Hawaii, where the general tendency has been for dispersal events from old to young islands along the chain, although back-dispersal has also occurred. Radiation of lineages on the big island of Hawaii has also been enhanced by the physical dynamism and variety of its environment. The importance of environmental change has been emphasized increasingly in the island evolutionary literature of late, particularly following recognition of the impact on island endemic bird populations of climatic fluctuations associated with El Niño events in the 1980s. Differential dispersal abilities, and evolutionary changes in dispersal powers, provide another key element to understanding the disharmony of island ecosystems and the context in which evolutionary change takes place in those lineages that do reach remote islands.

The final section of this chapter attempts to place the principal island evolutionary concepts and models into a simple framework of area versus island—or archipelago—isolation. It is suggested that different ideas come to prominence in different regions of this natural experimental factor space. Thus, for example, the taxon cycle may be an appropriate model to test in a not-too-remote archipelago, whilst adaptive radiation is more typical of the most isolated high island archipelagos. As effective dispersal range varies between (and within) taxa, different evolutionary patterns may emerge within the same archipelago by comparison of different types of organism. These geographical and taxonomic contexts provide a framework within which many of the ideas of island biogeography can be seen to be complementary rather than opposing theories.

7

Species numbers games

Area is the devils' own variable.

(Anon.)

Coastal islands are notorious for their accumulation, at all seasons, of a staggering variety of migrant, stray, and sexually inadequate laggard birds. In the absence of specific information on reproductive activity, it is therefore unwarranted to assume tacitly that a bird species, even if observed in the breeding season, is resident.

(Lynch and Johnson 1974, p. 372.)

7.1. Introduction

Chapters 7 and 8 of this book are centred on shorter time-scales and less isolated insular contexts than was Chapter 6. This change in emphasis might be distinguished by the label *island ecology*. Over the past 30 years or so much of the research in this field has been concerned, in essence, with species numbers games, or more explicitly with: the nature of species–area relations; the factors determining the number of species found on an island; and the rates of species turnover in isolates. This research agenda was opened up principally by the theory developed by Robert H. MacArthur and Edward O. Wilson, and first published in 1963 as 'An equilibrium theory of insular zoogeography' in the journal *Evolution*. They later expanded on the basic theory in their 1967 book *The theory of island biogeography*, in which they offered their framework as one not limited merely to zoogeography, but also applicable to the plant kingdom. Both works incorporate a wealth of ideas and theories, but the core concept is that of a **dynamic equilibrium**, a balance between immigration to an island (supplemented by evolution of new forms in some cases), and extinction from it of local populations, under the influence of island isolation and area, respectively. This elegantly simple core theory will be termed here the **equilibrium theory of island biogeography (ETIB)**, to distinguish it from the numerous additional theoretical elements and embellishments contained in their 1967 monograph. At the time of their collaboration, Wilson had already published seminal papers on character displacement and the taxon cycle. The island biogeography of MacArthur and Wilson was much, much richer than merely 'species numbers games'. The scope of their book, as they themselves saw it, is illustrated in Fig. 7.1, which demonstrates their vision of how their ideas might feed into an improved understanding of evolutionary change on islands. Much of the controversy which has surrounded their theory is, I believe, a result of a failure to distinguish between their core model and the broader body of ideas in their book. By highlighting the progress of their ideas and detailing attempts to test them, I hope to show by the end of this chapter how their equilibrium theory can be reconciled with other competing island ecological theories.

Island ecology has burgeoned since the 1960s and the result has been a radiation of lineages of ideas and models, with occasional episodes of hybridization and a few dead ends. The separation of issues concerned in essence with compositional structure into Chapter 8 reflects on what I happen to find a convenient way to understand and explain the development of ideas in this subject. The consideration of composition, assembly rules, and the structure of turnover are also, however, fundamental to a

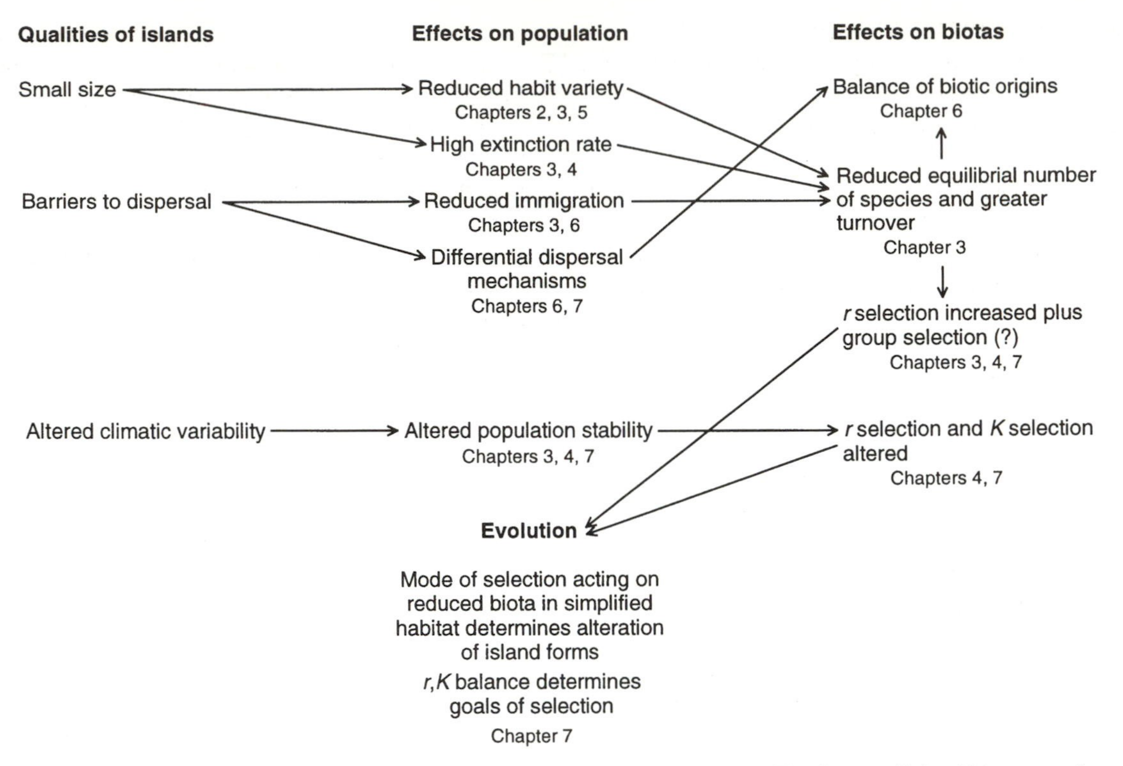

Fig. 7.1. The flow of ideas between chapters within MacArthur and Wilson's (1967) *The theory of island biogeography*, as set out by them in their Table 1. (Redrawn from MacArthur and Wilson 1967.)

full critique of the ETIB, which therefore is reserved to Chapter 8. Thus MacArthur and Wilson's theories provide a thread running through both Chapters 7 and 8. Their ideas have proven pivotal to debate in island ecology (and conservation biology) for several reasons. One is that their theory has focused attention on one of the most pervasive, fundamental themes of the subject: do ecological systems operate in an essentially balanced, equilibrial state (the balance of nature viewpoint), or do they exhibit non-equilibrium modes of behaviour? This dichotomy can be found in subfields of biogeography as widely spaced as diversity theory, succession theory, and savanna ecology, to name but a few. The island laboratories once again have been a testing ground for developing and refining ideas of broader relevance (Brown 1981, 1986), and the ETIB was the starting point. As the quotes above indicate, however, the path to enlightenment has not been an easy one.

7.2. The development of the equilibrium theory of island biogeography

Prior to the 1960s, as MacArthur and Wilson (1967) wrote in their monograph, there was little quantitative theory in island biogeography. The fundamental processes of dispersal, invasion, competition, adaptation, and extinction are, after all, quite difficult to study and understand. They are complex phenomena. Most research had been taxonomic in origin and historically orientated. The style of question asked was 'What was the ultimate origin of the Antillean vertebrate fauna?' Such questions were concerned with a limited number of higher taxa and, they contended, had not tended to encourage generalizations (but see Sauer 1969;

Williamson 1981). MacArthur and Wilson set out to formulate a biogeographic theory at the species level, relating distributional pattern to population ecological processes.

Several ingredients were combined in the ETIB (see account by Wilson 1995). It has long been known that there is a relationship between island area and species number. For instance, the zoogeographer Darlington (1957) offered the approximation that, for the herpetofauna of the West Indies, 'division of island area by ten divides the fauna by two'. However, the form of the relationship varies between cases, and in examining data for Pacific birds, MacArthur and Wilson (1963) noted that distance between the islands and the primary source area (i.e. island isolation) appeared to explain a lot of the variation. A second ingredient, commonly attributed to Preston (1948, 1962), was the discovery that there might be a connection between the form of the species–area curve and how individuals were distributed between species within an assemblage (the species–abundance curve). The third ingredient took the form of the turnover of species on islands, a force that Wilson (1959, 1961) had previously invoked on evolutionary time-scales in the taxon cycle, and which other biologists had described as occurring on ecological time-scales, for instance in analyses of plant (Docters van Leeuwen 1936) and animal (Dammerman 1948) colonization of the Krakatau islands. The other essential element was the idea that distance might have a crucial influence on turnover.

It has recently been pointed out that Dammerman (1948), in his monograph on Krakatau, had described several of the key elements of the eventual equilibrium theory, but had failed to put them into a mathematical framework (Thornton 1992). Another biologist, Eugene Munroe, went one better than Dammerman. In his studies of the butterflies of the West Indies, Munroe presented a formulaic version of equilibrium theory, remarkably similar to that independently developed by MacArthur and Wilson some years later. Munroe's formula appeared only in his doctoral thesis (1948) and in the abstract of a conference paper published in 1953. As Wilkinson (1993) noted, neither Dammerman nor Munroe promoted their ideas in the right form of packaging for immediate success and uptake by the wider scientific community. Munroe, in particular, deserves recognition for his discovery of the equilibrium theory, which was independent not only of MacArthur and Wilson, but also made prior to publication of Preston's analyses (1962) and of Dammerman's Krakatau monograph (see Brown and Lomolino 1989). None of this diminishes, in any way, the contribution made by MacArthur and Wilson themselves. They not only worked up the basic ideas, both descriptively and mathematically, but they had the insight to see the potential of these ideas for biogeography at that point in time and to develop and present the first modern, comprehensive theory of island biogeography. They also had a series of students following in their wake to carry forward the flame!

7.2.1. Species–area patterns

The relationship found between species and island area identified by earlier workers was explored further by MacArthur and Wilson (1963, 1967), who concluded that for a particular taxon and within any given region of relatively uniform climate, a fairly simple relationship often exists. It may be expressed by the **power function model**:

$$S = CA^z,$$

where S is the number of species of a given taxon on an island, A is the area, and where C and z are estimable parameters. Values for C depend on the taxon and biogeographic region: C is thus an empirically determined constant representing the biotic richness of an area. z is also a parameter, representing the slope of the diversity curve, and generally relates to the difficulty of reaching the islands. This simple formulation hides a wealth of opportunity for confusion for those (like the present author) who are not mathematically inclined, and even, it seems, for those who are (for problems associated with analysis of species–area curves see Connor and McCoy 1979; Sugihara 1981; Connor *et al.* 1983; Williamson 1988; Lomolino 1989; Rosenzweig 1995; Williams 1996).

Island biogeographers have typically used logarithmic transformations to represent species–area relationships as, very often, **log–log plots** (sometimes called Arrhenius plots, after O. Arrhenius who championed this approach in the 1920s) align the data along a straight line. The equation thus becomes:

$$\log S = z \log A + \log C,$$

which enables the parameters C and z to be determined using simple linear regression. In this equation, z describes the slope of the log–log relationship and $\log C$ describes its intercept.

From the data at their disposal, MacArthur and Wilson (1967) found that in most cases z falls between 0.20 and 0.35 for islands, but that if you take non-isolated sample areas on continents (or within large islands), z values tend to vary between 0.12 and 0.17. Thus, the slope of the log–log plot of the species–area curve appeared to be steeper for islands, or, in the simplest terms, any reduction in island area lowers the diversity more than a similar reduction of sample area from a contiguous mainland habitat. Wilson's (1961) own data for Melanesian ants (Fig. 7.2) was one of the data sets used in this analysis. While it is possible to find many examples that broadly support this island–mainland distinction (e.g. Begon *et al.* 1986, Table 20.1), there are also many exceptions. Williamson (1988) reviews surveys providing the following slope ranges: ordinary islands, 0.05–1.1.32; habitat islands, 0.09–0.957; and mainland samples −0.276–0.925. This particular island effect thus turns out to be a relatively weak one, susceptible to other influences (see below; Connor and McCoy 1979; Williamson 1981).

7.2.2. Species abundance distributions

In most plant and animal communities there are few species of many individuals and many more species of few individuals. This is one of the more robust patterns in community ecology. Two principal theories had been put forward concerning the distribution of abundance, they differ in relation to the abundance of the rarer species. Fisher *et al.* (1943), suggested that the largest class of species is of those that are individually rarest. This gives rise to the **logarithmic series of abundance**. Preston (1948, 1962) developed the alternative theory that species more typically fit a **log–normal series of abundance**, i.e. that the most numerous species were those of middling abundance. It was argued that insufficient sampling commonly had given rise to the apparent fit of the logarithmic model. This can be illustrated diagrammatically for a hypothetical community if the abundances of individual species are plotted on a log scale against number of species on an arithmetic scale (Fig. 7.3). If the sample is small, then the sparsest species of the log–normal distribution will not be sampled, and the abundance

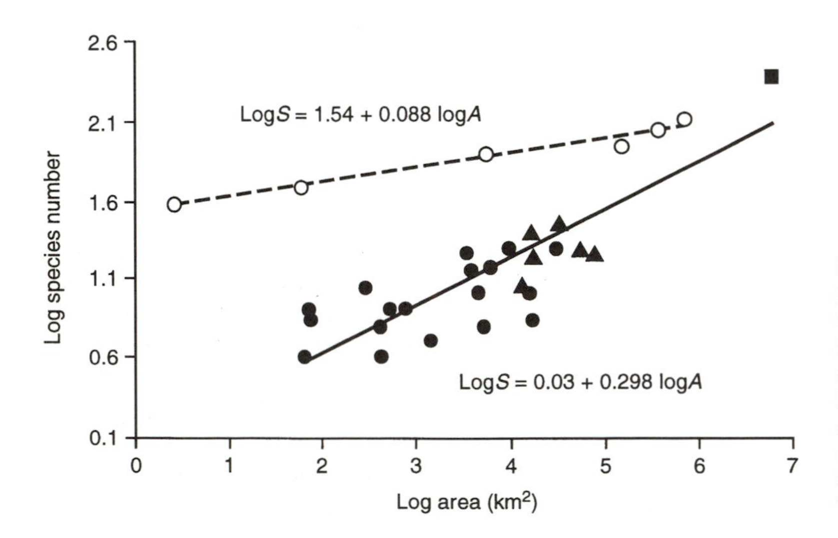

Fig. 7.2. Species–area relations for ponerine and cerapachyine ants in Melanesia. Solid dots represent islands; open circles, cumulative areas of New Guinea up to and including the whole island; triangles, archipelagos (not used in the regression); and the square, all of South-East Asia. (Adapted from Wilson 1961.)

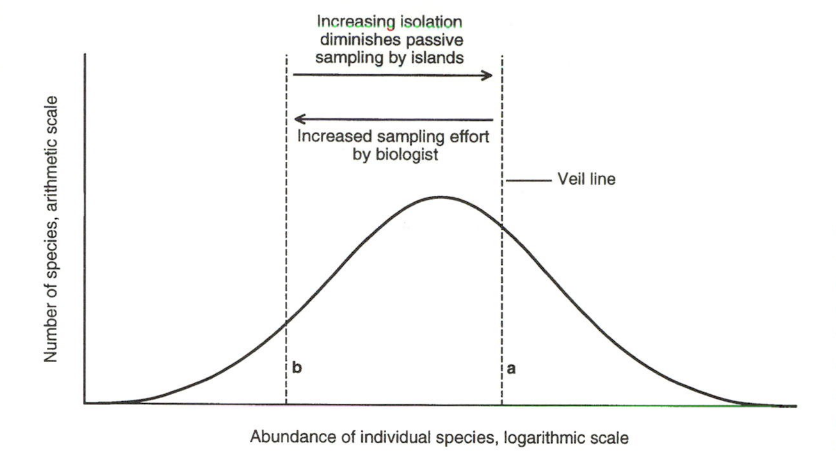

Fig. 7.3. Preston's log–normal species–abundance relationship. The portion of this hypothetical abundance distribution that is sampled may be a function of either active sampling effort (by the biologist) or passive sampling effort (by an island). Only the portion to the right of the veil line will be sampled.

distribution will be that shown to the right of the **veil line** drawn at point a (Preston 1948). On increasing the sampling effort, more of the rarer species of the system will be sampled, pushing the veil line to position b. By analogy, a small area veils the existence of all the species whose total abundance falls below a critical minimum.

To put this into a more familiar ecological formula, imagine an area delimited for study within a continuous woodland habitat, containing just one or two pairs of a bird species—the rare end of the distribution. Through chance effects, such as harsh weather or predation, the species might be lost locally, but be replaced almost immediately by other individuals from the surrounding woodland taking over the territory. Whereas, in an insular context, the species would be lost and might not be replaced by immigration for some years (e.g. Paine 1985). Average this across the whole island assemblage and the island will, at any one time, fail to sample a significant number of (principally) the rarer birds found in the mainland habitat patch. Theoretically, a log–normal series of abundance should give rise to z values of approximately 0.263, towards the low end of the range of values then available from islands and above those of continental patches. These differences therefore pointed to a role for isolation via population migration, and, in short, Preston's theorizing, combined with his analysis of island data, led him inevitably to a focus on turnover in his 1962 article, foreshadowing by a few months several of the elements of MacArthur and Wilson's thesis.

Williamson (1981) points out that there are theoretical reasons why neither the logarithmic nor the log–normal series can be exactly true. For example, the difference between species is discrete, and so cannot be matched perfectly by a continuous distribution such as the log–normal series. For present purposes these points of detail are not critical, yet it is of interest to determine which provides the better fit with real data. One way of establishing this is by examination of the form of the species–area relationship, as on theoretical grounds it is expected that a plot of S versus log A should produce a straight line where the logarithmic series applies. Whereas a log S versus log A plot should produce a straight line where the log–normal series applies. By reference to the argument in the previous paragraph, it might be anticipated that the former situation might be found to apply for large samples (or areas) and the latter to small samples (or areas). While noting exceptions, both Williamson (1981) and Rosenzweig (1995) conclude that empirical studies from islands show a fair degree of support for the Preston theory: the species–area plot is more usually linear on a log–log plot than in any other transformation. While this has a wider significance, for the moment it is sufficient to note that the position taken on species–abundance patterns by Preston, and by MacArthur and Wilson, was an eminently reasonable one.

7.2.3. The distance effect

MacArthur and Wilson (1963, 1967) predicted that the species–area curve would become steeper with increasing distance. Unfortunately, the increased impoverishment of island biotas with increasing isolation is confounded with variations in other properties of islands, particularly their area. The distance effect has turned out to be difficult to test, and the outcome of such tests as there have been remain equivocal. At the outset, MacArthur and Wilson (1963, 1967) produced a way of quantifying the effect for Pacific birds. First, they showed that there was a fairly tight, straight-line species–area relationship on a log–log scale for the closely grouped Sunda islands, but that an equivalent analysis for 26 islands and archipelagos from Melanesia, Micronesia, and Polynesia had both a much greater scatter and a tendency away from a straight-line relationship. Secondly, they reasoned that a line drawn through the points for near islands could be taken to represent a 'saturation curve', and the degree of impoverishment of islands below the line could then be estimated as a function of area by comparison with the saturation curve. In general, the degree of impoverishment, when corrected for area, was found to increase from islands 'near' to New Guinea (less than 805 km) to intermediate, to 'far' islands (greater than 3200 km). Williamson (1988) suggests that this demonstration is invalid because the example used archipelagos rather than single islands, and included archipelagos of different climate or deriving their avifaunas from different sources. He reports that, in fact, the line tends to be flatter the more distant the archipelago, such that (with some caveats) slopes of 0.28, 0.22, 0.18, 0.09, and 0.05 can be derived for birds in the progressively more isolated archipelagos of the East Indies, New Guinea, New Britain, Solomons and New Hebrides (Vanuatu) archipelagos, respectively. It appears that the assumptions made in analysing species–area curves in relation to isolation, e.g. whether to lump separate islands into archipelagos, whether to calculate isolation within or between archipelagos, whether to include islands of rather different climates, etc., are critical to the outcome of the analysis (compare Itow 1988; Williamson 1988; Rosenzweig 1995 (below)). In any case, the discovery since the 1960s of numerous 'new' bird species from subfossil remains on many of the islands of the Pacific (section 3.6; Chapter 10) renders this particular part of the MacArthur and Wilson analysis unsafe, as several points on their graph are below their 'natural' level. I have introduced their analysis of these data here principally because recognition of the distance effect formed an important part of their reasoning.

7.2.4. Turnover, the core theory (ETIB) and its immediate derivatives

It was these observations—of species–area patterns, species–abundance patterns, and distance—which were combined in the equilibrium theory, through the mechanism of the **turnover of species** on islands. The theory postulates that there are two ways in which islands gain species, by immigration or by evolution of new forms, and that these means of increasing species number will be balanced in the equilibrium condition by processes leading to the local loss of species from the island in question.

It is essential to have a clear definition of the processes involved prior to immersing ourselves in the detail of the model. MacArthur and Wilson (1967, pp. 185–91) developed the following formulations:

- **Immigration**. The process of arrival of a propagule on an island not occupied by the species. The fact of an immigration implies nothing concerning the subsequent duration of the propagule or its descendants.

- **Immigration rate**. Number of new species arriving on an island per unit time.

- **Propagule**. The minimal number of individuals of a species capable of successfully colonizing a habitable island. A single mated female, an adult female and a male, or a whole social group may be propagules, provided they are the minimal unit required.

- **Colonization**. The relatively lengthy persistence of an immigrant species on an island, especially where breeding and population increase are accomplished.

- **Colonization curve**. The change through time of numbers of species found together on an island.
- **Extinction**. The total disappearance of a species from an island (does not preclude recolonization).
- **Extinction rate**. Number of species on an island that become extinct per unit time.
- **Turnover rate**. The number of species eliminated (i.e. extinct) and replaced (i.e. immigrants/speciation) per unit time.

The forces for increase in species number (immigration + speciation) are likely to diminish as a fairly simple function with increasing distance from source pools; this is because immigration rates decrease with isolation. The increase in numbers by speciation is of increased proportional significance on remote islands but occurs too slowly to compensate in ecological time for the great reduction in immigration frequency. For the moment let us set aside the role of *in situ* evolution and concentrate solely on immigration. The local loss of a species population may be accomplished either by outmigration or by the death of the last representatives on an island, in either case leading to the local extinction of the species. The greater resource base of larger islands should mean that extinction rates are lower for larger islands than for smaller. Thus the ETIB postulates that the number of species found on an island represents a dynamic balance between immigration (I) and extinction (E), with immigration varying with distance from source pool and extinction with island area. This was presented in simple, accessible, diagrammatic form (Fig. 7.4)—possibly one of the keys to the broad uptake of the theory (Brown and Lomolino 1989).

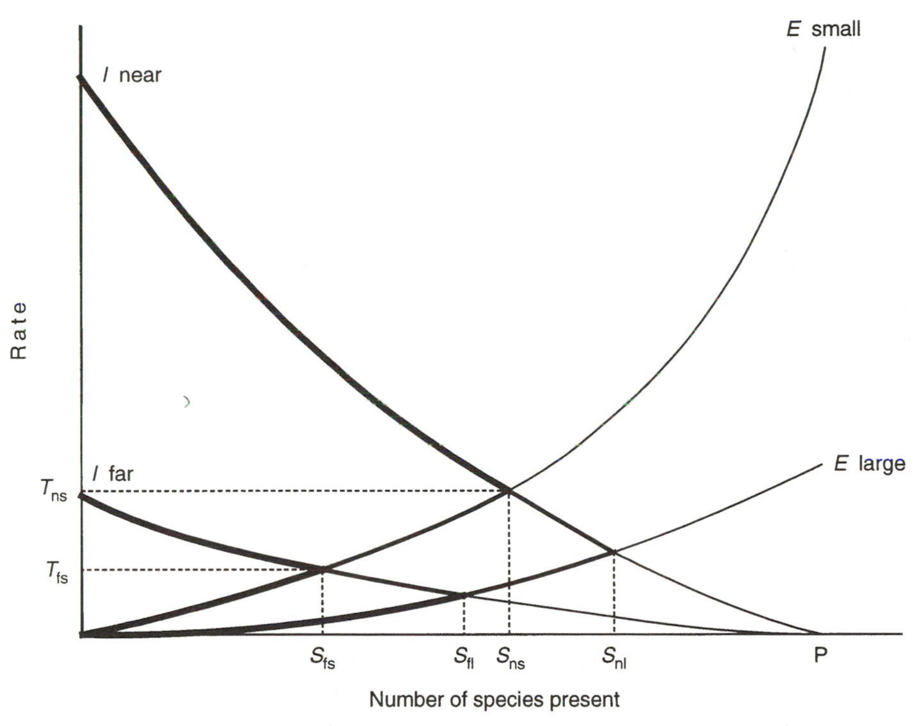

Fig. 7.4. A version of MacArthur and Wilson's (1963, 1967) equilibrium model (the core equilibrium theory of island biogeography), showing how immigration rates are postulated to vary as a function of distance, and extinction rate as a function of island area. The model predicts different values for S (species number), which can be read off the ordinate and for turnover rate (T) (i.e. I or E, as they are identical). Each combination of island area and isolation should produce a unique combination of S and T. To prevent clutter only two values for T are shown.

MacArthur and Wilson (1967) also considered the effects of slightly more complicated configurations of islands, for example where chains of islands ('**stepping stones**') are found strung out from a mainland, producing alterations in the expected immigration functions and thus in equilibrium values. However, the basic ideas may be most easily grasped without reference to such complicating features.

Figure 7.4 illustrates immigration and extinction as hollow curves, for the following reasons. When the number of species on an island is small relative to the mainland source pool, *P*, a high proportion of propagules arriving on an island will be of species which are absent from the island. As the island assemblage gets larger, fewer arrivals will be of species not already present, and so the immigration rate of new species lessens as *S* nears that of the total mainland pool. The hollow form of the extinction curve recognizes that some species are more likely to die out than others, and that according to the theories of species–abundance (above), the more species there are in a sample or isolate, the rarer on average each will be, and so the more likely each is to die out. At equilibrium, immigration and extinction rates should be approximately in balance and thus equal the rate of species turnover ($I=E=T$). Islands of different size or different isolation may have the same turnover, or the same species number, but they cannot have both the same turnover and species number. The figure thus shows how the theory predicts different combinations of *S* and *T* as a function of area and isolation. This reading of the figure uses it as a representation of space. The diagram may also be understood as a representation of change through time for a particular island, commencing with initial colonization by the first inhabitants, when immigration rate is high, species number low, and extinction rate low. The curves may be followed towards the equilibrial condition when species number is high and stable, immigration rate has declined and extinction rate has risen to meet it. If we take the case of a small, near island, species number at equilibrium will be S_{ns} and the rate of turnover at equilibrium will be T_{ns}. The portion of the *I* and *E* curves to the right-hand side of their intersection can, of course, be disregarded, as once they have met, neither rate should vary further to any significant degree; the island has reached its dynamic equilibrium point.

The equation for equilibrium is:

$$S_{t+1}=S_t+I+V-E$$

where S=number of species at time t, S_{t+1}=number at time $t+1$, I=immigration, V=additions through evolution (where applicable), and E=losses by extinction (it will be noted that as species are integers, continuous representations are approximations).

Integration of the rates of *I* and *E* as they vary over time produces what is termed the **colonization curve**, the temporal trend of species number on the island. This rises steeply initially, but ever more slowly as eventual equilibrium is neared (Fig. 7.5). How often can biologists be certain of recording all immigration and extinction events? Rarely, if ever, for real

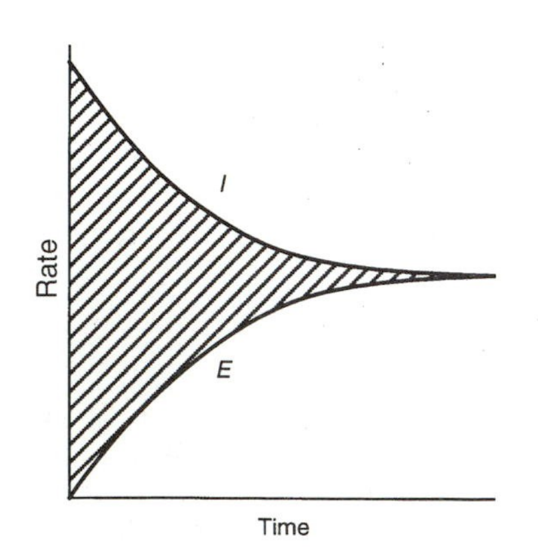

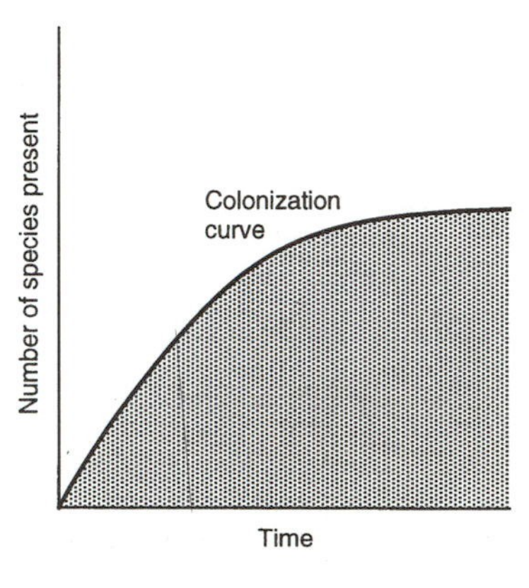

Fig. 7.5. Integration of immigration and extinction curves (left) should, theoretically, produce the colonization curve as shown (right). (Redrawn from MacArthur and Wilson 1967, Fig. 20.)

islands. MacArthur and Wilson (1967) recognized this but reasoned that the colonization curve might be used to back-calculate indirect estimates of what the real I and E rates might be at two stages of the colonization process. The first point is near the beginning of colonization, where I may be assumed to be close to the rate of colonization and E to be insignificant (although both assumptions appear rather bold when successional factors are considered: Rey 1985; Chapter 8). The second point is near the equilibrium condition. In this case they developed a proof for derivation of the extinction rate at equilibrium as a function of the species numbers on other similar islands already at equilibrium and the time taken to reach 90% of the equilibrial number of species on the target island. They detailed a number of studies in apparent support of this line of reasoning. The first of which was for bird recolonization of the Krakatau islands, which appeared to have reached a very dynamic equilibrium by the 1920s (MacArthur and Wilson 1963). This example is still cited in support of the theory (e.g. Rosenzweig 1995), but as MacArthur and Wilson (1963, p. 384) speculated might be the case, this was in fact a premature conclusion; indeed data had already been published which showed a further increase in species number (Hoogerwerf 1953).

The above constitutes what I view as the core theory, or ETIB, which occupies the second and third chapters of MacArthur and Wilson's (1967) monograph. It provides an essentially stochastic model of biological processes on islands, in which the properties of individual species get little attention. Its great virtue was its apparent testability. The predictive nature of the equilibrium theory is important. While it may have heuristic value without it, the contribution of the theory to biogeography was in large part to appear to provide a more rigorous scientific approach. As Brown and Gibson (1983, p. 449) put it:

> ... like any good theory, the model goes beyond what is already known to make additional predictions that can only be tested with new observations and experiments. Specifically, it predicts the following order of turnover rates at equilibrium: $T_{SN} > T_{SF} \equiv T_{LN} > T_{LF}$.

The next element of their thesis sketched out, in mathematical terms, a portrait of the biological attributes of the superior colonist. These ideas influenced Jared Diamond who subsequently developed the idea of **supertramp species** (see Chapter 8), highly ***r*-selected species** capable of high mobility and thus of reaching islands at an early stage, and of breeding rapidly to build large populations; but which are poor competitors and eventually lose out to slower dispersing and slower building ***K*-selected species** (naturally selected to maximize carrying capacity). Another chapter considered additional effects of stepping stones, allowing enhanced dispersal routes, and thus altering patterns of biotic exchange between different biogeographic regions. The later chapters also incorporated Wilson's ideas concerning niche shifts and the taxon cycle and radiation zones. MacArthur and Wilson (1967) thus built up their analyses from a highly simplified island ecological model, through the routes shown in Fig. 7.1 and into theories of evolutionary ecology. In parallel, there is a movement from an essentially stochastic model (the ETIB) to increasingly deterministic ideas. As they wrote on p. 121, 'A closer examination of the composition and behavior of resident species should often reveal the causes of exclusion, so that random processes in colonization need not be invoked.'

Lynch and Johnson (1974, p. 371), in their critical analysis of problems inherent in testing the ETIB, argued that for the theory to apply 'turnover should reflect the stochastic nature of an equilibrium condition (i.e. the turnover should not be attributable to some systematic bias such as ecological succession, human disturbance of habitats, introduction of exotic species, etc.)'. I have been at pains in this section to make it plain that MacArthur and Wilson's island biogeography did not simply end with a stochastic, simplifying, mathematical model of species turnover at equilibrium. But, this is a reasonable description of their core or ETIB model. It was this model that appeared to offer the prospect of a new scientific biogeography, complete with testable hypotheses. Lynch and Johnson's (1974) remarks that turnover must be stochastic to be consistent with the theory are thus unnecessarily restrictive in the broader terms of MacArthur and Wilson's vision for their theory, but at the same time are a fair assessment of the properties of the core model. This is indicative of a general problem in assessing the equilibrium theory: there is scope to define its limits in rather varied ways. Evidence that is claimed by one author taking a narrow (core model) view to be contradictory to theory, may thus be argued by a broad-view author to be consistent with MacArthur and Wilson. The approach taken here is that it is fair game to attempt to test and explore the limits of the core ETIB (a process MacArthur and Wilson themselves begin in their monograph), and that this should not be regarded as equivalent to impugning the wider body of ideas presented by its authors.

7.3. Spatial analyses of island species number

How, then, has the ETIB fared? Evaluation and refinement of the ETIB has taken two principal forms. The first can be characterized as attempts to test or explore the spatial manifestations of the theory. However, as one of the starting assumptions of the equilibrium theory is that there should be a close species–area relationship, demonstration of such a relationship is not of itself enough to provide a critical test of the theory (Gilbert 1980; Williamson 1988). The second, arguably more critical and more difficult form of test (Gilbert 1980) concerns turnover.

7.3.1. Area and habitat diversity

Assuming most islands to be at or near to equilibrium, the ETIB embodies particular roles for area and isolation in determining species numbers. MacArthur and Wilson (1967) employed area as a convenient starting point, indicating that the eventual theory which they envisaged being developed 'may not mention area because it seldom exerts a direct effect on a species' presence. More often area allows a large enough sample of habitats, which in turn control species occurrence.' (p. 8). Numerous studies have since been published which quantify, via regression, values for z, and the varying explanatory values of area, isolation, altitude, island age, and various other measures of habitat diversity. As a gross generalization, area is very often the most important explanatory variable.

Setting to one side the issue of turnover, two facets of the relationship between area and species number might be recognized: (1) a direct area or **area *per se*** effect, and (2) a **habitat diversity effect**. The former may be particularly relevant to very small patches of a particular and distinctive habitat, such as a small fringing salt marsh, or mangrove system, or the small patch of cloud-forest found around the summit of a small offshore island. It is in such cases that a true area effect, of insufficient space to fit in all the possible colonists, may be most apparent. The area involved will vary depending on the ecological characteristics of the group concerned. It may be very small for lichens growing on rock outcrops, or very large for top carnivores. Gibson's data (1986) for habitat islands of grassland plants within an Oxfordshire woodland (UK) indicated that an island biogeographical effect of this sort, if present in the herb flora, was detectable only in patches of less than 0.1 ha. With larger islands or reserves, it becomes increasingly difficult to separate the area *per se* effect from the habitat diversity effect. Depending on the size of the species pool, we can none the less appreciate that there may come a point where there is sufficient area to accommodate breeding populations of a large array of species, but where the lack of particular microhabitats prevents all species from colonizing. The formal testing of such ideas is of course difficult (but see Rafe *et al.* 1985; Martin *et al.* 1995).

An early paper on this theme, based on plant data from Kapingamarangi atoll (a Polynesian outlier in the west Pacific), revealed a discrepancy with the expected species–area relationship (Whitehead and Jones 1969). The findings suggested that area is not a simple variable in terms of how islands 'sample' the pool of available plants. The smallest islands, up to 100 m area, are effectively all strand-line habitats, their flora is therefore restricted to the limited subset of strand-line species. A little larger, and a lens of fresh water can be maintained, and the next set of plants can colonize, those intolerant of salt water. A little larger still and the island would be liable to human occupancy and would be topped-up with introduced species. These modifications of the area effect are not necessarily inconsistent with equilibrium theories.

7.3.2. The habitat-unit model

The notion that habitats have a distinctive, additive effect on species numbers was incorporated more directly in Buckley's (1982, 1985) **habitat-unit model of island biogeography**. The standard method used by most authors has simply been to include indices of habitat diversity, along with area, isolation, altitude, and other geographical factors, in multiple regressions of species number. Buckley took a different approach, dividing islands into definable habitats and floristic elements, and relating species richness to area and isolation independently for each unit. His analysis was based on a series of small islets in inshore waters off Perth, Western Australia. He distinguished three terrain units: limestone, white calcareous sands, and red sands, and constructed separate species–area relations for each. The values were then summed for the whole island. He found a significantly better fit for his habitat-unit model in predictions of actual species richness than derived from whole-island regression (Buckley 1982). He subsequently deployed the approach in a study of 61 small islets on bare hypersaline mudflats in tropical Australia, consisting of 23 shell, 19 silt, and 19 mixed-composition substrates (Buckley 1985). The flora comprised three elements: ridge species, salt-flat species, and mangrove species. In this second study, area was found to be the primary determinant of total species number. Moreover, elevation above mudflat was a better predictor of plant species richness than was substrate composition, probably through its effect on the soil salinity profile (cf. Whitehead and Jones 1969). Underlying these results, different floristic elements were found to behave differently as, not surprisingly, both island area and maximum elevation above mudflats were more important determinants of numbers of ridge species than of salt-flat species.

A similar approach to the habitat-unit model has been developed by Deshaye and Morrisset (1988). Working on plants in a hemiarctic archipelago in northern Quebec, they found that habitats worked as passive samplers of their respective species pools, but that at the whole-island level species number was controlled by both area and habitat diversity. They forcibly expressed the view that data on habitat effects are crucial to interpretations of insular species–area relationships (cf. Kohn and Walsh 1994). Rafe *et al.* (1985) provide another variant in their analyses of birds in reserves in Britain, showing that this sort of approach may also have value for zoological data (see also Martin *et al.* 1995). The methodological and conceptual developments discussed in this section appear promising and deserve wider application.

7.3.3. Competing hypotheses for the species–area effect

It may be helpful at this stage to codify some of these disparate observations into a framework of alternative theories concerning species–area effects. Some in fact pre-date the ETIB (Connor and McCoy 1979; Kelly *et al.* 1989).

1. The **random placement hypothesis** has it that if individuals are distributed at random, larger samples will contain more species. An island can be regarded as a sample of such a random community, without reference to particular patterns of turnover. Connor and McCoy (1979) term this passive sampling, and advocate its use as a null hypothesis against all alternatives (debated in Sugihara 1981; Connor *et al.* 1983).

2. The **habitat diversity hypothesis** regards the number of species as a function of the number of habitats: the larger the island the larger the number of habitats.

3. The **equilibrium hypothesis** postulates the number of species on an island as a dynamic equilibrium between immigration and extinction, dependent on island isolation and area. Only this hypothesis involves a constant turnover of species.

4. The **incidence function hypothesis** suggests that some species can occur only on large islands because they need large territories, others only on small islands where they can escape from competition (cf. Diamond 1974). This would affect species–area relations only if one or other type were more numerous.

5. The **small island effect hypothesis** is that certain species cannot occur on islands below a certain size, as suggested by the Kapingamarangi study (above). This effect may be more apparent in marine islands than non-marine isolates.

6. The **small island habitat hypothesis** postulates that small islands may be different in character because of their smallness, so that they actually possess habitats not possessed by larger islands, and thus sample an extra little 'pool' of species. This is analogous to the altitudinal zonation effect described in Chapter 2, in which vegetation zones are lower on small island peaks, thereby packing in representatives of an 'extra' pool of species.

7. The **disturbance hypothesis** postulates that small islands or 'habitat islands' suffer greater disturbance, and disturbance removes species or makes sites less suitable for a portion of the species pool. This effect was posited in relation to studies of small marine habitats by McGuinness (1984), who noted that although some earlier authors had proposed disturbance as an explanatory factor for the form of species–area curves, it had generally been neglected.

Other formulations have been put forward, but the above set, drawn from the review of Kelly *et al.* (1989) and from Simberloff and Levin (1985), parcels up the principal options. Kelly *et al.* (1989) made use of a rather underexploited feature of the ETIB: that given a knowledge of the species pool, it should be possible to predict the proportion of species in common between islands, or *between equal-sized sample areas* on the islands. Kelly *et al.* (1989) have used the latter approach to evaluate the first three of the above hypotheses for 23 islands within Lake Manapouri, New Zealand. It had previously been established that there was a statistically significant species–area relationship for the whole flora, with some indication that habitat diversity might be the cause. In their study, they sampled for richness of vascular plant species in two vegetation types (beech forest and manuka scrub) with a fixed quadrat size. This should eliminate the effect of island area on observed species richness if either the random placement or the habitat diversity hypothesis is correct, but should leave an effect of island area if the equilibrium hypothesis is correct. The equilibrium hypothesis failed this particular test. As their sampling design was also intended to eliminate the random placement and habitat diversity hypotheses, they suggested that other ideas, such as those labelled 4–6 (above) might have explanatory value. In particular, they favoured the small island habitat hypothesis, although it should be stressed that in such studies there may often be a combination of several contributory factors, notwithstanding efforts to control and simplify data collection.

Rosenzweig (1995) has expressed some doubts as to the meaning of this refutation, arguing that the samples were too small, that the restriction to single habitat types was inappropriate, and that the islands were insufficiently isolated. The plots used by Kelly *et al.* (1989) were of 100 m^2, which is not only a fairly standard size for use in New Zealand forests of this character (J. B. Wilson, personal communication), but is also respectable in relation to islands varying from 242 m^2 to 2.675 km^2 in total area. The restriction of sampling to particular forest types also seems a reasonable procedure for distinguishing between the three hypotheses as stated above, but it should be recognized that this assumes the narrow definition of the core ETIB model. MacArthur and Wilson (1967) are very clear that they used area in their modelling largely because good data on habitat diversity were lacking. The point about isolation, although it does not invalidate the test, is a fair one. Unfortunately for ETIB enthusiasts, similar problems apply to many other island ecological studies. Arguably, this study exemplifies the point I will develop later, that the ETIB holds relevance to far more restricted geographical circumstances than many of its advocates have envisaged (Haila 1990). But it is also evident that the test depends on the narrow or core definition of the equilibrium theory. While some progress in distinguishing area and habitat diversity effects is possible by tests of this sort (e.g. Kohn and Walsh 1994), to really get at the crux of the matter we need data on turnover.

7.3.4. Area is not always that important

Not all studies identify area as the prime variable. Connor and McCoy (1979) speculate that non-significant correlation coefficients between species number and area are probably published less often than discovered, because they may be perceived by authors or reviewers to be uninteresting. One example of a null relationship is provided by Dunn and Loehle (1988) who examined a series of upland and lowland forest isolates in south-eastern Wisconsin. Their data failed to fit the hypothetical species–area curve, the slopes for both cases not differing statistically from zero. Their isolates were not real islands, nor was there an enormous range of areas in their data set. Factors other than size, such as disturbance and forest-edge effects were argued to be influences on plant species number. Their result was consistent with the indirect nature of the control exerted by area, but also with the obvious point that meaningful depiction of species–area effects requires the selection of a wide range of island areas (Connor and McCoy 1979).

Power's (1972) study of bird and plant species on the California Islands incorporated 16 islands or island groups with a broad range (0.5–347 km^2) of areas. The results are presented in a path diagram showing the relationships among variables as indicated by stepwise multiple regression (Fig. 7.6). While area was an important determinant of plant species number, it was a relatively poor predictor of bird species numbers. The role of climate in mediating plant species richness was indicated by the role of latitude. Power suggested that near-coastal islands with richer and structurally more complex floras tended to support a larger avifauna. The analysis thus indicates the importance of hierarchical relationships between plants and birds in determining island species numbers, a line of argument which may also be developed in a successional context (Bush and Whittaker 1991).

7.3.5. Species–energy theory

In a review paper on diversity theory in 1981, Brown discussed progress made following Hutchinson's seminal 1959 paper 'Homage to Santa Rosalia', in which the question 'why are

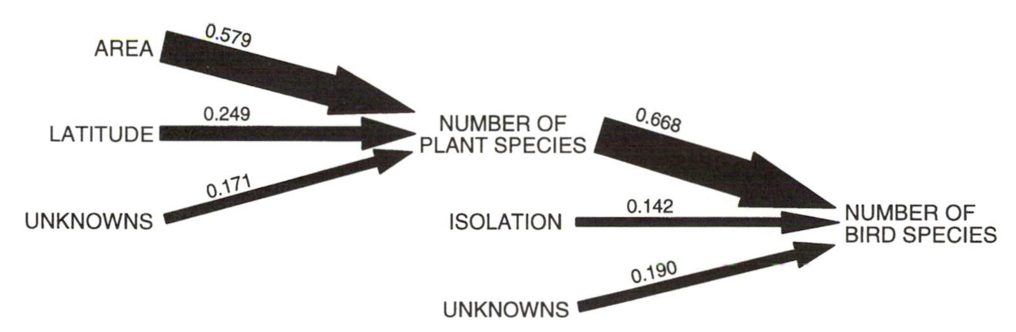

Fig. 7.6. Path diagram showing relationships among variables in explanation of species numbers of plants and birds on the California Islands (modified from Power 1972). The coefficient associated with each path is the proportion of the variation in the variable at the end of the path explained by the independent variable at the beginning of the path, while holding constant the variation accounted for by other contributing variables.

there so many kinds of living things?' was posed. Brown pointed to a divide that had opened up within (mostly) American ecology, between two schools of thought: one influenced by Eugene Odum and concerned with energetics at the expense of the diversities and roles of individual species, and the other following MacArthur and concentrating on the ecological and evolutionary interactions between species to the general neglect of energetics. Brown called for a renewed effort to examine the role of energetics in species diversity theory, noting that total productivity must (together with habitat diversity) be one of the prime controlling factors limiting the number of species that an island may support (see also Schoener 1989*b*).

The effort was made by Wright (1983), who took the seemingly simple step of replacing 'area' with 'available energy' in the models. Ideally, available energy should be measured in units of energy available to the trophic level in question per unit time. For plants he used actual evapotranspiration (AET), a commonly used, although not ideal, parameter (see O'Brien 1993), which estimates the amount of water used to meet the environmental energy demand. As with area, available energy does not estimate directly the variety of resource types present on an island, but it is likely to be correlated with it. Wright argued that, particularly for islands of variable per-unit-area resource productivity, the species–energy model should hold greater applicability than the original. His approach was to substitute energy (correcting for island area) into Fig. 7.4, making the equivalent assumption that available energy should often have a negligible effect on immigration rates. This assumption is debatable, in that for instance, frugivorous birds or fruit bats may be preferentially attracted to islands that possess a ready supply of fruit over those that do not, as appears to be demonstrated by data from the recolonization of the Krakatau islands (e.g. Docters van Leeuwen 1936; Dammerman 1948). Consideration of the example of successional development on these islands (section 8.5) illustrates that as a representation of temporal change, rather than spatial variation at equilibrium, this particular modification of the ETIB does not escape the limitations of the original. But, as a model of spatial variation assuming effective equilibrium, it may be more realistic, and it is here that the value of the approach probably lies. While AET is not the ideal variable for describing the effective energy regime, Wright argued that it provides a usable first approximation. Wright's first test was on a data set of 36 islands, ranging in size from Sumba (11 000 km^2) to the island continent of Australia (7 705 000 km^2). Using AET as the index of energy available for plants and total net primary production (TNPP) for birds, it was found that, respectively, 70% and 80% of the variation in log species number could be accounted for by log 'energy'; a result that was consistent with species–energy theory. However, islands at higher latitudes tended to have slightly lower species numbers than low-latitude islands, suggesting that consideration of energy alone may not be sufficient. Water availability might be another important variable to factor into this approach (cf. O'Brien 1993). Williamson (1988) has noted that one reason why MacArthur and Wilson (1967) thought species–area curves were steeper on more distant archipelagos, was that the archipelagos that they used to represent distant areas tended to have drier and less favourable climates for birds than those selected for less distant areas.

Wright's (1983) empirical test was conducted on a series of islands ranging from large to beyond the size allowed by many as the largest island. It is perhaps unsurprising that climate should be found to be important across such a range of areas. Studies on gamma (regional) scale patterning in diversity within large land masses have reported that climate can provide a first-order, albeit partial, explanation of species richness (O'Brien 1993). On first principles, it might be argued that scale will be important to the applicability of species–energy theory. We know that the species richness of two samples will be dependent on how large an area is sampled. Comparison of a 1 m^2 patch of chalk grassland with a whole hillside of the same habitat will inevitably find the whole hillside to be richer. It follows that in searching for causal explanations of diversity patterns between differing locations, area should be held constant. On large spatial scales (comparing areas of tens to hundreds of kilometres), climate is one of the more prominent sets of variables influencing plant species richness. On much smaller scales, factors such as geology, slope, and grazing regime are likely to be proportionally more important. That is to say that different theories of diversity may hold relevance to different (albeit overlapping) spatial scales (Wright *et al.* 1993). Island species–energy theory builds both energy and area in to the ordinate, as island position on the energy axis is corrected for total island area. Thus it is likely that where island sizes span several orders of magnitude, deviations from the best-fit relationship may have different causes for different portions of the island size continuum. A similar argument might be deployed in relation to the simpler species–area approach, where, however, it might be more easily appreciated (see below).

Wright (1983) notes that species–energy theory is consistent with several long-established patterns. For instance, the energy requirements higher up the trophic pyramid are such that smaller areas may be unable to support top predators, and large-bodied animals are often among the first to be lost from areas newly turned into isolates. Wylie and Currie (1993*a*) have applied the species–energy approach to mammals (excluding bats) on land-bridge islands from across the world. They made use of a slightly different set of explanatory variables, but the approach was otherwise very similar. They found, as Wright had for birds and angiosperms, that island energy explained more of the variation in mammal species richness than did area. However, the improvement was much less marked for mammals. The range of island sizes in the mammals data set was $0.4\,km^2$ to $741\,300\,km^2$, and it is possible that the greater range, and in particular, the extension of range down to much smaller islands, may have given rise to a stronger relationship to area in their data. Latitude was also found to have a stronger effect on mammal species number than Wright had found. Of 100 log–log species–area regressions published in Connor and McCoy (1979), the average explanatory value of area was less than 50%. In the three data sets analysed by the species–energy method, the figures ranged from 70 to 90%.

Species–energy theory is an interesting development which, surprisingly, does not appear to have generated many direct tests to date. It will be particularly interesting to find how well the approach scales down. Wylie and Currie (1993*b*) followed up their global-scale study with a local application, attempting to predict the mammal species richness of isolated nature reserves in the Australian wheatbelt. They found that the global, island-based model underestimated the total richness in the reserves, but accurately predicted the number of species that forage exclusively within the reserves and not on surrounding farmland. The first empirical tests of the approach are thus encouraging, in suggesting that the theory offers improved predictive models for species richness of isolates, but more studies will be necessary, over a wider range of situations, before this can be confirmed.

7.3.6. Isolation and distance

The depauperate nature of isolated archipelagos has long been recognized (Chapter 3). A number of studies have shown that isolation explains a significant amount of the variation in species number once area has been accounted for in stepwise multiple regression (e.g. Fig. 7.6; Power 1972; Simpson 1974). However, isolation is not just a question of distance, but of the nature of the land, or seascape, or atmospheric systems, separating isolates. An example relating to the colonization of Mona Island by butterflies blown by the predominant wind system from Puerto Rico rather than the equidistant and larger Hispaniola was given in section 2.4.4. Some modellers have gone to considerable lengths to develop elegant proofs concerning the effects of particular island–mainland geometries (e.g. Taylor 1987). Such attempts may seem futile in the absence of data on current systems, basic ecological attributes of the source pool populations, and migration routes. However, they do serve to illustrate that it is possible to improve the realism of the mathematical models, which is to be welcomed.

Several authors have made use of the power function model as indirect evidence for the presence or absence of equilibrium. Connor and McCoy (1979) caution against such procedures, arguing that the slope and intercept parameters should be viewed simply as fitted constants, devoid of specific biological meaning. However, in particular cases, it may be possible to construct highly plausible scenarios by such means. Studies of non-volant mammals on isolated mountain tops in western North America by Brown and colleagues, provide an illustration (e.g. Brown 1971; Brown and Gibson 1983).

The mountains of the Great Basin have been isolated from comparable habitat since the onset of the Holocene (*c.* 8000–10 000 years). Brown assumed that immigration across the intervening desert habitats was not able to occur. It follows that the system dynamics must be dominated by extinction events, i.e. they are 'relaxing' towards lower diversity, ultimately to zero species. Small islands would have had higher extinction rates and have already lost most of their species, whereas large islands still retain most of theirs (Brown 1971). This explained the much steeper z values found for the species–area curve for mammals than for birds. For birds, the isolation involved was not sufficient to prevent saturation of the habitats. Thus, the mammals provided a non-equilibrium pattern, and the birds an equilibrium, although not quite to the ETIB model, as it appeared that on the rare occasion that a species went extinct, it tended to be replaced not by a random draw from the mainland pool, but by a new population of the same species (Brown and Gibson 1983). It is possible to test such models by independent lines of evidence, if for example, subfossil remains can be found of species once found on the mountains but which have become extinct since isolation. Some such evidence was found in the Great Basin. The approach has been applied to a number of other North American mountain data sets, for instance by Lomolino *et al.* (1989), again showing how palaeo-ecological data can be combined with reasoning based on species–area relationships to derive plausible models of the dynamics of the insular systems. In this case, however, the montane isolates in the American South-west (Arizona, New Mexico, and southern Utah and Colorado) were argued to be influenced by post-Pleistocene immigrations as well as extinctions.

The assumptions involved in such studies are open to question. For example, Grayson and Livingston (1993) observed one of the mountain species, the Nuttall's cottontail (*Sylvilagus nutallii*) within the desert 'sea', suggesting that immigration across these desert oceans may occasionally be possible. Lomolino and Davis (1997) provide a re-analysis of data for both the Great Basin and the American South-west archipelagos. Their approach is predominantly statistical. They conclude that while the most extreme desert oceans (southern Great Basin) represent in effect a barrier to dispersal, in less extreme cases (the American South-west) they represent immigration filters, allowing inter-mountain movements of at least some forest mammals on ecological time-scales. The scenarios constructed are thus more complex than the original, invoking both equilibrium and non-equilibrium systems and varying contributions for immigration and extinction. While plausible and informative, these deductions ultimately demand verification by independent lines of evidence.

7.3.7. Species–area relationships in remote archipelagos

It was recognized in the ETIB that species number could rise through two means, immigration (I) and evolution (V). In ecological time, only immigration occurs at measurable rates, but in remote islands the evolution of new forms may be more rapid than the immigration rate from mainlands; the inclusion of V was thus necessary for the theory to extend to such systems. Wilson (1969) subsequently published a modified 'colonization' curve, showing how species number might rise above the value first achieved via a process of ecological assortment (e.g. successional effects) and, over long time-spans, rise again via evolutionary additions. No means of quantifying the time-scale or amplitude of the adjustment was made, making the adjusted theory difficult to falsify.

None the less, several authors have applied equilibrium thinking to evolutionary biotas on remote islands (e.g. MacArthur and Wilson 1967; Juvig and Austring 1979). If the islands within a remote archipelago exchange species moderately frequently, species–area relationships may be detectable in biogeographical patterns at the archipelago level. Moreover, taxa capable of rapid rates of speciation, may perhaps so rapidly saturate even geologically young islands that a clear positive correlation could result between species richness and island area. This appears to apply to the highly volant Hawaiian *Plagithmysus* beetles, but not to the flightless Hawaiian *Rhyncogonus* weevils, which instead show a pattern of increasing species number with increasing island age, and a decline with island area (the largest Hawaiian island being the youngest) (Paulay 1994). Juvig and Austring (1979) have undertaken an analysis of species–area relationships within the Hawaiian honeycreeper-finches, which produced results they claimed were consistent with equilibrium interpretations. However, their analysis was undertaken before the extent of the historic extinctions (section 6.3.1) became clear. The full distribution of this family prior to Polynesian colonization is unknown, casting doubt upon interpretations of such species–area calculations (but see the attempt by James (1995) to correct for extinctions). Moreover, the history of introductions to remote islands suggests that while native species may lose out, the total number of species on such islands commonly rises faster than species are lost, suggesting that the islands in question are effectively unsaturated and therefore non-equilibrial (e.g. Groombridge 1992, p. 150). This is perhaps a weak line of argument because carrying capacities have undoubtedly been altered by human alterations of habitats and this, in tandem with the anthropogenic increase in immigration rate, will set a new equilibrium point for the island in question. However, this does not mean that a new dynamic equilibrium will be achieved. Gardner (1986) has shown that the lizards of the Seychelles do not produce a significant species–area relationship, and that there are unfilled niches on some islands. The patterns of distribution of endemic and more widespread species were found to be best explained by reference to historical events, such as periods of lower sea level in the Pleistocene, and to non-equilibrium hypotheses (see also Lawlor 1986).

The applicability of equilibrium ideas to remote archipelagos may ultimately depend on the effective response rate of particular taxa and the scale at which they perceive the island environment. It is not sufficient to view demonstration of a species–area relationship (and associated spatial distributional patterns) as proof of equilibrium in those island groups in which evolutionary phenomena are dominant over ecological phenomena (cf. Williamson 1981, 1988; Lawlor 1986).

7.3.8. Scale effects

The details as to which variables best explain variations in S in a given data set hold fascination principally in their specific context. They are difficult to generalize. This is in large part because each group of islands (real or habitat) has its own unique spatial configuration. Studies of plant diversity where maximum distance is of the order of 500 m, but where elevational range encompasses distinct habitats, vertically arranged, and where area varies by several orders of magnitude, are unlikely to show a significant isolation effect, but should have both area and habitat effects. This is a description of Buckley's (1982) Perth islets data set, in which the purpose of the study actually required that isolation be effectively controlled. As a number of authors have pointed out, it is important to take account of scale factors in analysing species–area relations (e.g. Martin 1981; Lomolino 1986; Rosenzweig 1995; Sfenthourakis 1996). The relative significance of such key factors as area and isolation will depend in part on how great the range (how many orders of magnitude) the sample islands encompass for each variable. However, as I have argued in an evolutionary context (Chapter 6), it is not just a matter of how wide the range is but what the upper and lower limits of the variables are relative to the ecology of the taxon under consideration. This can be appreciated in terms of habitat, wherein the addition of a patch of mangrove habitat will not do the same for plant diversity on a tropical island as the addition of an area of montane cloud-forest.

In terms of isolation, the effective mobility of different taxa (e.g. birds versus terrestrial mammals) and of different ecological guilds (e.g. sea-dispersed versus bird-dispersed plants) can differ greatly, which may also serve to shape the form of species–area comparisons across archipelagos. For a moderately dispersive group, it might be assumed that very small distances have no impact on colonization and thus on species richness. At the other extreme, beyond a certain limit, in the order of hundreds of kilometres, the group might be absent, and thus further degrees of isolation again have no impact. Similarly, Martin (1981) has shown that where the range of areas spans small island sizes, slopes are liable to be lower than for larger areas. It is thus not surprising that comparative studies show a range of forms for species–area effects between taxa and islands, with different explanatory variables to the fore (e.g. Sfenthourakis 1996). The collective value of such studies is the promise of refinement of, for instance, the dispersal limits for particular taxa, and thus of insights into the dynamics of particular insular systems (e.g. Lomolino and Davis 1997).

One factor that complicates the study of isolation effects is that migration may occur between islands within an archipelago. To avoid this effect, Adler and Rosenzweig analysed the birds of the tropical Pacific not on an island-by-island basis, but using grouped data for whole archipelagos (results in Rosenzweig 1995). When this was done, they found that distant archipelagos did have higher z values than near-mainland archipelagos, albeit with a considerable degree of scatter in the relationship (compare with the alternative analysis reported in section 7.2.3). Rosenzweig (1995) offers re-analyses of two further studies of this effect: first, data for birds from isolated paramo blocks in the equatorial Andes; and, secondly, for plants on British islands. In each case the data could be deployed to demonstrate the effect, but in each case the clarity of the finding was diminished, it was argued, by the complicating effects of within-archipelago rather than island–mainland distance effects.

While there is no doubt that the ETIB has stimulated much of the research on species–area relationships, Williamson (1988) concludes that the direct contribution of the theory to their understanding turns out to be rather limited. He favours an explanation based on an improved appreciation of environmental heterogeneity—opening up a rather different form of scale effect. His thesis is that **environmental variation has a reddened spectrum**. The term 'reddened spectrum' is derived from an analogy with the properties of light, which we need not go into. Put in its simplest terms, the argument is that for many environmental variables, the further apart measurements are made in space or time, the more different they will be. What the reddened spectrum does is to put this simple observation into mathematical form using spectral analysis. Williamson adds a second ingredient, by suggesting that the pattern of environmental variation is to some extent fractal, i.e. that as you increase the magnification by which a landscape is studied, you see more detail in it. The relevance of fractals and the reddened spectrum of environments to a fuller understanding of species–area effects has yet to be explored either empirically or theoretically.

The viewpoint expressed by Williamson plays down the significance of turnover. Others continue to take a more dynamic view. For instance, Rosenzweig (1995) follows a similar line of reasoning as Preston (1962) and MacArthur and Wilson (1967) in arguing that, within an area of mainland, species populations can be found which are essentially **sink populations**, sustained only by a degree of inward migration from **source populations** occurring in more favourable habitats (source–sink ideas are discussed further in Chapter 9). For this reason, the slopes of intra-island species-area plots are flatter than those of inter-island species–area plots, whereas comparisons between different regions, or provinces (where evolutionary processes dominate), reveal the steepest of slopes.

The differences in perspective offered by different authors are at first sight confusing.

However, there is consensus that there are scale effects interwoven into species–area patterns, and for the approximate form that they generally take. Thus, for instance, the conclusion that inter-provincial curves are steeper is consistent with the reddened spectrum of environmental variation. The issue remains, to what extent do the patterns observed reflect dynamic equilibria of the form proposed in equilibrium theory? It is thus essential, as Gilbert (1980) advocated, to give close consideration to species turnover.

7.4. Turnover

According to equilibrium thinking, it is a reasonable assumption that most islands are at, or close to, their equilibrium species number most of the time. The ETIB postulates that this should be a dynamic equilibrium, approached by means of monotonic alterations in rates of immigration and extinction. Each of these postulates warrants examination. Prior to doing so, it is important to consider the nature of the evidence involved. Whether or not these postulates are well founded, there is another important issue that must be considered, which is: when turnover does occur at measurable rates on ecological time-scales, is it homogeneous (across all species) or heterogeneous (involving particular subsets of species)? This question is of fundamental importance to applications of island thinking to conservation (Chapter 9).

7.4.1. Pseudoturnover and cryptoturnover

The distinction between colonization rates and immigration rates made above, is a critical one, as they are not one and the same thing. It is equally important to recognize that it is exceedingly difficult to measure *real* immigration and extinction rates (Sauer 1969; Abbott 1983). Most studies of faunal or floral build-up are based on lists of species found at different points in time. The rates presented are perforce observed rates of *changes in lists*: despite commonly being labelled immigration and extinction rates (e.g. Whittaker *et al.* 1989), they are really rather different from the terms as originally defined.

For an extinction to occur, the species must first be present. Unfortunately, if we accept that most extinctions are likely to be of the rarer species, it follows that it will be difficult to determine which have really colonized in the first place. Only 'proper' colonists should be counted in determining the extinction rate. Examination of the definitions of terms given above reveals a certain fuzziness at the centre of this business. The distinction between immigration and colonization is indeterminate, as it relies on what constitutes a propagule for a particular taxon (actually it may vary from species to species) and by what is meant by 'relatively lengthy persistence'. Lynch and Johnson (1974) argue that persistence (for a colonist) or absence (for an extinction) through at least one breeding cycle should be the minimum criteria. In practice, most island biogeographical studies have lacked adequate data on these phenomena, relying (as already noted) principally on lists of species recorded for different islands or different points in time, without adequate knowledge of breeding status, etc (Fig. 7.7).

These problems have plagued island turnover studies, and have led to some pointed exchanges in the literature. The term **cryptoturnover** was coined for the situation when island surveys are rather irregular, wherein species might become extinct and re-immigrate within the inter-survey period (or vice versa), thus depressing turnover rate estimates (Simberloff 1976). A different form of error, **pseudoturnover**, occurs when censuses are incomplete, and information on breeding status inadequate, leading to species appearing to turn over when they have actually been residents throughout, or alternatively when they have never properly colonized. For both these reasons, estimates of turnover (and of I and E) are dependent on census interval (Diamond and May 1977; Whittaker *et al.* 1989) as well as data quality (Lynch and Johnson 1974).

Fig. 7.7. The strand-line on Anak Krakatau, 1989. An area of 5 × 10 m yielded this collection of anthropogenic flotsam. Over 200 varieties of plastic water bottle from a wide region of origin have been recorded from Krakatau in recent years (T. Partomihardjo, personal communication), but despite the regular occurrence of many in repeat surveys, there is no evidence that any of these varieties have established breeding populations on the islands: Krakatau thus represents a population sink for plastic bottles. See text for discussion of problems of assessing residency in plants and animals. Many species of plants have colonized Krakatau by sea-dispersal. (Photo: RJW 1989.)

The study that prompted Lynch and Johnson's critique and which purported to demonstrate turnover at equilibrium was that of Diamond (1969), who censused the birds of the Channel Islands, off the coast of southern California, 50 years on from an earlier compilation. Diamond found that among the nine islands 20–60% of the species had turned over. However, the interpretation of this turnover was disputed by Lynch and Johnson (1974), who, having first pointed to problems of pseudo-turnover, went on to argue that most of the thoroughly documented changes in the birds could be attributed to human influence. This included the loss of several birds of prey due to pesticide poisoning, and the spread into the islands of European sparrows and starlings—both part of a continent-wide process of range expansion. They concluded that the occurrence of natural turnover at equilibrium had not been established by the study. Notwithstanding these data-quality problems, Hunt and Hunt (1974) pointed out an interesting feature in the data for one of the islands, Santa Barbara: the predatory land birds were represented by tiny populations, generally just one or two pairs, and this level of the trophic hierarchy thus underwent greater fluctuations in species representation than birds of lower trophic levels.

Nilsson and Nilsson (1985) conducted an interesting experiment by re-surveying the plants of a series of small islets within a Swedish lake. Even with consistent survey techniques, they achieved at best only 79% efficiency per survey. As different species are missed on different occasions, a significant proportion of the recorded turnover could be attributed to pseudoturnover. Having assessed the rate of pseudoturnover, they were then able to calculate a best estimate of real turnover. However, such procedures cannot be applied with any confidence to surveys by different teams many years apart, wherein the expertise, experience, special taxonomic interests or biases, methods, and time spent in surveying are scarcely known or quantifiable. If the islands in question are large or otherwise difficult to survey, the problems are multiplied. It follows that larger, richer islands will have greater rates of apparent turnover because the degree of error in surveying them is liable to be at least as great as for small islands, and will therefore involve larger numbers of species. In circumstances of this sort, resort has to be made to special pleading in order to justify rate estimates and theory verification (e.g. Rosenzweig 1995, pp. 250–8). Such data may tell interesting ecological stories, but scarcely allow unequivocal tests of the ETIB to be made (Whittaker *et al.* 1989).

In his 1980 review, Gilbert argued vigorously that turnover at equilibrium, a key postulate of the ETIB, and one that set it aside from other competing theories, had not been convincingly demonstrated; the best empirical evidence coming from Simberloff and Wilson's mangrove arthropod experiments (below). Simberloff (1976) had acknowledged the same point in an earlier article, and in the terms demanded by Lynch and Johnson (1974) one is hard-pressed to find in the scientific literature convincing evidence of stochastic, equilibrial turnover to this day.

7.4.2. When is an island in equilibrium?

MacArthur and Wilson (1967) understood that measuring the real rates of I and E in the field is exceptionally difficult. They therefore suggested a means of testing turnover using the **variance–mean** ratio. They argued that if a series of islands of similar area and isolation were to be selected, it is reasonable to suppose that the variance of the number of species on different islands will vary with the number of species present, i.e. variance should be a function of the degree of saturation. At the earliest stages of colonization the variance should be close to 1, declining to about 0.5 as islands become saturated. This provides an approach to assessing whether a series of similar islands (i.e. similar in isolation, area, etc.) might be viewed as equilibrial (e.g. Brown and Dinsmore 1988), but does not constitute a particularly precise tool.

Simberloff (1983) argued that for equilibrium status to be judged it is necessary for researchers to spell out how much temporal variation they will accept as consistent with an equilibrium condition. Only by adopting a specific colonization model, and then testing it against the data can the equilibrium hypothesis be considered falsifiable. He undertook a study using a simple Markov model, in which extinction and immigration probabilities were estimated independently from data for passerine birds of Skokholm Island, land birds of Farne Island, and birds of Eastern Wood (a habitat island in southern England). By comparison of the simulations with the observed data series of between 25 and 35 years, he found a poor fit with the equilibrium model. The data did not show the regulatory tendencies expected if species interactions cause species richness to be continuously adjusted towards an equilibrium.

7.4.3. Propagules and pools

As defined by MacArthur and Wilson (1967) in their glossary (above), I, E, and T rates concern change per unit time, and take no account of the size of source pools (although they do elsewhere in that publication). A number of analyses of these phenomena have used formulations in which rates are expressed as a function of the size of the biota on an island during the course of the study, and/or as a function of the size of the mainland pool (e.g. Thornton *et al.* 1990). Unfortunately, defining the mainland pool can be an extremely problematic question. For this particular purpose, and despite some attention to the topic, I would not venture to offer what the mainland pool might be for the Krakatau flora, there being insufficient mainland data to make the calculation.

Other researchers have made better progress in this direction. For example, Graves and Gotelli (1983) provide an approach to the study of neotropical land-bridge avifaunas that builds a bridge between the traditions of equilibrium theory and habitat determinism. They applied organism-based methods for delimiting source pools, habitat availability, and geographic ranges of source pool species. By comparing 'total' and 'habitat' pools for their islands, they found that habitat pool was a superior predictor of species richness in each family they examined.

Where P is unknown, or unknowable, then evaluation of the theory must follow one of the other routes provided by MacArthur and Wilson (1967). Particular formulations may be optimal for different contexts (e.g. early periods of build-up, or near-equilibrium condition) and arguably should be given a different nomenclature (see the useful discussion in Thornton *et al.* 1990).

7.4.4. The rescue effect and the effect of island area on immigration rate

The basic ETIB model promises a workable, testable model because extinction rate is influenced by area (or its correlates) and immigration rate by isolation. If, however, it can be shown that area also influences immigration, and isolation influences extinction, then the model shown in Fig. 7.4 is in some jeopardy (Sauer 1969).

It will be recalled that immigration rate is the term given to the arrival of species not already present on an island. It is logical to assume that a near-shore island for which there is a high immigration rate will also continue to receive additional immigrants of the species which *are* present. This might be termed **supplementary immigration** to distinguish it from immigration proper. An island population declining towards extinction might be rescued by an infusion of new immigrants of the same species, thereby lowering extinction rates on near-shore islands compared with distant islands. This effect of supplementary immigration was termed the **rescue effect** by Brown and Kodric-Brown (1977), who censused the number of individuals and species of arthropods (mostly insects and spiders) on individual thistles growing in desert shrub-land in south-east Arizona. Their findings conformed quite well with the predictions of ETIB, with the exception that turnover rates were higher on isolated plants than on those in close proximity to others. This was argued to be because of increased frequencies of immigration events for species present on the thistles at time 1, either preventing extinctions or perhaps effecting recolonization prior to subsequent survey at time 2. Either way, each census was likely to record the species as present on the less-isolated thistles. It has been pointed out that this study was based on rather a notional form of island (individual thistles) and was not based on breeding populations. Since the phrase was coined, the rescue effect has been invoked by a number of authors in explanation of their data (e.g. Hanski 1986; Adler and Wilson 1989); often, as in the original, this seems to be based on its good sense rather than on any direct proof (but see Lomolino 1984*a*, 1986).

Similarly, immigration rate may be affected by area. A large island presents a bigger target for random dispersers. A simple study of sea-dispersed propagules on 28 reef islands in Australia showed that the fraction of the dispersing propagule pool intercepted by each island is proportional to beach length (Buckley and Knedlhans 1986). Larger islands also present a greater range of habitats and attractions to purposeful dispersers, such as some birds, elephants swimming across a stretch of sea, or mammals crossing ice-covered rivers in winter (Johnson 1980; Lomolino 1990). Williamson (1981) points out that MacArthur and Wilson recognized the possible influence of area on immigration in their 1963 paper but discretely altered their position in their longer book.

At least one study has reported both the above 'problems'. Toft and Schoener (1983) undertook a two-year study of 100 very small central Bahamian islands, in which they recorded numbers of individuals and species of diurnal orb-weaving spiders. They found that the extinction rate was related positively to species number and to distance, and was negatively related to area. Immigration rate was positively related to area and negatively related to species number and distance. In consequence, the turnover rate showed no strong relationship to any variable, thus the net result of these interactions is that the effects of area and distance on turnover tend towards the indeterminate. Put simply, the ETIB does not work for this particular insular system.

7.4.5. The path to equilibrium

One of the first and only rigorous tests of the process of colonization and development of equilibrium was provided by Simberloff and Wilson's classic experiments on mangrove islets in the Florida Keys (e.g. Simberloff and Wilson 1969, 1970; Simberloff 1976). These experiments involved, first, a full survey of all arthropods on individual isolated mangrove trees (11–18 m in diameter); secondly, the elimination of all animal life from islets; and thirdly, the monitoring of the recolonization. Although seemingly a satisfactory demonstration of the achievement of equilibrium and one in which turnover was occurring, there were some problems. One concerned the handling of certain species groups which were regarded as not treating the trees as an island. Some of these were included in the analysis and others excluded. Simberloff also had problems determining which species were proper immigrants as opposed to transients (Williamson 1981). One interesting features was that a degree of overshoot occurred, suggesting that the islands could support more than their 'equilibrium' number of species while most species were rare, but as populations approached their carrying capacities, competition and predation eliminated the excess species. As Simberloff (1976, p. 576) described it: 'more highly co-adapted species sets find themselves by chance on an island and persist longer as sets'. This development was termed an **assortative equilibrium**.

Demonstration of equilibrial turnover demands the most stringent data. The clearest evidence inevitably comes from 'islands' which are relatively small and simple, and for organisms with fast life cycles. The mangrove 'island' experiments are the classic, first test and they fit this description. These tiny patches of mangrove consist of a single habitat type (albeit not to an insect), the arthropods concerned have very short generation times, and can be expected to respond very rapidly to changing opportunities and to exhibit relatively little community development. Such features may be atypical of islands in general. It is also notable that although turnover rates were indeed high in the early stages of the experiment, a couple of years on, once the assortative process was completed, and pseudoturnover or transient species removed from the analysis, the rates were found to have slowed greatly. This is not expected if area and isolation remain constant. Simberloff, co-author with and student of Wilson, had become rather guarded about the success of the theory by the time of his 1976 article.

A less well known but similar study was conducted by Rey (1984, 1985). His islands were patches of the grass *Spartina alterniflora*, varying in size from 56 to 1023 m^2, but structurally simpler than the mangroves. Once again, fumigation was used and arthropod recolonization was monitored. The study relied on species presence in weekly surveys and thus it corresponds only loosely with the requirements of a stringent test. Initially, the colonization rate was slow because extinction rates were high. As the assemblages built up, populations persisted longer and extinction rates fell. Rey makes a useful distinction between two extremes in turnover patterns. If all species are participating equally in turnover, such that each species is involved somewhere in the archipelago in extinction and immigration events, he terms the pattern **homogeneous turnover**. If, in contrast, only a small subset of the species pool is involved in turnover, this would be **heterogeneous turnover**. If strongly heterogeneous, the turnover would be inconsistent with the ETIB. Rey's data were intermediate between these two extremes, neither totally homogeneous, nor excessively heterogeneous.

Why is heterogeneous turnover problematic for the ETIB? The most basic equilibrium model of MacArthur and Wilson (1967) contains the assumption that each species has a finite probability of becoming extinct at all times. The rate of extinction would then depend on how many species there were. This model assumes entirely homogeneous turnover. However, their favoured model took into account that the more species there are, the rarer each is (on average) and hence an increased number of species increases the likelihood of any given species dying out. This results in the curved function for extinction. This is clearly correct under stochastic birth and death processes, provided you assume that the more species there are on an islands, the smaller, on average, each population is. As already established, this provides us with the hollow forms of immigration and extinction curve and the expectation that equilibrium species number should be approached by a colonization curve of negative exponential form. It also generates the expectation that, on the whole, the rarest species on a particular island are most likely to become extinct. This view, although basically stochastic, is at least part way along the axis towards a model of structured turnover. Now, if, in general, turnover at equilibrium involves a subset of fugitive populations, while another, larger group of species, mostly of larger populations, is scarcely or not at all involved in turnover, then the turnover is highly heterogeneous. Both Williamson (1981, 1983, 1989*a,b*) and Schoener and Spiller (1987) have concluded that most empirical data point to turnover mostly being of such fugitive, or ephemeral species. Whether this is consistent with the ETIB or not is an arguable point: at best, if this really is an accurate description of turnover at equilibrium, then it suggests that the theory is 'true but trivial' (Williamson 1989*a*).

How is equilibrium approached? Can you assume that the more species an island has, the smaller the populations? Do the curves for immigration rate and extinction rate follow the trends described by MacArthur and Wilson of smooth decline and increase, respectively? Actually, as they put it themselves (1967, p. 22): 'these refinements in shape of the two curves are not essential to the basic theory. So long as the curves are monotonic, and regardless of their precise shape, several new inferences of general significance concerning equilibrial biotas can be drawn'. This acknowledges that it is the monotonicity of the rate changes which is critical. It is possible to test whether, indeed, immigration and extinction trends are monotonic. The answer in at least two cases appears to be 'no'. Rey's study illustrates this in a simple system. Observations from the Krakatau islands illustrate much the same point in a complex system.

The Krakatau islands were effectively sterilized in volcanic eruptions in 1883. Recolonization of the three islands in the group commenced shortly after. It started in a limited number of places and gradually claimed more of the island area. Niche space and carrying capacity increased as the system developed. Successional processes kicked in, and habitat space waxed and waned as grasslands developed and then diminished—overwhelmed by a covering of forest. Bush and Whittaker (1991) demonstrated that, if the lists of species present are taken at face value, the trends in immigration and extinction from the lists (not the real, unmeasurable I and E on the islands) behave non-monotonically for birds, butterflies, and for plants. Bush and Whittaker's (1991) bird data were later shown to be incomplete for the latest survey periods, but even after correction (Thornton *et al.* 1993), the non-monotonicity of trends remained (Bush and Whittaker 1993). It therefore appears that for complex islands, there is more ecological structure to the assemblage of an island ecosystem than is allowed for in the ETIB. Precisely how these successional processes structure the colonization on the Krakatau islands will be discussed in Chapter 8.

7.4.6. What causes extinctions?

What causes extinctions, and thus turnover, is one of the crunch issues connected with the equilibrium theory. MacArthur and Wilson (1967) were not explicit. There is no doubt that they regarded it as, in general, a function of population size, but to what extent does their theory assume it to be random (and homogeneous), as opposed to a deterministic outcome of species–species interactions, particularly competition? The 'narrow' answer is that the ETIB is essentially a stochastic formulation (above), but the 'broad view' answer is that their text acknowledges the importance of competition. For instance, the data for the recolonization of Krakatau by plants (up to 1932) failed to fit their expectations. One of the two explanations they offered invoked a successional pattern in which 'Later plant communities are dependent on earlier, pioneer communities for their successful establishment. Yet when they do become established, they do not wholly extirpate the pioneer communities...' (MacArthur and Wilson 1967, p. 50). This invocation of competitive effects is a community-wide, rather than one-on-one, form of competition, and is sometimes given the label **diffuse competition**. Wilson (1995, p. 265), writing retrospectively about his field reconnaissance prior to the mangrove islets experiments, recollects 'For ants the pattern was consistent with competitive exclusion. Below a certain island size, the colonization of some species appeared to preclude the establishment of others...'.

Rosenzweig (1995) invokes both competition and predation as reasons why extinction rates should rise with increasing size of the assemblage. The data he cites for a direct role of predation are not entirely convincing of a general effect, as they come (with one exception) from predators introduced by humans to remote islands, the biota of which had evolved without them. The exception was for data for orb spiders on very small islands where they were preyed upon by lizards. The support in this case is not entirely unequivocal (Toft and Schoener 1983) and, indeed, Spiller and Schoener (1995) have shown that the effects of the lizards on spider densities within these islands vary significantly over time as a function of rainfall variations (data on turnover was not given in the latter study).

Additional evidence for predators causing rising extinction rates have come from Thornton (1996) and colleagues' studies of the colonization of the newly formed island of Anak Krakatau (the fourth Krakatau island, which emerged from the sea in 1930). As the island began to support appreciable areas of vegetation in the 1980s, frugivorous and insectivorous birds were able to colonize in greater numbers, but following the establishment of raptors on the island, several species suffered considerable reductions. In two cases at least, their disappearance from the island was attributed to predation. This is an intriguing finding, as in other ecological contexts predators have often been invoked as preventing competitive exclusion and thereby enhancing diversity at lower trophic levels (Caswell 1978; Begon *et al.* 1986). In the mid-1980s Anak Krakatau was able to support only a single pair of oriental hobbies, which in 1989 were replaced by a pair of peregrine falcons, suggesting the position of top predator to be rather marginal. Anak Krakatau's avifauna is set in the context of three nearby islands, and individuals found on the island may well be only partially dependent on it for food supplies. For instance, Thornton (1996) has noted that the home range of the other avian predator on Anak Krakatau, the barn owl, is sufficient to take in all four Krakatau islands. In such cases, a predator may be able to remove one element of its *in situ* food supply while supplementing its diet on prey from other nearby islands (equally, new arrivals or stragglers may be killed by resident predators on small islands, cf. Diamond 1974). In this fashion, Anak Krakatau may be functioning not as a closed but as an open system (cf. Caswell 1978).

The role of predators may also be scale dependent, such that development of the island's ecosystems to a greater biomass might diminish the ability of predators to cause the total loss of a prey species. This suggestion is consistent with Diamond's (1975*a*) observations of contrasts between Long Island and the more depauperate Ritter, both being islands recovering from volcanic disturbance. Caswell (1978) makes a similar suggestion on the basis of his non-equilibrium modelling of predator–prey relationships in open systems.

In order to demonstrate that the extinction curve is indeed a concave, rising form produced by species interactions, Rosenzweig (1995) had to resort to a complex line of reasoning, based on species–area relationships from two data sets, each of island archipelagos created at the end of the last ice age and presumed to be undergoing relaxation. This constitutes a rather indirect line of 'proof', reliant upon a series of assumptions being taken as read. It is intriguing that he had to work so hard to find support for such a basic element of the ETIB. The generality of the concave, rising trend of extinction does not seem to be founded on much empirical evidence yet. Data can also be found that, at face value, contradict this expectation (section 7.4.5).

There is no reason to assume that there is a single path to extinction. In the process of a species declining (perhaps in fluctuating fashion) to extinction its population must, for some period, be small. The small size of the population may be a good predictor of extinction, but why is the population small in the first place? The explanation could lie in the trophic status of the organism; in predation, disease, or disaster; in resource shortage; or in competition with other species of its trophic level. It is, as Williamson (1981) notes, remarkable that the equilibrium theory was widely accepted with almost no evidence on the distribution of probabilities of going extinct and how these vary (cf. Pimm *et al.* 1988). It is arguable that, in many cases, habitat availability and stability, and food web connections (and continuous availability of resources), are more important than competition in determining extinction probabilities (Whittaker 1992).

7.4.7. Forms of equilibria and non-equilibria

We are not dealing with a well-stabilized situation of long standing. This may account for some of our difficulties in fitting existing situations into a theory which concerns itself only with final equilibrium.

(Preston 1962, p. 429)

In the previous section it was shown that the path to equilibrium may not be quite as the

ETIB suggests it to be. Moreover, islands may reach an asymptote, a point where species number is stable, but not always with the form of turnover originally envisaged. What then? Equilibrialists assume that once at the asymptote, it should be maintained. If island area remains constant, and island isolation remains constant, so should *S*. There are, however, situations where island carrying capacities and isolation do not remain constant. This can be shown both for low- and high-amplitude events.

Many accounts of species turnover rates have treated ocean waters as a constant barrier to dispersal. Scott's (1994) observations of an invasion of San Clemente Island, 80 km from the Californian mainland, led him to question this assumption (see also Diamond 1975*a*). In the late summer of 1984, 26 black-shouldered kites (*Elanus caeuleus*) took up temporary residence on the island for a period of several months, although previously only one or two kites had been recorded during 19 autumn/winter cycles of bird observation. Scott suggested that the 1984 irruption may have been a consequence of an unusual pattern of Catalina eddies—which are seasonal cyclonic winds—possibly linked to an El Niño event. Each pulse of kites arriving on the island coincided with the first or second day of a Catalina eddy. It appears that the eddy system is like a door to San Clemente Island that irregularly opens and closes. The same El Niño influenced carrying capacity for Darwin's finches in the Galápagos (Chapter 6). From such simple empirical observations, it might be reasonable to postulate that fluctuations in the weather might produce variations in rates of *I*, *E*, and *T*. None the less, if the fluctuations are not of great amplitude and the response time of the system is swift, then most of the time an island could indeed be close to equilibrium: theory does not demand a precise fixed value (Simberloff 1976).

One study that appears to demonstrate that equilibrium values do change in response to modest, non-catastrophic changes in island environments on time-scales of years and decades is that of Russell *et al.* (1995). They found that a non-equilibrium model provided improved predictions of observed turnover of birds for 13 small islands off the coast of Britain and Ireland. They suggested that turnover could be viewed as operating on three scales: first, year to year 'floaters' (trivial turnover); secondly, on a time-scale varying between 10 and 60 years, an intrinsic component equivalent to that envisaged in the ETIB was observable; thirdly, most islands show a change in numbers over time due to so-called extrinsic factors, such as habitat alteration.

The glacial/interglacial cycles of the Quaternary have involved high-magnitude effects, the boundaries of biomes being shifted by hundreds of kilometres, sea levels falling and rising by scores of metres. These changes have resulted in at least two types of relictual pattern studied by island biogeographers. The first is that of **land-bridge islands**, those formerly connected to larger land masses (their mainlands), which gained at least some of their biota by overland dispersal prior to becoming islands (e.g. Crowell 1986). The second well-studied context is where habitat 'islands' have become isolated in a similar fashion within continents (e.g. Brown 1971). In both contexts, data often support a non-equilibrium interpretation.

The ETIB would have it that slopes of species–area curves should be steeper (higher *z* values) for distant (oceanic) island faunas than for nearer (often land-bridge) islands in the same archipelago. This is due to the combined effects of extinction and low colonization rates on very isolated islands, and because the impact of both is greater on small islands (Fig. 7.8a). However, as Lawlor (1986) noted in his review of terrestrial mammals on islands, large remote islands tend not to attain the species richness predicted for them on the basis of their area. This is because immigration is very slow to distant oceanic islands, and is little influenced by island size in such remote contexts, and thus the islands tend to remain undersaturated, i.e. non-equilibrial. Species–area slopes for distant archipelagos calculated in this way are thus often found to be relatively flat. Land-bridge islands, in contrast, may have steeper species–area curves because of the predominance of extinction, as over long time-periods, members of the assemblage stranded by island formation

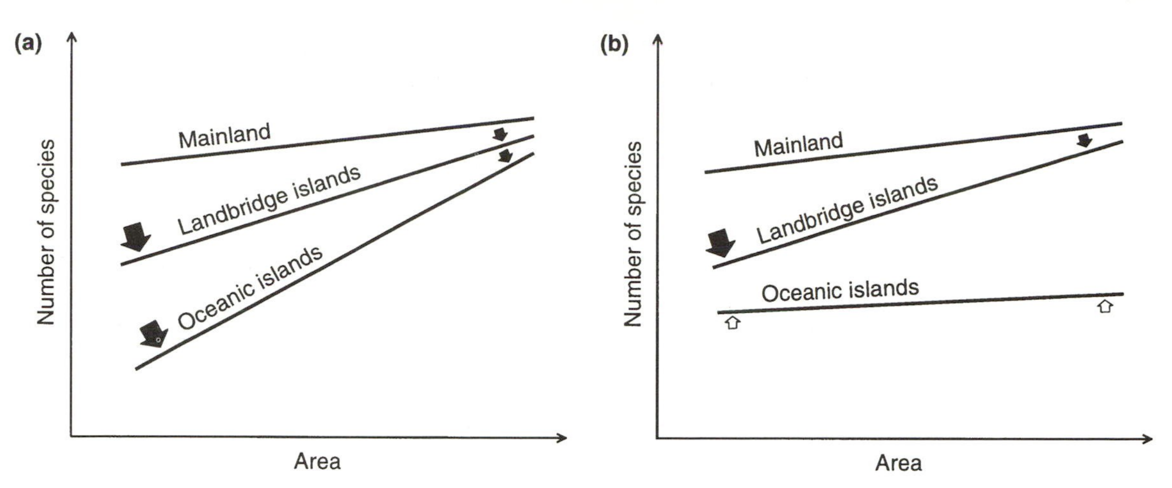

Fig. 7.8. Equilibrial and non-equilibrial representations of species–area relationships for a hypothetical archipelago, as summarized by Lawlor (1986). (a) In the equilibrial view, the effects of colonization and extinction produce steepening of the species–area curves with increasing distance, as extinction (arrows) has greater relative impacts on remoter islands. (b) In the non-equilibrial model, oceanic islands may be undersaturated, as immigration (open arrows) is too slow to 'fill' the islands (extinction is relatively unimportant), whereas land-bridge islands undergo 'relaxation', i.e. their richness patterns are extinction dominated (solid arrows).

suffer occasional attrition, supplemented by little immigration (Fig. 7.8b). In such cases, the fauna at the point of isolation has been termed **supersaturated**, i.e. it contains more species than at true equilibrium, thereafter, the dominance of extinction over immigration results (in theory) in **relaxation** of species number, declining towards eventual, hypothetical equilibrium. Classic illustrations of this effect come from non-volant mammals occupying montane habitats in the Great Basin of western North America (section 7.3.6). These studies have been selected for their illustrative value, and by no means all non-equilibrial systems are products of Quaternary climate change (Brown and Gibson 1983). The distinction between true oceanic islands and land-bridge islands is, as Lawlor (1986) has shown, a crucial one for understanding species–area relations of non-volant mammals of remote islands, and analyses which fail to distinguish the two groups (e.g. Lomolino 1984*b*), are thus likely to be misleading.

We have seen that there is an important distinction between equilibrium and non-equilibrium models. I think there is an equally important distinction between dynamic and static models of island ecology. These notions provide for a variety of combinations, shown conceptually in Fig. 7.9, in which I have highlighted four extremes: dynamic equilibrium, static equilibrium, dynamic non-equilibrium, and static non-equilibrium. A similar framework was drawn up by Rey (1984), although he recognized only three of the extremes. I believe the failure to recognize and distinguish these four extremes has provided something of a hindrance to the understanding and development of island ecological theory.

The **dynamic equilibrium hypothesis** corresponds to the ETIB, the properties of which have already been considered at length. While the core ETIB model is essentially stochastic ($ETIB_1$ in Fig. 7.9), MacArthur and Wilson (1967) recognized in their book that some structuring of turnover commonly occurred, hence their fuller theory could be assigned a position part way along the ordinate ($ETIB_2$ in Fig. 7.9). By either view, their ideas are strongly associated with the dynamic equilibrium corner. Figure 7.9 recognizes a continuum from purely stochastic (homogeneous) turnover, through increasingly heterogeneous turnover to the ideas of habitat determinism with minimal turnover. This end of the ordinate describes the static

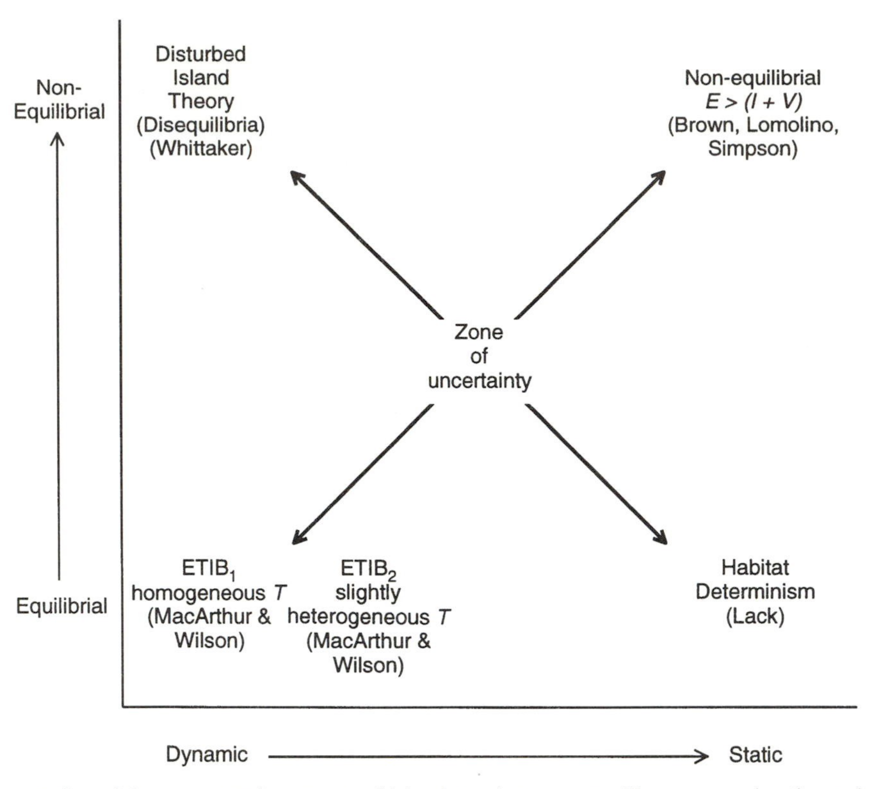

Fig. 7.9. A representation of the conceptual extremes of island species turnover. The terms and authors given in the diagram correspond to studies cited in the text. Any author finding him- or herself in the untenable position of the zone of uncertainty, is liable to ascribe their system to one of the four extreme positions, while usually acknowledging (often emphasizing) that the data do not fully agree with such a simplification. Within a single island (or archipelago), different taxa may occupy different positions in this conceptual space. *T*, turnover; *I*, immigration; *E*, extinction; *V*, species increase via evolution. $ETIB_1$ represents the core equilibrium model of MacArthur and Wilson (1967), and $ETIB_2$ the broader position adopted by their book.

equilibrium hypothesis, which may be characterized crudely as describing situations where the key controls of species presence/absence appear to be habitat controls (as Martin *et al.* 1995) and where species turnover measured on a time-scale of generations is insignificant. It should be noted that it is possible to reconcile ideas of habitat determinism with the occurrence of systematic variation in species richness related to area and isolation, if habitat structure itself is significantly influenced by the area and isolation of an island (Martin *et al.* 1995). Hence, once again, the pattern of turnover is central to distinguishing between alternative hypotheses. The emphasis on habitat controls has often been associated with David Lack, who based his analyses on studies of island birds (e.g. Lack 1969, 1976). He argued that the failure of birds to establish on islands comes from a failure to find the appropriate niche space rather than from a failure to disperse to the island. He therefore played down the relevance of turnover. Others have supported this viewpoint in so far as saying that 'most genuine turnover of birds and mammals seems attributable to human effects' (Abbott 1983) or that turnover, where it does occur, mostly involves subsets of fugitive species and is thus ecologically trivial (Williamson 1981, 1988, 1989*a*). A similar conclusion was reached in a study of newly isolated Canadian woodlots. Once successional effects were removed from the analysis, turnover within the vascular flora, if occurring at all, was at a very slow pace (Weaver and Kellman 1981).

A similar distinction, between the dynamic and the static, can be made within non-equilibrium ideas. First, we may recognize disturbed islands: those in which equilibrium is reached only rarely because environmental dynamics outpace the response times of the biota (cf. the disturbance hypothesis of McGuinness (1984)). The biotas may track changing equilibrium points through time, but always remain a step or two out of phase. This idea has been given various labels (e.g. Heaney 1986; Bush and Whittaker 1993; Whittaker 1995), but within the scheme developed here it is identified as the **dynamic non-equilibrium hypothesis**. We may distinguish this notion from that introduced by Brown (1971) for relictual assemblages dominated by extinction. In Brown's system, although isolates may be losing species on a millennial time-scale, and are thus in a non-equilibrium condition, on ecological time-scales the system of isolates appears effectively static, i.e. they fit the **static non-equilibrium** condition.

The difference between the dynamic non-equilibrium hypothesis and the dynamic equilibrium hypothesis (or ETIB) is of degree, relating to the causation of turnover and the biotic response times (Whittaker 1995). In fact, MacArthur and Wilson (1967) acknowledged the implications of environmental disturbances for particular island systems. For instance, they recognized the following phenomena: the occurrence of successional turnover; the influence of storms on immigration rates; and that much extinction, especially that resulting from storms, drought, and invasion of new competitors, is accompanied by a severe reduction in the carrying capacity of the environment. None the less, they tended to downplay the general significance of disturbance, arguing that only equilibrium models are likely to lead to new knowledge concerning the dynamics of immigration and extinction. But, as others have noted, where turnover is the product principally of abiotic forcing, turnover patterns may be poorly predicted by equilibrium models (Caswell 1978; Heaney 1986; Whittaker 1995; Morrison 1997).

I have argued elsewhere that disturbed islands are more abundant (both naturally and when including anthropogenic effects) than generally recognized in the discussions of equilibrium theorists (Whittaker 1995). In illustration, hurricanes are frequent, devastating events throughout the region 10–20° north and south of the equator (which covers a lot of islands) (e.g. Nunn 1994). For the Caribbean, between 1871 and 1964 an average of 4.6 hurricanes per year have been recorded, the island of Puerto Rico having a return period of just 21 years (Walker *et al.* 1991*a*). Hurricanes are high-energy weather systems, involving wind speeds in excess of 120 km/h, which may have paths tens of kilometres wide, and which can have huge ecological impacts on island ecosystems (e.g. Walker *et al.* 1991*b*; Elmqvist *et al.* 1994). Such phenomena require consideration in island ecological modelling.

Turnover patterns in response to disturbances of this and other kinds can be anticipated on theoretical grounds to be dependent upon taxon choice (e.g. Diamond 1975*a*, p. 369). Schoener (1983) established empirically the fairly obvious point that percentage species turnover declines from 'lower' to 'higher' organisms, as follows (approximate figures): 1000%/year in protozoa, 100%/year in sessile marine organisms, 100–10%/year in terrestrial arthropods, and 10–1%/year in terrestrial vertebrates and vascular plants. Moreover, across this range of organisms, turnover declines approximately linearly with generation time. Given which, and restricting consideration only to islands subject to such environmental dynamics, the attainment of a dynamic (biotically balanced) equilibrium in certain *K*-selected (long-lived, only moderately dispersive) taxa may be postulated to be comparatively rare. The full development of this line of thinking demands consideration of successional dynamics, and this will follow in Chapter 8.

In my view we should no longer give equilibrium models primacy in island ecology. Non-equilibrium models are now much more advanced than they were in the 1960s. Their application has led to improved understanding of ecological systems (e.g. Caswell 1978; Weins 1984; Williamson 1988; Russell *et al.* 1995; Villa *et al.* 1992; Whittaker 1995). Incorporating environmental disturbance into island ecological models and modelling presents significant but not insurmountable difficulties. An example is provided by Villa *et al.* (1992). They constructed a model island consisting of a habitat map of cells containing individuals, a list of the species involved and their life history characteristics. The model involved colonization of the island, and varying degrees of perturbing events that caused the mortality of individuals. The model was highly simplified, but none the less allowed the testing of ideas. One intuitively appealing result was that the global attainment of equilibrium within the simulations was strongly dependent on the rate of increase of the species involved. This demonstrated that equilibrium for slow-growing species (i.e. those with a low intrinsic rate of population increase) should be tested over longer periods, a point equally clear from Schoener's (1983) review of empirical data. Another interesting feature was that in the presence of disturbance, turnover was mostly accounted for by ephemeral populations (arguably trivial ecologically). This held even in equilibrial simulations, a finding again in concordance with empirical evidence, in the form of bird data from Skokholm (Williamson 1989*a*,*b*).

Figure 7.9 is thus offered as a device for organizing the ideas on turnover, and is thus a restatement of island ecological theory, in which the 'narrow' view of ETIB is represented in the broader subject space against other 'narrow' view alternatives. The response times, dispersal abilities, and trophic status of different taxa may mean that comparative studies of different taxa from the same set of islands adopt differing positions in the subject space (Bush and Whittaker 1993).

The four extreme positions on turnover are thus: dynamic equilibrium, static equilibrium, dynamic non-equilibrium, and static non-equilibrium. In practice, they may be difficult to distinguish with the data available. It is anathema to most of us to admit to being uncertain as to the meaning of our studies (and it goes down badly with journal referees), and so we tend to try to label our system as supporting one of the more definite positions in the diagram. However, most authors, including those identified in Fig. 7.9, do not take up the absolute extremes, and many do recognize more than one outcome. For instance, Crowell (1986), in discussing the mammalian fauna of a set of land-bridge islands, aired the notion that it may be possible to recognize a continuum, from isolated land-bridge islands having non-equilibrial faunas, to those of intermediate isolation with both relict species and those in equilibrium, to islands so near, and/or so small, that all species are in equilibrium. The options sketched out in this diagram are similar to the distinctions drawn about 20 years ago by Caswell (1978) (see also Weins 1984; Schoener 1986; Case and Cody 1987). The scale of Caswell's application is rather different from those generally under discussion here, but the parallel is apparent none the less. He set out to develop models of predator (or disturbance) mediated co-existence. He distinguished between open systems (involving migration between 'cells') and closed systems (no migration), and between equilibrium and non-equilibrium systems, concluding that it would not do to suggest that communities as a whole are entirely equilibrium or entirely nonequilibrium systems. 'Perhaps a community consists of a core of dominant species, which interact strongly enough among themselves to arrive at equilibrium, surrounded by a larger set of nonequilibrium species playing out their roles against the backdrop of the equilibrium species' (Caswell 1978, pp. 149–50). We can only go so far towards evaluating such ideas by consideration of species numbers. Next we must look at species composition.

7.5. Summary

This chapter has reviewed numerous tests and developments of the equilibrium theory of island biogeography (the ETIB). Do the numbers add up? Views on the success of the ETIB vary from the damning (Gilbert 1980), or largely dismissive (Williamson 1989*a*) to the enthusiastic: 'The theory of island biogeography holds up well…' wrote Rosenzweig (1995, p. 263), taking the broad view of a torch-bearer! James Brown (1981, p. 882) wrote perceptively that the theory had contributed significantly to explaining organic diversity, but that its value has been largely heuristic: 'It has been tested repeatedly, often rejected, and not yet to my knowledge proven to be both necessary and sufficient to account for the diversity of a single insular biota ...'. The initial hopes that the theory provided the prospect of general and fairly simple laws for the understanding of island species richness have been largely disappointed. Debate and controversy has been fuelled by the frustrating problems of obtaining sufficiently rigorous data for the various tests attempted and by the deceptive complexity of superficially simple variables such as area. None the less, the nature of species–area effects and probable influences on them are now much better understood than they were 30 years ago. The *apparent* testability of the ETIB provided the crucial catalyst, and one of lasting effect. Models and theories for the effects of isolation, habitat determinants, energetics, scale factors, and so forth have been promulgated. In some cases these have been tested and refined, and, in other cases, such as Williamson's ideas on fractals and the reddened spectrum of environmental variation, they have yet to be fully specified and explored.

MacArthur and Wilson (1967) recognized that not all island systems were at equilibrium, and yet postulated that turnover might be detectable in non-equilibrial systems. Subsequent work bears this out. The occurrence of non-equilibrial systems does not of itself contradict the ETIB, as the non-equilibrium situation can be argued to be merely a stage in the development of equilibrium. Yet, if equilibrium is actually reached only rarely because of insufficient time (producing in ecological time an apparently static non-equilibrium) or irregular environmental conditions (dynamic non-equilibrium), there is a danger that the equilibrium position may have become effectively irrefutable while being generally inapplicable. For those who prefer their science traded in theory refutation, and in the 'narrow sense', the ETIB has been refuted enough times for it to be dead and buried. Arguably, however, it retains relevance to many insular systems, it is just that its domain in terms of predictive ability is much more limited than its proponents and others had hoped (see also section 8.7). Intriguingly, for a theory published in the journal *Evolution*, it has less explanatory value for the remote archipelagos on which evolution is dominant than for near-shore continental-shelf islands. There is thus no single answer to the equilibrium–non-equilibrium dichotomy. It is not actually a dichotomy but a continuum. Neither is there a simple answer to the static–dynamic debate.

The emphasis on habitat determinism by Lack and others has been opposed by many supporters of dynamic island biogeography. They see it as returning the subject to a narrative form. The presence of many endemics on very isolated islands clearly demonstrates that even among a taxon of generally good dispersers, e.g. birds, there comes a point where isolation is sufficient to render immigration and colonization a rare and essentially probabilistic process. However, there is no need to draw such a stark distinction between dynamic and static views of islands. Studies reviewed in this chapter suggest that is possible to build bridges between these differing positions and that very often the determinants of species number and composition lie somewhere between the essentially stochastic, dynamic equilibrium and more 'static' behaviour, and require a more complex explanation of community organization on islands than provided for by the ETIB. Some more of this complexity will be considered in the next chapter.

8

Community assembly and dynamics

... studies of island biogeography show that a community is not simply a collection of all those who somehow arrived at the habitat and are competent to withstand the physical conditions in it ... a community reflects both its applicant pool and its admission policies ... We need to formulate a new generation of population and community models with the transport processes explicitly built in ...

(Roughgarden 1989, pp. 217–19, with thanks to Thornton 1996.)

8.1. Introduction

This chapter follows a different but overlapping agenda to that on species numbers. In Chapter 7, species were treated largely as exchangeable units. Such an approach ultimately has its limits. This chapter is therefore concerned with those aspects of island biogeographic theory dealing with compositional pattern and how it is assembled (mostly) in ecological time. The forces structuring island assemblages, as opposed to determining how many species they contain, are evidently much the same; it is merely the response variable that differs. However, by considering the much more complicated response variable of 'composition' it is possible to cast a clearer light on to some of the problems left over from the species numbers games (Worthen 1996). Pattern interpretation in this chapter will again invoke forces such as competition and predation, but will also pay greater attention to species autecology, dispersal attributes, and succession.

Colonization and ecosystem development of a not-too-distant island arguably constitutes just a special case of ecological succession, a process (realistically many processes) that in complex ecosystems, dominated by long-living organisms, may be evident over hundreds of years. This is where I must declare my own special interest, if it is not already apparent. I have developed my position on island ecology through a long-term involvement with several colleagues in the study of the recolonization of the Krakatau islands: doubtless it has biased my judgement. The rationale for giving special attention to the Krakatau story is, however, a reasonable one. It was, after all, the first data set MacArthur and Wilson turned to having formulated their equilibrium theory (Wilson 1995) and it illustrates many of the features that appear to be important in the present context. However, it is the theories of Jared Diamond which provide the starting point for this chapter, just as MacArthur and Wilson provided us with the framework for the previous chapter.

8.2. Island assembly theory

In my school days, assembly rules were that you all stood quietly in lines, class by class, and behaved yourselves while certain school rituals were observed. Of course, these behavioural rules were broken daily, furthermore, the precise order and number of individuals in each line varied, but on the whole the system approximated to the form laid down. The casual observer looking in might, however, have failed to recognize this on occasion. This analogy might apply to island ecosystems, but here detection of the rules, if they exist, requires a great deal more application. This particular branch of island ecology was given its impetus by an extensive series of studies of the avifauna of New Guinea and its surrounding islands by Diamond (e.g. 1972, 1974, 1975*a*). The impetus was lost as the approach became embroiled in a technical dispute concerning the formulation and testing of hypotheses, and controversy concerning the relative significance of competition in pattern formation. Yet what Diamond (1975*a*) presented amounted to a fairly comprehensive island biogeographical theory in its own right: to distinguish it the term **island assembly theory** will be given to it here.

8.2.1. Assembly rules

Diamond adopted the working hypothesis, subsequently challenged, that through diffuse competition, the component species of a community are selected, and co-adjusted in their niches and abundances, so as to fit with each other and to resist invaders. In his studies he identified the following patterns or **assembly rules** (quoted from Diamond 1975*a*, p. 423):

1. If one considers all the combinations that can be formed from a group of related species, only certain ones of these combinations exist in nature.
2. Permissible combinations resist invaders that would transform them into forbidden combinations.
3. A combination that is stable on a large or species-rich island may be unstable on a small or species-poor island.
4. On a small or species-poor island, a combination may resist invaders that would be incorporated on a larger or more species-rich island.
5. Some pairs of species never coexist, either by themselves or as part of a larger combination.
6. Some pairs of species that form an unstable combination by themselves may form part of a stable larger combination.
7. Conversely, some combinations that are composed entirely of stable subcombinations are themselves unstable.

These assembly rules were drawn up on the basis of distributional data. The elements that went into Diamond's analyses will first be described, prior to an examination of the debate about their validity which followed. Diamond made use of a variety of biogeographic circumstances and patterns, including: incidence functions; recolonization of defaunated islands; checkerboard distributions; and recognition of guild structure. The assembly rules were the descriptions and interpretations of these emergent patterns.

8.2.2. Incidence functions and tramps

Incidence functions are an essentially simple idea, reliant on the availability only of good data on species distributions across a series of islands. Sometimes incidence functions have been calculated as a function of island area. However, as originally employed by Diamond (1975*a*), they take the form of a plot of island species number, S, versus the incidence, J, of a given species on all islands of that value of S. Unless a huge number of islands is available, it is necessary to group together islands of a given range of values of S in order to obtain enough observations in each class. Figure 8.1 shows the functions calculated by Diamond for a subset of his bird species. The legend provides a labelling of the different distributional patterns, high-S species, A-tramp, and so on. In order to interpret these labels some background is needed.

The region lies near the equator, with the predominant natural vegetation cover being rain forest. The New Guinea bird species pool consists of about 513 breeding non-marine species. Offshore, there are thousands of islands of varying sizes and degrees of isolation, for many of which bird data were available, in cases including instances of successful and unsuccessful colonization. Some land-bridge islands can be assumed to have gained much of their avifauna at times of lowered sea level, and were thus regarded as supersaturated. Others, having

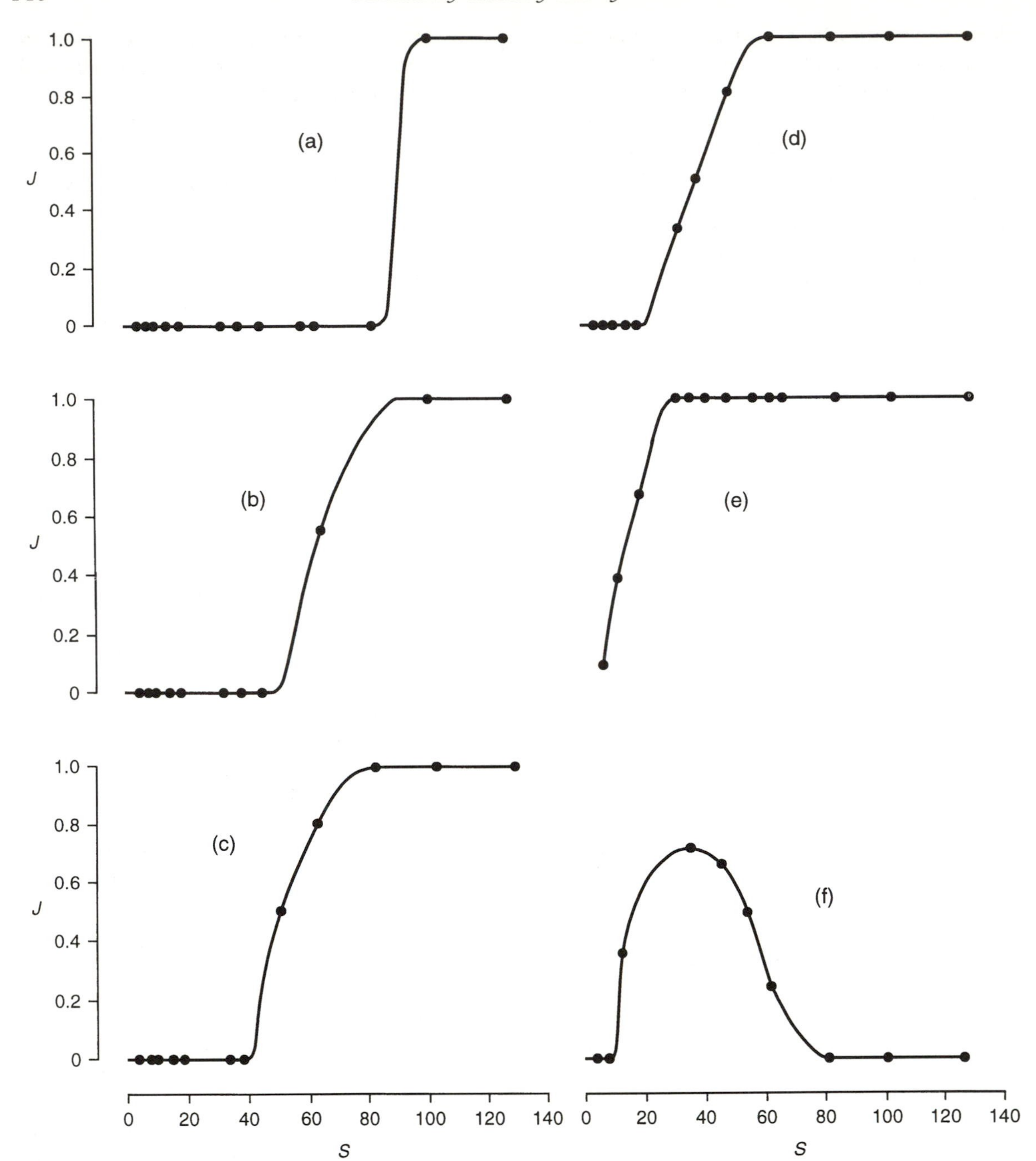

Fig. 8.1. Incidence functions for birds of the Bismarck Archipelago. S = number of species on an island, but the points represent grouped data for a narrow range of values of S from between 3 and 13 islands per point, except the two largest values, which each represent a single island. The index $J = 1.0$ if the species occurs on all islands, and 0.0 if the species occurs on none. (a) *Centropus violaceus* (cuckoo), a high-S species; (b) *Diacaeum eximium* (berrypicker), an A-tramp; (c) *Pitta erythrogaster* (pitta), a B-tramp; (d) *Ptilinopus superbus* (pigeon), a C-tramp; (e) *Chalcophaps stephani* (pigeon) a D-tramp; (f) *Macropygia mackinlayi* (pigeon), a supertramp. See text for explanation of categories. (Redrawn from Diamond, J. M. (1975) In *Ecology and evolution of communities* (ed. M. L. Cody and J. M. Diamond). Copyright © by the President and Fellows of Harvard College. Reprinted by permission of Harvard University Press.)

undergone disruption by volcanic action (e.g. Long and Ritter islands), were regarded as displaced below equilibrium. Others were regarded as equilibrial. Thus, within the 50 islands studied, although a significant species–area relationship was recorded, there were statistical deviations, which appeared intelligible in relation to historical and prehistorical events, and which were interpreted in terms of varying long-term trajectories in avifauna size and composition. Some species of birds, such as *Centropus violaceus*, *Micropsitta bruijnii*, and *Artamus insignis*, occur only on the largest, most species-rich islands. These were termed **high-S** (or **sedentary**) species (Diamond 1974), the precise usage being species found on the four richest Bismarck Islands (81–127 species) plus not more than one from the next richest category (43–80 species). At the other extreme, a small number of species are absent from the most species-rich islands and are concentrated in the smallest, or most remote, and most species-poor islands. An example of this **supertramp** category is the pigeon *Macropygia mackinlayi*. In between, Diamond differentiated arbitrarily four other categories of lesser **tramps**, for those species occurring on some or all of the most species-rich islands and, depending on the category of tramp, on other islands of varying size and species richness within the Bismarck Islands: A-tramps 3–9 islands; B-tramps 10–14 islands; C-tramps 15–19 islands; D-tramps 20–35 islands. The smallest or most species-poor islands generally have representatives of both tramp and supertramp categories.

The first conclusion reached was that the avifaunas represented highly non-random selections of the species pool. As species number varied with factors such as area, isolation, and elevation, and as incidence functions were drawn up on the basis of island species number, this simple analysis appeared to indicate a systematic bias in how species were drawn from the mainland pool by the island systems. Diamond (1975*a*), in discussing the relationship between incidence functions and island area, also invoked roles for: the absence of suitable habitat; island sizes being smaller than minimum territory requirements; seasonal or patchy food supply; historical factors such as land-bridge connections; disturbance and changes in carrying capacity; and interactions of these factors with population fluctuations.

8.2.3. The dynamics of island assembly

In addition to being found predominantly on the smaller islands, the supertramps were also characteristic of the recently recolonized volcanic islands (Diamond 1974), suggesting that these species are excellent colonists but poor competitors. The more widespread tramp species, being found in relatively high frequencies on all but the smallest islands must, in contrast, be both capable colonists and good competitors. The species restricted to large islands only, might by definition be assumed to be relatively weak colonists, but this could be either because of poor dispersal or some other form of unsuitability to small islands. As this high-S set includes species of larger body size (e.g. herons), or requiring larger home ranges (e.g. large hawks), at least in some cases the second category of explanation appeared important. Diamond concluded that the high-S and A-tramp categories actually included a fairly heterogeneous mix of species, some limited by dispersal and some limited by other factors (the term high-S is thus preferred to his alternative label 'sedentary'). It should also be noted that about 56% of the high-S species are endemic to the Bismarcks at the semispecies or species level, such that the patterns under discussion retain at least some (arguably a lot of) signal from evolutionary time-scales.

Some species distributions were interpretable in relation to habitat requirements. For example, as only five islands in the data set exceed 3500 feet (1067 m) elevation, the 13 montane bird species are restricted to these five islands. Of the five islands, one is very small and another was recently defaunated by volcanic explosion: 12 of the 13 montane birds are lacking from both of these islands. Similar fits to apparent habitat controls could be found for other Bismarck bird groups, as indeed have been described in other island biogeographical studies (Chapter 7). The endemic species, which are mostly also high-S species, are mostly lowland forest or mountain species. The tramp species confined to scarcer lowland habitats were not found to have differentiated beyond subspecies level. This was thought to be indicative of the greater frequency of dispersal needed to maintain their lineage in a scattered system of habitat patches across the archipelago. The supertramp species differ again, in that where they occur on an island, they tend to occur in a greater variety of habitats, and Diamond (1975*a*, p. 381) characterized their ecology as 'Breed, disperse, tolerate anything, specialize in nothing.' According to this view, the lack of supertramps on an island is a product of competitive exclusion and not of dispersal constraints. The gradient between the supertramps and the high-S species can be viewed as equivalent to that between *r*-selected and *K*-selected species, or pioneering and late successional plant species. In each case, the gradient represents an approximation of a more complex reality, with varying degrees of explanatory power from one case study to another (Begon *et al.* 1986; and for a critical evaluation of *r*–*K* selection see Caswell 1989).

In some respects there are parallels between Diamond's ideas and the taxon cycle devised by Wilson for the Melanesian ants (Chapter 6). Both theories seek to explain distributional patterns by means of competitive effects, habitat relationships, and evolutionary considerations. It is only by a matter of degree that the taxon cycle is viewed more as an evolutionary model, and Diamond's an ecological model. Diamond (1975*a*) himself suggested that the parallel was quite a close one, although he noted that while in the taxon cycle ants colonizing a species-poor island initially occupy lowland non-forest habitats and subsequently undergo niche shifts to occupy the interior, many colonizing birds already occupy lowland forests. This suggests that a key factor separating the postulated biogeographical behaviours of the two taxa is the superior average dispersability of the birds.

As noted in earlier chapters, there can be great differences in dispersal abilities within taxa. This can be deduced for New Guinea birds by reference to those species absent from non-land-bridge islands. On the assumption that a dispersive species should somewhere have found one or more suitable islands in the non-land-bridge set, only 191 of the 325 lowland bird species have shown a capacity to cross water gaps greater than 5 miles (8 km); of the remaining 134, just four are flightless (Diamond 1975*a*). In illustration of apparently limited dispersal ability, the flycatcher, *Monarcha telescophthalmus*, occurs on all the large land-bridge islands around New Guinea, but none of those lacking a former land bridge. More direct evidence of dispersal abilities can be derived from the recolonization of islands such as Long, Ritter, and Krakatau, that are known to have been defaunated, or by direct observations of birds moving between islands (e.g. Diamond 1975*a*; Thornton 1996). These observational elements were also a point of difference between the taxon cycle and the dynamics of Diamond's assembly model. In effect, his was a successional model of island assembly (below).

8.2.4. Checkerboard distributions

In developing island assembly theory, the incidence functions provided the first step in the reasoning. The second step relied upon a different form of distributional patterning, the so-called **checkerboard distributions**, found for pairs of congeneric bird species in the Bismarcks. A checkerboard is where two (or more) species have mutually exclusive but interdigitating distributions across a series of isolates, such that each island supports only one species. Several examples were given by Diamond (1975*a*). One example is provided by two species of flycatcher. *Pachycephela melanura dahli* is found on 18 islands, and its congener *P. pectoralis* on 11, but the two do not occur together on any island. Similar patterns were found for two species of cuckoo-doves of the genus *Macropygia* (Fig. 8.2). *Macropygia mackinlayi* has a weight range of 73–110 g and mean of 87 g, *M. nigrirostris* has a range of 73–97 g and mean of 86 g. Given these values, they might reasonably be expected to have similar resource requirements. The interpretation offered for these distributional patterns was that either chance determined first arrival, or that

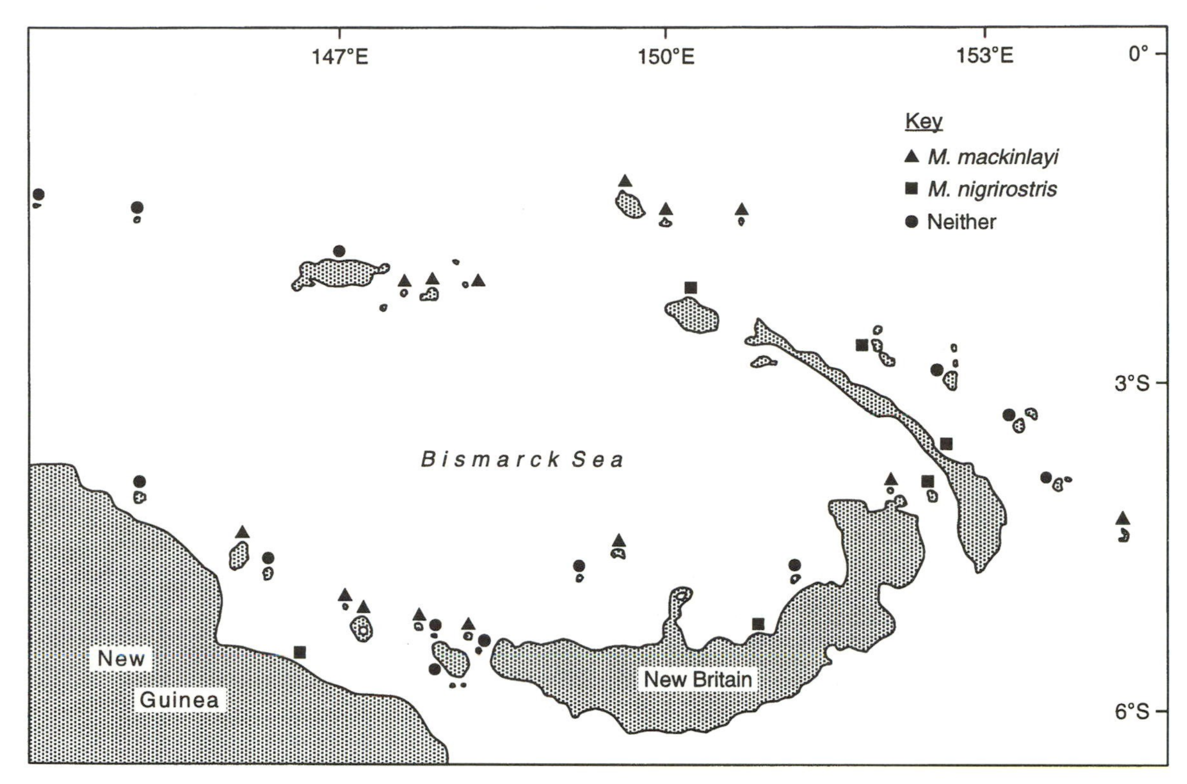

Fig. 8.2. Checkerboard distribution of cuckoo-doves (*Macropygia*) in the Bismarck region. Of those islands for which data are available, most have one, but no island has both species. (Modified from Diamond, J. M. (1975) In *Ecology and evolution of communities* (ed. M. L. Cody and J. M. Diamond), Fig. 20. Copyright © by the President and Fellows of Harvard College. Reprinted by permission of Harvard University Press.)

slight ecological advantages favoured one species over the other in a particular island. Once established, the resident would then be able to exclude the congener either indefinitely, or at least for a long time.

Checkerboards and incidence functions are merely different facets of the same distributional data sets. In the particular case of the *Macropygia*, as already noted, *M. mackinlayi* was one of the supertramps identified by the analysis of incidence functions. Its congener occurs on all the larger land masses and only a few of the small islands. Again, Diamond invoked competitive exclusion to explain the failure to find both species resident on a single island. It will doubtless have occurred to the reader that as some islands lack both these species of cuckoo-dove, and as the identification of *M. mackinlayi* as a supertramp involves analysis of entire assemblages, not just other cuckoo-doves, there must be more to explaining the distributions of such congeners than simply their relationships to one another. The data are certainly consistent with a role for competitive interactions, but other factors and other biogeographic scenarios cannot be ruled out on the basis of Fig. 8.2. One means of testing the competition hypothesis would be provided in the event of one of the pair of congeners colonizing an island 'held' by the other. According to the interpretation of the incidence functions, there ought to be an asymmetry in their competitive interactions, and *M. mackinlayi* should generally be the loser.

8.2.5. Combination and compatibility—assembly rules for cuckoo-doves

Macropygia mackinlayi, although not overlapping with *M. nigrirostris*, does occur on islands with one of two larger species of cuckoo-dove: namely

M. amboinensis (average weight 149 g) or a representative of the *Reinwardtoena* superspecies (average weight 297 g). This suggests that a certain degree of niche difference may be necessary to enable coexistence, a weight ratio of 1.5–2.0 between members of the same guild being generally found to be sufficient (Diamond 1975*a*). Diamond undertook a more formal analysis of the species combinations to ensure that they were indeed non-random, and thus evidence for competitive effects in island assembly. As there are four Bismarck cuckoo-dove species (represented by the letters AMNR), there are 15 possible combinations in which they might occur on any island. Only six combinations were actually observed on the 26 (out of 50) islands which had one or more cuckoo-dove species. With such a small sample size relative to the number of permutations, it is not expected that each of the 15 would occur. The analyses thus relied upon a modelling exercise, built on top of the simple incidence functions, and designed to test whether the empirical data exhibited a significant degree of departure from that which would occur were the combinations the product of chance or random processes of assembly.

The interpretation of the combinational patterns begins with the incidence rules, which determine that certain species combinations are 'forbidden' (i.e. do not occur). The second form of rule was termed a **compatibility rule**. The allopatric *M. mackinlayi* and *M. nigrirostris* exemplify this rule, as they are so ecologically similar in their resource requirements (if not in their incidence characteristics) that they are presumed to be unable to co-occur other than on the most temporary basis. Diamond postulated that the two are probably the product of fairly recent speciation from within a former superspecies. Thus the combinations MN, MNR, AMN, and AMNR cannot (i.e. do not) occur. The final rule type, **combination rules**, in effect mops up the unknowns. For instance, Diamond calculated on the basis of the incidence functions that the combination AMR should occur frequently on islands with a medium species number. It is allowed by the compatibility rules which were deduced, yet it was not found to occur within the data set. This was taken to imply additional combinational rules which were preventing apparently reasonable combinations occurring with the frequency to be expected by chance. One further possibility, reliant again upon the incidence and compatibility rules, was that certain combinations which might otherwise be possible, could simply be difficult to assemble from other permitted combinations. That is, as most changes to an island's avifauna are likely to be produced by single species additions or losses, recombinations requiring two or more gains and/or losses simultaneously (to avoid a 'forbidden' combination) might be anticipated to have low **transition probabilities**.

Diamond undertook similar statistical analyses for the guilds of gleaning flycatchers, myzomelid sunbirds, and fruit-pigeons. The niche relationships among the species of *Ptilinopus* and *Ducula* fruit-pigeons provide a classic example of resource segregation. Diamond first established the form of community structure within the 'mainland' lowland forests of New Guinea. The larger pigeons forage preferentially on bigger fruits, providing one element of resource segregation. If large and small species do feed within individual trees, the lighter pigeons are able to feed on the smaller, peripheral branches, better able to support their weight. Thus niche separation can be maintained within a guild of co-occurring species. Yet, within a single locality in New Guinea, no more than eight species will be encountered, forming a graded size series, each species weighing approximately 1.5 times the next smaller species. There are thus eight size levels filled by combinations drawn from 18 species. On satellite islands off New Guinea, subsets are drawn in such a way that on smaller or more remote islands, size levels are emptied as the guild is impoverished according to a consistent pattern. Level 1 (the smallest birds) empties first, followed by levels 2 and 5, followed by level 8. It is intuitively sensible that the smallest birds should be excluded first as they will be restricted to the smallest fruit types, for which they would have to compete with slightly larger pigeons that can also make use of larger fruits. This form of structuring could also be identified on those islands regarded as being supersaturated, or undersaturated (as discussed above), suggesting that the time required to adopt such structure is shorter than that required for equilibration of species number on islands. This observation resonates with subsequent analyses of ecosystem build-up within the flora and at least some elements of the fauna of Krakatau (e.g. Whittaker *et al.* 1989; Zann and Darjono 1992; Thornton 1996).

The above constitutes the bulk of Diamond's species assembly theory as it concerns us here. However, he also applied the ideas to non-insular communities, a context in which many of the recent developments in the formulation and testing of assembly rules have come (e.g. Wilson and Roxburgh 1994; Wilson and Whittaker 1995; Weiher and Keddy 1995). Diamond's studies incorporated distributional data at the island level in the first instance, but also involved: data on habitats in which particular species were observed or trapped; body weights; feeding relationships; observations of bird movements between islands; and historical records of species colonizations and extinctions, in cases providing apparent evidence of the unsustainability of certain combinations of species.

8.2.6. Criticisms, 'null' models, and responses

Diamond's papers appear to provide a cohesive theory of the biogeography of the New Guinea island birds. This island assembly theory could be seen as a development of several of the lines of argument to be found in MacArthur and Wilson's (1967) *Theory of island biogeography*. Yet, the approach, and particularly the assembly rules, drew criticism from Daniel Simberloff and colleagues (Simberloff 1978; Connor and Simberloff 1979). The problems were sufficiently troublesome, or intractable, as to hamper subsequent progress, and the approach has arguably failed to receive the attention that its intrinsic interest warrants (Weiher and Keddy 1995).

One element of the logic employed by Diamond that might be criticized is the implicit acceptance of the equilibrium thinking at the heart of the ETIB itself. Thus, for instance, he writes, 'The fact that the larger land-bridge islands around New Guinea still have double their equilibrium species number [of birds], more than 10 000 years after the land bridges were severed, emphasizes for what long times many species can escape their inevitable doom on a large island' (Diamond 1975*a*, p. 370). As noted in Chapter 7, observation of such non-equilibrium situations is frequently argued to be consistent with the equilibrium model, yet if it takes in excess of 10 000 years to reach an equilibrium, the effect would appear to be a relatively weak one, potentially or actually to be overridden by other factors. The inevitability of their doom, the fate ultimately of all living things, as the fossil record reveals, is not necessarily in practice to be the consequence of their island systems moving closer to a *biotically regulated* equilibrium. Simberloff (1978) made a similar point when he argued that the statistical analyses of distribution were not uniquely interpretable as a function of diffuse competition.

The second critical paper, was more strident. It claimed to show that 'every assembly rule is either tautological, trivial or a pattern expected were species distributed at random' (Connor and Simberloff 1979, p. 1132). The rules derived in Diamond's analyses were cast in terms of 'permitted' and 'forbidden' combinations (as rule 2, above) and 'resistance' to invasion. Thus, while they were derived empirically mostly from distributional data, competition was invoked within their formulation. One element of the critique concerned the great difficulty of demonstrating a particular role over time for competition in determining present-day distributional patterns (Law and Watkinson 1989; Schoener 1989*a*). Simberloff (1978) argued that the role of competition had been over-played by several authors, citing Diamond's studies in illustration. His argument focused largely on the methods by which non-random distributions are calculated. He argued that confidence limits were not given on the combinational rules, which were therefore difficult to evaluate, and that in any case there were too few islands involved to allow rigorous statistical assessments. He argued that the **null hypothesis** should be formulated in terms of species being considered essentially as similar units, i.e. without differing dispersal abilities and ecologies, and that these species should be modelled as bombarding islands entirely at random. The different islands should be accepted as having different properties, however. While accepting that the first of these assumptions is biologically unrealistic, Simberloff's argument was that it provided a baseline by which to examine biogeographical distributions. When the data are not consistent with this null model, ecological phenomena, such as competition or dispersal differences can then be examined for explanations. Simberloff also argued that the biological characteristics of each species should be examined for explanatory value as being a more parsimonious explanation than competition. Thus one species might be a superior colonist with respect to another because it has better dispersal ability and/or better persistence once immigration occurs (as a function of per capita birth and death rates).

Thus Simberloff, whose own formative experience was the mangrove islet study (a fast, near, dynamic system), adopted a position which is arguably closer to the stochastic, dynamic ETIB than did Diamond in his seminal study of New Guinea island avifaunas (a slow, more isolated, and less dynamic system with a strong evolutionary signal). This difference in view was recognized by Simberloff, who noted that botanists and invertebrate zoologists have most often sought distributional explanations in responses of individual species to physical factors, whereas vertebrate zoologists commonly invoke competition. The generality of competitive effects was in Simberloff's view largely unproven, and the reason was that appropriate statistical tests had yet to be applied to sufficient data sets—indeed such data sets were, by the nature of the statistical requirements, bound to be difficult to obtain. This line of argument was extended by Connor and Simberloff (1979) who contended that the patterns deduced by Diamond as the outcome of competition could be produced by random processes providing that the following three constraints were accepted: '(1) that each island has a given number of species, (2) that each species is found on a given number of islands, and (3) that each species is permitted to colonize islands constituting only a subset of island sizes' (Connor and Simberloff 1979, p. 1132). They based their claims on data for New Hebridean (Vanuatu) birds, West Indian birds, and West Indian bats, which they analysed for evidence of Diamond's assembly rules. The statistical part of their analysis failed to support the existence of assembly rules in their data.

The quote given earlier referred to tautologies and trivial features within the assembly rules. These arguments are difficult to summarize but amount to an attack on the line of reasoning used to construct the assembly rules. In illustration, the third rule 'A combination that is stable on a large or species-rich island may be unstable on a small or species-poor island' was argued to be reducible to 'A combination which is found on species-rich islands may not be found on species-poor islands' because, first, species number is used operationally by Diamond in lieu of island size and, secondly, because stability is largely dependent on the combination of species in question having been observed (relatively little species–time data being available). As large islands contain on average more species, it would be expected by chance alone that some combinations found on rich islands will not occur on species-poor islands and that the number of these combinations will exceed those found only on species-poor islands. By this line of reasoning, the third rule was deemed to be a trivial outcome of the definitions used and otherwise a probabilistic outcome.

Another important line of criticism concerned the checkerboard distributions, which they argued could represent on-going cases of geographic speciation without re-invasion, i.e. could be cases of divergence between allopatric or parapatric lineages, rather than being the outcome of competitive exclusion. The position they adopt at the end of their paper is clear. The assembly rules were flawed because at no time was a simple null hypothesis framed and tested, competitive effects were invoked without proof and where simpler explanations were available. None the less, Connor and Simberloff stressed that they were not implying that island species are distributed randomly and that interspecific competition does not occur, just that neither non-random distributions nor the primacy of competition in determining distributions were convincingly demonstrated by Diamond's analyses.

Not surprisingly, these criticisms drew a response, in the form of consecutive articles in the journal *Oecologia* by Diamond and Gilpin (1982) and by Gilpin and Diamond (1982), and indeed generated a flurry of articles and comments within the literature of the late 1970s and early 1980s. Perhaps the core of Diamond and Gilpin's (1982) response is a challenge to the primacy of Connor and Simberloff's conception of a null hypothesis. Diamond and Gilpin (1982) take the reasonable position that biogeographical distributions may be influenced by many factors, including competition, predation, dispersal, habitat, climate and chance, and that the best mix is liable to vary with the group of organisms considered and with spatial and temporal scale. Further, it follows that no single theory, hypothesis, or conceptual position has logical primacy, or special claim to be the most parsimonious null hypothesis (they signify this by placing 'null' in quote marks throughout). Each should be regarded as a competing or contributory hypothesis to be evaluated on its merits. This is a pragmatic position adopted generally in the present book, that different answers may be found to similar questions posed in different island biogeographical contexts. However, there is, of course, more to the debate than this. Diamond and Gilpin in fact took the opportunity to respond to a series of papers by Simberloff, Connor, Strong, and others, employing null hypotheses. The problem with null hypotheses, i.e. hypotheses of 'no effect', is the assumptions that are made in their formulation (Colwell and Winkler 1984; Shrader-Frechette and McCoy 1993). Diamond and Gilpin argue that in generating their null distributions Connor and Simberloff (1979) took the following inappropriate steps:

1. They diluted the data for particular target guilds with irrelevant data from the whole species pool, whereas only *within guilds* is it reasonable to expect to find evidence of the competition proposed within the assembly rule theory.

2. They incorporated hidden effects of competition into their constraints. The assumptions employed in their null models build in occurrence frequencies and incidences of species which can be shown to be influenced by competition by reference to occurrence frequencies of members of the same guild in different archipelagos (Bismarck, Solomon, and New Hebrides). These data show that the fewer competing species of the same guild that share an archipelago, the higher the occurrence frequency of particular species, often by a factor of 10 or more.

3. They produced a statistical procedure incapable of recognizing a checkerboard distribution.

4. Simulation procedures used were inadequate to provide realistic simulations of the matrices.

This is an incomplete listing of the objections raised by Diamond and Gilpin (1982), but the general position they adopt is summed up as follows: 'how can one pretend that one's 'null model' is everything-significant-except-X, when it was constructed by rearranging an observed database that may have been organized by X?' (Diamond and Gilpin, 1982, p. 73). The problem thus afflicts the proponents on both side of the argument, and is one that has since re-appeared in other applications of null models (papers in Strong *et al.* 1984; Weiher and Keddy 1995). All null models involve assumptions, the trick is to recognize what they are, the disagreements come over which are the most realistic and appropriate set to use.

In their second paper, Gilpin and Diamond (1982) developed a test that they regarded as an improvement over the Monte Carlo modelling used by Connor and Simberloff. They favoured a log-linear model for generating expected probabilities of occurrence, but were able to demonstrate non-random distribution patterns by the application of several different modelling techniques. Comparing results for the Bismarck Islands and the New Hebrides, they found non-random patterns in each system, but that the richer Bismarck Islands had clearer evidence of competitive effects. The New Hebrides have a smaller avifauna (56 compared with 151 species) and thus negative associations (as would be produced by competition) did not exceed a random expectation, neither were checkerboard distributions evident for the New Hebrides. In contrast, they re-affirmed the existence of negative associations and checkerboard distributions amongst ecologically similar pairs of Bismarck species. They found that some pairs of species have more exclusive distributions than expected by chance, and invoked as likely controlling factors, competition, differing distributional strategies, and different geographical origins. Examples of the first two have been given above. In illustration of the third category, the hawk *Falco berigora* has spread from the west, whereas the parrot *Chalopsitta cardinalis* has spread from the east, but in each case to a limited degree. They are not members of the same guild, obviously, and their currently exclusive distributions relate to their differing geographical origins rather than competition. Other species pairs were found to have more coincident distributions than would be expected by chance alone, and this they interpreted in relation to shared habitat, single-island endemisms, shared distributional strategies, or shared geographical origins (Table 8.1).

As Gilpin and Diamond (1982) point out, much of the information about non-random co-occurrences is actually contained in the incidence functions—the first element in the whole analysis—which are much more intuitively accessible sources of information than are the more sophisticated modelling exercises. Equally, the checkerboard distributions provided unambiguous patterns, most simply explained by invocation of competitive effects: in such cases restricting analyses to members within a single guild is clearly necessary (Diamond 1975*a*). Thus, where observed within guilds of pigeons for which there is excellent evidence of colonizing ability across moderate ocean gaps, and where an otherwise widespread species is absent from large, ecologically diverse islands offering a similar range of habitats to those occupied by the species elsewhere, but which happen to be occupied by the alternative species, competition is the obvious answer.

Other authors have since joined in the debate. Stone and Roberts (1990) agreed with Diamond and Gilpin's criticism that Connor and Simberloff's null hypotheses were flawed in that they smuggled in species-interaction effects. They have re-analysed some of the data employed in Connor and Simberloff's (1979) article, but by means of a new form of test which they termed the *C*-score, a means of quantifying the degree of 'checkerboardedness' (a term that hopefully will not catch on). Their analysis attempted to avoid some of the problems of Connor and Simberloff's methods while maintaining the constraints that they assumed. In contrast to Connor and Simberloff, their re-analyses led to rejection of the null hypothesis of the compositional patterns being of a form as might be generated by random colonization, at significance levels of $P < 0.001$ (New Hebrides) and $P < 0.04$ (Antilles) (see also Roberts and Stone 1990). Thus, they found support for checkerboard distributions in both faunas, while stopping short of invoking competitive explanations for their findings. In a further analysis, Stone and Roberts (1992) examined the degree of negative and positive relationships within confamilial species of the New Hebridean

Table 8.1

Factors invoked in explanation of non-random co-occurrence by Diamond and Gilpin in their articles on assembly rules

Negative	Positive
Competition	Shared habitat
Differing distributional strategies	Shared distributional strategies
Differing distributional origins	Shared geographical origins
	Single-island endemics

avifauna. In this case they found evidence not for checkerboard tendencies, but instead for a greater degree of aggregation than expected by chance, i.e. confamilial species tended to occur together. This they interpreted as evidence for similarities of ecology within a family overriding competition between confamilial members. However, as they did not restrict analyses to particular ecological guilds, this 'within-family' result is both unsurprising and by no means contradictory to Diamond's (1975*a*) arguments.

Such debates are rarely settled by the triumph of one viewpoint to the complete exclusion of the other, but, as in this case, can sometimes result in the re-examination and refinement of the original models and positions (an example of counter-adaptation in island terms!). Arguably, Connor and Simberloff's (1979) intervention led to increased attention to the problems of statistical analyses in island assembly structure, which was their stated intent. While Gilpin and Diamond (1982) reaffirmed the original assembly rules, the restatement gives greater prominence to alternative deterministic explanations than Diamond's 1975*a* synthesis. In that sense, the apparent primacy of competition in island assembly theory was diminished (Table 8.1). It is perhaps unfortunate that the original statement of assembly rules gave such prominence to competition, as there are clearly many other factors (e.g. predation) that can produce non-random species assembly, and there are circumstances where these may be dominant and competition insignificant as an assembly force. However, Gilpin and Diamond's position remained that stochastic processes could not explain many of the distributional patterns recorded, that deterministic processes were important, and that particular lines of evidence from the Bismarcks did indeed provide strong evidence for competitive effects *within guilds* in some cases. For the record, Connor and Simberloff (1983) continued the debate, and a further series of exchanges can be found in Strong *et al.* (1984): they add little that cannot be found in the papers cited above (but see Steadman (1997) for recent data on human modifications of bird distributions in Polynesia). Further evidence for structure in island assemblages follows in the next two sections of this chapter.

8.2.7. Other studies of assembly structure

One of the problems of evaluating island assembly theory as detailed above is that the biogeographical context of the island avifaunas under discussion is in each case quite complex and it is difficult to isolate particular factors. This section details further related ideas and studies, some of which constitute simpler systems, enabling particular factors to be isolated.

Non-volant mammals on systems of small, near-shore islands have provided some simpler configurations. Hanski (1986) discusses occurrence and turnover of three species of shrews on islands in lakes in Finland, a system in which individual species populations could be studied in some detail. The largest species, *Sorex araneus*, was found to have a low extinction rate except on the smallest islands, of less than 2 ha, on which demographic stochasticity becomes important. The smaller *S. caecutiens* and *S. minutus*, in contrast, were absent from many islands in the 2–10 ha range, which Hanski suggested could have been due to environmental stochasticity. Temporary food shortages, or 'energy crises' might be more problematic for the smaller species than for the larger *S. araneus*, thus potentially explaining their lower frequency of occurrence. Arguably, interspecific competition might also have a role to play, but Peltonen and Hanski (1991) subsequently reported no evidence for a significant effect of competition on extinction rates. Both studies conclude that the size of species has bearing on their occurrence on islands; an idea which might be tested on other island shrew assemblages. As the larger species do not appear to be affected by interactions with the smaller species within the guild, they predict that they should occur consistently on islands greater than a minimum size set by demographic stochasticity. Small species should have a more erratic distribution, because of their increased susceptibility to environmental stochasticity. This interpretation combines elements of the incidence function and small island effect hypotheses introduced in Chapter 7.

Indeed, Hanski (1992) shows how a simple model of incidence functions can generate realistic estimates of colonization and extinction rates for this system, given a knowledge of minimum population sizes.

Incidence functions, relying as they do on species richness or island area, are relatively simple, crude tools. An alternative approach was developed by Schoener and Schoener (1983) who constructed occurrence sequences from species presence/absence data and a large set of independent variables. Their approach depended largely upon the use of univariate analyses to establish which variables had the best explanatory power. Adler and Wilson (1985) provided a more sophisticated multivariate approach, using multiple logistic regression to estimate the probabilities of individual species occurrences and which of 11 explanatory variables might influence them. Their system consisted of records of nine species of small terrestrial mammals for 33 coastal islands off Massachusetts, and their explanatory variables provided various measures of island area, length, isolation, habitat, and source pools. Principal components analysis was used as a dimension-reducing tool to generate two composite environmental variables, termed 'size' and 'isolation'. These composite variables were then used together with a categorical variable representing the dominant habitat type of each island ('domhab'), in multiple logistic regressions. For all but one species, statistically significant functions were obtained. In four cases (species of *Scalopus*, *Peromyscus*, *Microtus*, and *Zapus*), occurrence on islands was positively related to increasing island size, and in four cases (*Sorex*, *Blarina*, *Tamias*, and *Clethrionomys*), to decreasing island isolation. The habitat variable, domhab, was also included in the functions of two species (*Peromyscus* and *Microtus*). Their study thus showed another means of quantifying relationships between species occurrence and a suite of environmental variables: their probability functions allowing the relative significance of different variables to be tested within a common framework. One other feature of interest in the present context was that they obtained population density data by means of live-trapping over a 2–4 year period. These data demonstrated a positive relationship between population density and insular distribution, thus the most widely distributed species within the islands are also those which reach the highest population densities.

The importance of both constancy of resource availability and of isolation was shown by Adler and Seamon's (1991) study of the population fluctuations in just a single species, the spiny rat *Proechimys semispinosus*, on islands in Gatun Lake, Panama. No other species of mouse- or rat-like rodents were captured on the islands. Their data revealed that colonization and extinction of the rat occurs regularly to and from the islands. Larger islands with year-round fruit production, regardless of isolation, have persistent populations of rats. Small, isolated islands do not, or only rarely, have rats, because they lack a year-round food source and because immigration is rare. Small, near islands frequently have rats when fruits are present. These observations thus reveal how in some circumstances (of size, isolation, etc.) fluctuations in island carrying capacities influence species composition and turnover. However, it should be stressed that the degree of isolation of these islands was in all cases fairly slight, no island being more than 455 m from another land mass.

A particular land or aquatic system may provide dispersal filters which differentiate between different members of a source biota, and indeed different species within a single taxon. For instance, as shown by Lomolino (1986), particular functional groups of small mammals may be differentially affected by the same immigration filters. In regions with season ice cover, small coastal, in-shore, or riverine islands may have their populations supplemented by mammals moving across the snow-covered ice in winter. In some cases, such movements have been recorded by tracking studies. In his studies of mammal movements across the snow-covered St Lawrence River, Lomolino found that larger, winter-active species were good colonizers, whereas the smaller species and hibernators seldom travelled across the ice, and if they did so, then only for limited distances. In related studies of incidences of species across eight mainland and 19 insular sites, it was found that species that were known to utilize the ice cover had significantly higher insular rankings. Furthermore, within a particular functional guild, such as the insectivores, the smaller species were found to have lower insular rankings.

While these data are strongly suggestive of immigration effects, species interactions may of course be operative. It is a logical corollary of the ETIB that in addition to species numbers being a result of interactive effects of immigration and extinction, so also should species composition. Lomolino (1986) seeks to draw a distinction between combined (additive) effects of factors influencing immigration and extinction, and interactive or **compensatory effects**. The latter appears to be, in essence, merely another name for the rescue effect of Brown and Kodric-Brown (1977), whereby supplementary immigration of species present but at low numbers on an island, enables their survival through to a subsequent census. Thus, where immigration rates in the strict sense are high, so too will be rates of supplementary immigration. As the distinction between the two forms of population movement is effectively abandoned in this study, it might be termed instead the **arrival rate**. Species may thus have a high incidence on islands either where low arrival rates (poor dispersers and/or distant islands) are compensated for by low extinction rates (good survivors and/or large islands) or where high extinction rates are compensated for by high arrival rates. Thus species are common on those islands where their rate of arrival is high relative to their rate of loss. Of itself this is a trivial observation, but when combined with other ecological features it provides another option for understanding species compositional patterns, and one that stresses the dynamism inherent in MacArthur and Wilson's (1967) ideas.

Figure 8.3 summarizes five hypothetical insular patterns for the distribution of particular species. First, a species may be distributed randomly (b) within an archipelago. Secondly, it may have minimum area requirements and not be dispersal-limited within the archipelago, in which case it will be present on virtually all islands over a critical threshold size (c). Thirdly, if it is a poor disperser which has low resource requirements relative to insular carrying capacities, it may occur only on the least isolated islands (d). Fourthly, it may depend on both island size and isolation, but not exhibit compensatory effects, thus resulting in the block pattern (e). Finally, it may exhibit the compensatory relationships between immigration and area and thus show the diagonal pattern of part (f) of the diagram. Lomolino (1986) postulated that compensatory effects should be more evident for archipelagos with a large range of area and isolation relative to the resource requirements and vagility (mobility) of the fauna in question. If it is further assumed that within an ecologically similar group of species, larger species have both greater resource requirements and greater vagility than smaller species, an interesting prediction arises from this compensatory model. On the least isolated archipelagos, the smaller species will have a higher incidence on the smaller islands. But, as isolation of a hypothetical archipelago increases, the low persistence of the smaller species combined with their poorer vagility means that larger species should be the more frequent inhabitors of smaller classes of islands, i.e. their incidence relationships 'flip-over' as a function of isolation. In short, if recurrent arrivals and losses are important in shaping the composition of the islands in question, the incidence of a species as a function of area will also vary as a function of immigration.

Lomolino developed a form of multiple discriminant analysis which enabled him to distinguish between the block effects and compensatory effects of Fig. 8.3. He applied this procedure to 10 species of mammals, each occupying at least two of 19 islands studied in the Thousand Islands region of the St Lawrence River, New York State. Circularity of argument in using distributional patterns to infer causation could be avoided as independent lines of evidence on species movements, body size, and other features of their autecology were available.

It was found that none of the 10 species exhibited a maximum isolation effect (Fig. 8.3d), which was not surprising given the relative lack of variation in isolation in the data set. Only one species, *Microtus pennsylvanicus*, a small but

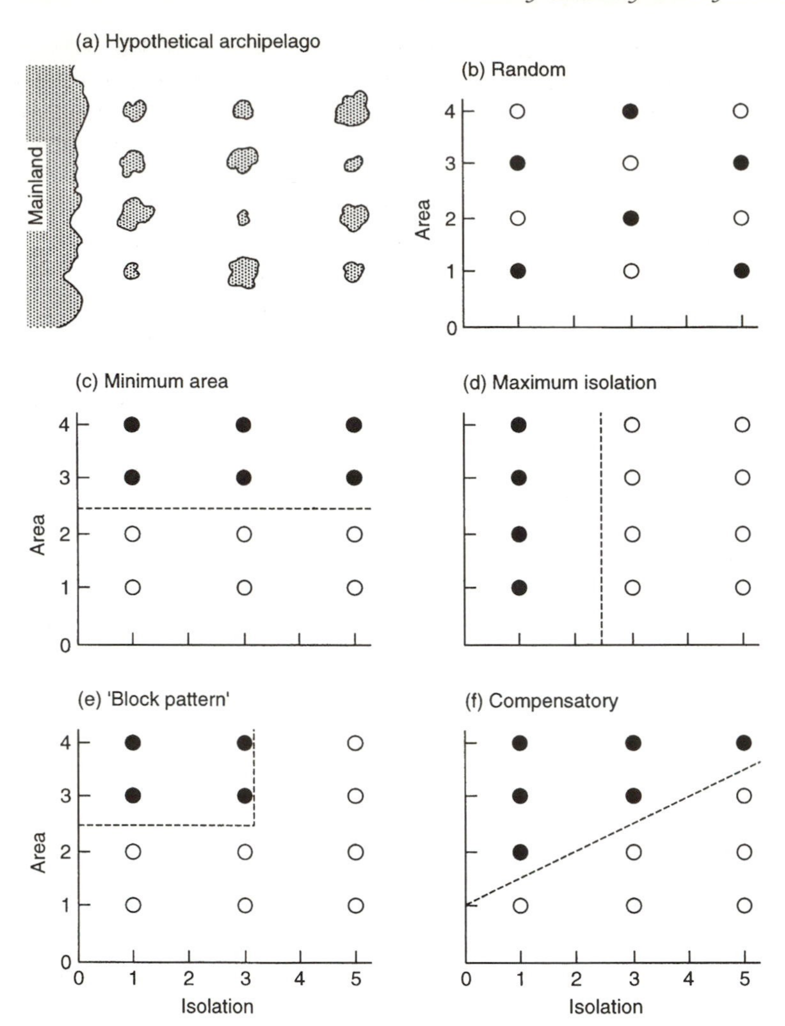

Fig. 8.3. Patterns of insular occurrence of a species on islands of a hypothetical archipelago (a). Presence and absence in (b)–(f) are indicated by filled and open circles, respectively. The units for isolation and area are arbitrary. (Redrawn from Lomolino 1986, Fig. 1.)

dispersive generalist herbivore, showed no area or isolation effect. Six species exhibited minimum area effects (Fig. 8.3c) and the remaining three species exhibited compensatory effects (Fig. 8.3f). *Blarina brevicauda* is an example of the latter group, which is consistent with independent evidence that it is a relatively poor disperser. Lomolino (1986) applied his discriminant analyses to additional data sets for the islands of Lake Michigan, Great Basin mountain tops, and islands in the Bass Straits. He found general support for the model he had developed, the relevance of scale of isolation to the incidence patterns, and, in appropriate cases, for patterns predicted by the compensatory model (see also Peltonen and Hanski 1991).

Microtus pennsylvanicus and *Blarina brevicauda* in the Thousand Islands region provide an illustration of another important effect structuring island communities, namely predation. The carnivorous shrew *Blarina brevicauda* preys mainly on immature stages of the vole *Microtus pennsylvanicus*. In the absence of the shrew on more remote islands, *Microtus* was found to undergo ecological release, occurring in habitats atypical of the species (Lomolino 1984*a*). When introduced into islands, *Blarina* brought about drastic declines of *Microtus* densities, restricting it to its optimal habitat and, in at least one case, causing its extinction. As the vole is the better disperser, the two species exhibit a negative distributional relationship (particularly in terms of population densities) across the islands studied. Lomolino's studies thus demonstrate the significance both of recurrent arrivals and losses and of ecological controls such as predation in structuring the mammalian communities of relatively poorly isolated islands. As the islands in question undergo seasonal variation in their immigration filters, such that at times they are more akin to habitat islands than real islands (cf. the metapopulation models of Chapter 9), care must be taken in transferring such findings to other systems with different scales of isolation.

Haila and Järvinen (1983) and Järvinen and Haila (1984) examined the bird communities of the Åland Islands, off Finland in the Baltic. There are thousands of islands in this group, ranging from only a few square metres to the main island of 970 km^2. Isolation of the islands tends to be slight, the main island being only 30 km from Sweden, and 70 km, via a stepping-stone chain, from Finland. Twenty species of land birds found in the same latitudes in Finland and/or Sweden do not occur on the main island. These form three groups:

1. species for which habitat is available on Åland, and for which absence is believed to be due to lack of over-water dispersal (e.g. green woodpecker, marsh tit, and nuthatch);
2. species for which historical explanations can be offered (e.g. collared turtle dove and tree sparrow have recently expanded their mainland distributions but have yet to reach the island); and
3. species lacking because of absence of suitable habitat (e.g. great grey shrike).

They also noted that small, uninhabited islands in the archipelago lack those species of birds that are associated either directly or indirectly with humans. They found that species numbers and density of the bird communities tended to be similar where habitats were of highly similar structure, and tended to differ when habitats were not directly comparable. Their overall conclusion was that, in most cases, habitat variation between islands or between islands and mainland provided explanations for the assembly of the bird communities of the Åland Islands.

An interesting application of null models to neotropical land-bridge island avifaunas is provided by Graves and Gotelli (1983). They identify three problems that afflict many other analyses. First, source pools are often poorly known or specified, and often arbitrary geopolitical units are used. Secondly, habitat preferences and availability are not explicitly incorporated in null models. Thirdly, the use of incidence functions to estimate colonization potential may be unsatisfactory if habitat effects and distance effects are not controlled in the analysis. They put forward what they termed 'organism-based' methods for constructing geometrically standardized source pools, habitat availability, and geographic ranges of source-pool species. The (former land-bridge) islands analysed were Coiba, San Jose, Rey, Aruba, Margarita, and Trinidad and Tobago. The three procedures were as follows. Their **source pools** were constructed by drawing circles of various diameters on maps. Circles of 100 km produce mainland source pools lacking many species found on the islands; increasing the size to 500 km radius would include many mainland habitats lacking from the islands. They settled, arbitrarily, on a radius of 300 km from their target islands, and specified the mainland source pools as all mainland species found within that line. They classified **habitat types** into forms such as mangroves, palm savanna, and lowland rain forest, and considered a habitat present if the total area on an island exceeded 50 ha. The **pool of colonists** available to an island they represented in two ways: first, the whole pool of species and, secondly, those species of the mainland pool for which the island possessed appropriate habitat—this they termed the habitat pool. The final element was to use range size as a surrogate for colonization potential, dividing species into widespread and restricted species.

Their analyses showed that, with exceptions, land-bridge islands could be viewed as a random subset of the mainland 'habitat' pool when judged at the family level. They used families because they felt unable to assign species to guilds, and because species within a family are usually ecologically and morphologically similar. While acknowledging that this is not ideal—families do not necessarily represent units of interspecific competition—they contend that non-randomness of island avifaunas might reasonably be expected to be detectable at the family level. Using a simple model assuming all source-pool species to be equiprobable colonists, they found that out of 40 families, three are unusually common on land-bridge islands: pigeons (Columbidae), flycatchers (Tyrannidae), and American warblers (Parulidae), and only one, the puffbirds (Bucconidae), was present less often than expected. Thus, while in absolute terms many species and families are absent from the islands, the proportional representation of most families is consistent with mainland source pools. Comparison of the whole pool of species with the habitat pool showed that, as expected, the habitat pool is a superior predictor of species richness in each family. Examination of the mainland range of the species in the pools showed that those species with widespread mainland ranges are disproportionately common on islands. This effect was stronger when using the whole mainland pool, but was also found when using the habitat pool only.

Graves and Gotelli interpret their findings as showing that habitat availability and area of a species range are important factors in explaining differences between island and mainland communities. They stress that their study does not provide insight into the extent to which interspecific interactions might be important in determining habitat use on islands. For this, additional forms of data would be required. The finding that availability of habitat is important in explaining island assemblage structure (as argued by Lack 1969, 1976) led Graves and Gotelli (1983) to question one of the assumptions commonly made in studies of relaxation (sometimes termed faunal collapse) on land-bridge islands. They point out that if available habitat were not important, the total pool would have served just as well in predicting species richness within families of birds. As it does not, models of relaxation that imply that habitat changes since the Pleistocene have not affected extinction (or recolonization) may be flawed. Graves and Gotelli's study is interesting, but it is by no means obvious that their assumptions are more biologically realistic than those they criticize in their paper. For instance, it might be contended that the definition of source pools is crude, and that habitat use on islands may differ from habitat use on the mainland. Perhaps more troubling is how to interpret the relationship to size of mainland range. As the islands concerned are land-bridge islands, it is not clear whether the presence of particular bird species on the islands reflects long-term persistence of a population established when the island was much less isolated, or its colonization under present-day conditions of isolation (cf. Lack 1969). Size of mainland range could thus be indicative either of colonization potential or persistence ability. Notwithstanding these concerns, the patterns depicted appear fairly robust. The study illustrates the potential of analyses combining use of autoecological data with biogeographical distributions in understanding island assembly structure.

It is generally difficult to assess the importance of competition in the process of community assembly on islands. One reason for this is the difficulty of observing immigration events. The purposeful introduction of land birds to remote oceanic islands provides a form of inadvertant experiment on the role of competitive effects in structuring insular avifaunas. The immigration rate of land birds to such islands can be assumed to be negligible, but the rate of introductions of species by people appears, for certain islands, to have been both quite high and well documented (Lockwood and Moulton 1994). Only a proportion of introduced species succeeds in establishing and maintaining populations over lengthy periods. In tests of data for introductions to the Hawaiian islands, Tahiti, and Bermuda, it has been found that the surviving species exhibit what is termed morphological overdispersion, i.e. they are more different from one another in characteristics that are believed to relate to niche separation than chance alone would predict. These results, while controversial, appear to support the hypothesis that interspecific competition shaped the composition of these communities (see analyses and references in Lockwood and Moulton 1994).

8.3. Nestedness

If island biotas are in essence randomly drawn from a regional species pool, as per the ETIB, then an archipelago of islands should exhibit differences in composition from one island to the next and the degree of overlap should be predictable on the basis of a 'null' or random model. The assembly rules provide one form of departure from such an expectation, another form is nested distributions. The phenomenon has been recognized for some time (e.g. Darlington 1957), but only recently have statistical tests been developed for its formal detection. A **nested distribution** describes the situation where smaller insular species assemblages constitute subsets of the species found at all other sites possessing a larger number of species (Patterson and Atmar 1986). Perfect nestedness would indicate an extremely deterministic structuring of island assemblages and would also require no statistical test in its proof but, given what we have already established about island ecology, can be anticipated to be unlikely to occur in other than the smallest of insular matrices. The question thus becomes one of developing appropriate statistical tests to describe whether and to what extent a series of insular assemblages are indeed nested. When the indices were developed, it was found that a high proportion of systems examined were indeed structured in this way (Patterson and Atmar 1986; Patterson 1990; Blake 1991; Simberloff and Martin 1991). In instigating this line of analysis, Patterson and Atmar (1986) took the view that nestedness was most likely to occur in extinction-dominated systems. But, as we will see, it now appears that this may not be the case, and that interpretation of the meaning of nestedness, as with other such tools, requires independent lines of evidence.

The nestedness concept, as applied by Patterson and Atmar and as generally followed by others, is based on size of fauna or flora. This notion is thus independent of island area in its calculation. Some authors, however, have used the same term when ordering the data matrix not by species richness but by island area (e.g. Roughgarden 1989; Lomolino and Davis 1997). Although area and species numbers are often strongly correlated, they are rarely perfectly correlated and so this second type of application differs (Simberloff and Martin 1991) and ought to be distinguished by a different nomenclature: I suggest **area-ordered nestedness** rather than the original **richness-ordered nestedness**.

There are similarities between incidence functions and nestedness calculations, as both involve the examination of species occurrences as a function of species richness (or area) of a series of islands. It will be recalled that incidence functions describe the frequency of occurrence of a species in islands of differing richness, requiring the use of an incremental series of richness classes. In contrast, the first step in nestedness analysis (at least as described above) is to test the degree to which all the species across a series of islands form a nested series. Once this has been done, it is actually much more illuminating to identify which particular species exhibit departure from nestedness. For instance, supertramp species might a priori be expected to display departure from nestedness, as their incidence functions denote them to be species found mostly on islands of low species richness. Patterson and Atmar (1986) therefore produced a procedure for calculating a nestedness index for individual species, N_i. This describes the 'errors' for each species, operationalized as the number of times it fails to occur in faunas richer than the smallest assemblage from which it is recorded. Thus, on an individual species basis, the incidence function and the index N_i are merely differing properties of a data matrix in which is specified the occurrence of each species across a series of islands.

Diamond (1975*a*) intended his assembly rules approach to be deployed on groups of ecologically related species, or guilds, whereas the analyses of nestedness (like incidence functions) provide descriptive tools which may be applied to either narrow or broad groups of species. It might thus be possible, at least hypothetically, for assembly rules to be detectable within a particular guild of birds, whereas the avifauna analysed as a whole exhibits a tendency towards nestedness.

Patterson and Atmar's (1986) index of (richness-ordered) nestedness for the whole biota, N, is derived by summing the separate values of $N_{i.}$. Departure from randomness may then be assessed by Monte Carlo simulation. The index N should not be used for direct comparison of multiple data sets as it tends to increase in value with the size of the matrix (number of species and/or islands). It may also be influenced by the assumption that extinctions have shaped species distributions, as it counts all absences from biotas of greater species richness as unexpected absences (Cook and Quinn 1995). This was deemed an appropriate assumption by Patterson and Atmar (1986), indeed it was specifically tailored for 'relaxing' faunas, because it was believed that extinctions were chiefly responsible for the distribution of mammals in their primary data set—the montane habitat islands of the Southern Rocky Mountains (USA). The analysis revealed a significant degree of nestedness. The overall composition of the biotas showed progressive changes with species richness, such that carnivores and mesic forest herbivores were over-represented in richer faunas and under-represented in more depauperate ones, while the converse held for xeric forest herbivores. The individual species nestedness scores revealed that of the 26 species in the 28 'islands', 12 were perfectly nested. Two of these occurred in all but two islands, perfect nestedness being unsurprising, and two occurred only in one island, leaving eight perfectly nested species occurring on between three and ten islands. The remaining 14 species exhibited departure from perfect nesting in the range 9–120%. The general interpretation of the results was that nestedness was due chiefly to selective extinction of species in the millennia since the switch to postglacial climates isolated the faunas.

Patterson and Atmar (1986) also found significant nestedness in secondary data for mammals on maritime land-bridge island archipelagos, offering a similar interpretation of the results, i.e. that they were the product of selective extinction from islands only rarely influenced by colonization events. They recognized none the less that nested structure might also result from deterministic patterns of colonization (Fig. 8.4), if immigration rates were high enough to dwarf differences in extinction rates.

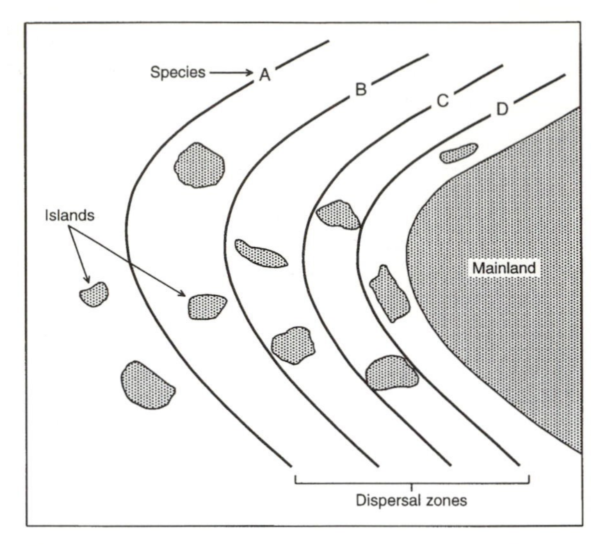

Fig. 8.4. Hypothetical scenario for the production of nested subsets via differential colonization. From Patterson and Atmar (1986, Fig. 4), who contended, none the less, that extinction is a more potent structural determinant in the majority of insular systems.

The relative importance of differential extinction and colonization was investigated by Patterson (1990), following the same methods, and considering studies at three differing temporal scales: palaeo-ecological studies spanning the entire Holocene; historical or long-term ecological studies; and short-term ecological studies (Table 8.2). Data for the Great Basin mammals and birds were each found to be nested, and while the former were assumed to be extinction-dominated, in the latter case, there was evidence for a role for recurrent colonization as well as extinction. Non-nested distributions within the 'Holocene changes' systems appeared characteristic of the most distant oceanic island systems (but see below), again with some variation interpretable in relation to dispersal abilities of the taxon concerned. In the data sets structured by processes operating over historical time-scales, extinction appeared to be

Table 8.2
Summary of significant nested subset structure (according to Patterson 1990), in assemblages affected over various time periods by: mainly extinction; mainly colonization; or both processes in concert

Processes	Nested	Non-nested
Holocene changes		
Extinction dominated	Southern Rocky Mts mammals New Zealand land-bridge birds Baja land-bridge mammals Baja land-bridge herptiles Great Basin mammals	
Colonization dominated	Baja oceanic birds	New Zealand oceanic birds Baja oceanic mammals Baja oceanic herptiles
Both processes	Baja land-bridge birds Great Basin birds	
Long-term changes		
Extinction dominated	Australian wheatbelt mammals North American park mammals São Paulo sedentary birds Bass Strait mammals	Australian wheatbelt lizards
Colonization dominated		All São Paulo birds
Both processes	Penobscot Bay mammals	
Short-term changes		
Extinction dominated		Mangrove area reduction
Colonization dominated	Dispersing shrews and voles Weeds of young lots	Breeding shrews and voles Weeds of old lots
Both processes		Mangrove defaunation

The study was based on the nestedness index, N, of Patterson and Atmar (1986), which has been argued to be suboptimal for the analysis of systems affected by colonization (see discussion in Cook and Quinn 1995).

more important in producing nestedness. Indeed, within the São Paulo bird data set, drawn from three fragments of Atlantic forest, nestedness detectable in the distributions of sedentary bird species disappeared when the transient species were added to the analysis. On the shorter timescales, the mangrove island experiments of Simberloff and Wilson (Chapter 7) were reported to be non-nested. Indeed of the 'short-term changes' data sets considered by Patterson, only those for invasion of weeds on cleared urban lots, and data for dispersing (as opposed to resident) populations of shrews and voles, produced significant nestedness. On the whole, the analysis appeared to support the views expressed by Patterson and Atmar (1986) that systems dominated by extinction are more liable to exhibit nestedness than those in which colonization dominates.

Simberloff and Martin (1991) caution that the preference for a leading role for extinction over immigration in producing nestedness is not intuitively obvious, and indeed point out that Darlington (1957), made exactly the opposite suggestion. They also criticize the assumption that particular systems are in fact dominated by extinction (e.g. land-bridge islands, some mountain-top mammal studies) or by colonization (e.g. Baja California islands), pointing out that most of the evidence for such events is indirect or inferential. This is, as ever, a problematic point in island studies. Cook and Quinn (1995) have also suggested that, in theory, nestedness may result not only from selective extinction or selective colonization, but also from other mechanisms, such as nested habitat. Passive sampling is another possibility, but has yet to be demonstrated convincingly (Worthen 1996).

Cook and Quinn (1995) examined the significance of these differing causes, by means of an analysis of 52 data sets, for native resident species and including data for mammals, birds, reptiles, amphibians, insects, and plants. They adopted a different index of nestedness, the *C* index of Wright and Reeves (1992). *C* is derived from the index N_c, which counts the number of times that a species' presence at a site correctly predicts its presence at richer sites, and sums these counts across species and sites. This index is a scaled metric, and so is not affected by the size of the matrix. Neither is it based on any underlying assumption of causality, unlike Patterson and Atmar's (1986) *N* index. The relationship of the *C* index to incidence functions is even closer than that of *N*, because species richness is regarded as fixed in the analysis, and therefore the observed values of N_c are in effect determined by the species' incidence totals.

Within their 52 data sets Cook and Quinn (1995) found that the vast majority have a highly significant degree of nestedness, with only 3 of 16 land-bridge systems and 3 of 22 oceanic biotas showing no significant tendency to nestedness. The use of the *C* statistic allowed Cook and Quinn to make direct comparisons of different taxa from within the same archipelago, with interesting results. They compared groups of non-endemic species differing in dispersal ability, for four different land-bridge or intracontinental island archipelagos and for two oceanic island archipelagos. They found that in 18 out of 20 pairwise comparisons, the taxa of superior dispersal abilities exhibit a greater degree of nestedness than poorer dispersers (e.g. birds > reptiles > mammals). In contrast to the earlier studies, Cook and Quinn (1995) concluded that frequent colonization enhances rather than obscures patterns of nestedness among most island biotas. They also found some evidence for a role for habitat nestedness in explaining patterns of nestedness for birds on islands within the Sea of Cortéz.

The ability of differential immigration to produce nested patterns may best be examined by studies of the early phases of colonization. A good example is Kadmon's (1995) study of seven islands created by the filling of the Clarks Hill Reservoir, Georgia, USA. The islands were logged and cleared of woody plants prior to their separation from the mainland. Thirty-four years later species number was found to be inversely related to island isolation. The woody plant floras were found to be nested when ranked by richness, or isolation, but not by area. These results point to the significance of differential immigration in structuring the data set. Only a small proportion of the species included in the analysis contributed to the observed nestedness, principally being those that lacked adaptations for long-range dispersal. Wind-dispersed species showed no evidence for nestedness. This study thus pointed to the significance of plant dispersal attributes in structuring the rebuilding of insular assemblages, a feature also of the recolonization data from Krakatau, as we shall shortly see.

It is now well established in the literature that nestedness is a common feature of many insular biotas. It is equally clear that different species and taxa from the same types of island system exhibit differing degrees of nestedness: a finding that is unsurprising from a knowledge of the earlier literature on incidence functions, but which, none the less, holds its own interest. The extent to which such patterns are the product of extinction as opposed to colonization arguably remains to be firmly established (Lomolino 1996). The general preference for extinction expressed by Patterson and Atmar (1986) and Patterson (1990) has been countered by more recent studies (Simberloff and Martin 1991; Cook and Quinn 1995; Nores 1995). Indeed, Cook and Quinn (1995) reach an almost diametrically opposite conclusion: that the prevalence of selective extinction as a natural phenomenon has yet to be established, and will be difficult to detect in most cases, because of incompleteness of knowledge of prior distributions and because of the positive effect that colonization appears to have on nestedness.

To end on a cautionary note, it seems likely that there is still some way to go in the development of nestedness indices, and indeed James Sanderson (pers. comm.) has suggested that there has been a general over-estimation of the degree of nestedness in the literature to date due to properties of the metrics used. As was the case with earlier assembly rules literature, it may prove difficult to reach a consensus on how the null models for nestedness should be specified and the resulting statistics interpreted. At the time of writing, the array of nestedness and related indices continues to grow (Cutler 1991; Atmar and Patterson 1993; Cook 1995; Kadmon 1995; Lomolino 1996). Several of the developments and applications are concerned primarily with problems in conservation biology, and so the applied relevance of nestedness will be discussed in Chapter 9. Consideration of nestedness in this chapter has led to the point where questions concerning the colonization process require more detailed attention.

8.4. Successional island ecology: first elements

Both island biogeography and succession theory go back a long way. Biologists have been studying them in combination since the nineteenth century (e.g. references in Whittaker *et al.* 1989), although not necessarily within the same frameworks we use today. Ridley may well have been influenced by the early Krakatau data when he wrote in the preface to his classic text on plant dispersal (Ridley 1930, p. xii):

> An island rises out of the sea: within a year some plants appear on it, first those that have sea-borne seeds or rhizomes, then wind-borne seeds, then those borne on the feet and plumage of wandering sea-fowl, and when the vegetation is tall enough, come land birds bringing seeds of the baccate or drupaceous fruits which they had eaten before their flight.

This provides the first elements of a dispersal-structured or **successional model of island assembly**, which it is the purpose of this section to develop.

Successional effects were discussed in relation to equilibrium theory by MacArthur and Wilson (1967), but in a fairly simplified fashion and with little empirical data. Following their work, one of the more interesting contributions was provided by the data for the two volcanically disturbed islands from Diamond's (1974) study of New Guinea islands. Ritter was sterilized in 1888, and Long was devastated about two centuries prior to Diamond's survey of the two islands in 1972. Vegetation recovery on each island was found to be incomplete, Ritter supported *Pandanus* (screw-pines) up to 12 m high on its gentler slopes, but was bare in steeper areas, succession being retarded by landslides, strong prevailing winds, and erosion. Long's recovery was further advanced, but its forests were more open and savanna-like than on the older volcanic (control) islands nearby. The species–area plot in Fig. 8.5 reveals that both Ritter (four species) and Long (43 species) remained depauperate as a function of area relative to the control islands (Diamond 1974). These deficits were postulated to be a product either of slow colonization by birds, or of the incomplete recovery of the habitat. Diamond favoured the successional explanation, noting that the arrival on Ritter of species that could not establish resident populations in the sparse vegetation was confirmed both by direct observation and by feathers at the plucking perch of a resident peregrine falcon. Diamond compared his data with a survey from 1933, calculating by the means explained in section 7.2 that its species number was approximately 75% of the equilibrium value. He found that species numbers had increased only slightly in the interim, with a degree of turnover occurring—a result he attributed to the slow pace of forest succession 'arresting' the progress of the Long avifauna towards its equilibrium value. As turnover was

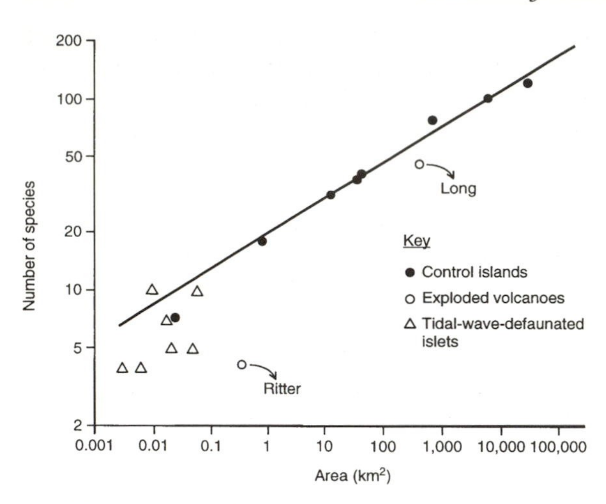

Fig. 8.5. The relationship between log-species number and log-island area for resident, non-marine, lowland bird species on the Bismarck Islands (redrawn from Diamond 1974). The line of best fit is for the so-called control islands. The explosively defaunated islands of Long and Ritter remain significantly ($P<0.001$) below the line of best fit, which Diamond interpreted as demonstration of incomplete succession and failure to attain equilibrium in the period since their volcanic disturbance.

not greater than on older islands he termed it a quasi-steady state. This hierarchically determined pattern provides an interesting parallel with subsequent studies of the build-up of bird numbers from the Krakatau group (below).

Diamond's line of argument was of much broader biogeographical scope than just the islands of Ritter and Long, yet arguably these two islands held a key place in the overall theory of the dynamics of the New Guinea islands. Hence, I feel that his theory can be described as, in effect, a successional model. Diamond's reasoning was as follows. Small islands, on which extinction rates should be high, will have a high incidence of the best dispersing species, i.e. they have a high incidence of tramps. Larger defaunated islands will first be colonized by the supposedly *r*-selected supertramps. Other, less dispersive tramps arrive subsequently and eventually crowd out the less competitive supertramps. Prior to their loss, however, the island acts to supply the surrounding area with its surplus population, and the species therefore will survive by having found another defaunated island prior to exclusion from the first. Diamond calculated that for some supertramps the availability of Long island will have enabled the quadrupling of the entire population of the species. Long and Ritter thus hold considerable interest. Yet, relatively little species–time data are available for either island. Krakatau is set in a less well developed biogeographical context but provides data for the developmental process over a uniquely long period, thus complementing the earlier studies of Diamond.

8.5. Krakatau—succession, dispersal structure, and hierarchies

8.5.1. Background

Krakatau was the first site used to test the ETIB's predictions of an approach to a dynamic equilibrium. The natural recolonization of the islands by birds between 1883 and 1930 appeared consistent with the ETIB: species number reached an asymptote and further turnover occurred (MacArthur and Wilson 1963, 1967). However, the plant data from the islands revealed a poorer fit, as species numbers continued to rise. This was postulated to be due to one of two effects: either that the pool of plant species was sufficiently large to prevent a depletion effect on the immigration rate, or successional replacement of the pioneer communities was incomplete, thereby leading to a reduction in extinction rate for a time. The latter effect was favoured, and illustrated in diagrammatic form, but the ETIB was not modified formally to account for the data.

The Krakatau group has undergone repeated phases of volcanic activity. In 1883 it consisted of three islands, arranged in a caldera which resulted from prehistoric eruptive activity. The largest island of the group became active in May of that year, commencing a sequence that ended on 27 August in exceptionally destructive eruptions. Two-thirds of the island disappeared, vast quantities of ejecta were thrown into the atmosphere, and a series of huge waves (tsunami) generated in the collapse resulted in an estimated 36 000 human casualties in the coastal settlements of Java and Sumatra flanking either side of the Sunda Strait some 40 km or so from Krakatau. The meteorological and climatological effects of the eruption were observable across the globe and the islands excited considerable and lasting scientific interest from a number of disciplines (Thornton 1996). Each of the three Krakatau islands were entirely stripped of all vegetation. The main island, now known as Rakata, lost the majority of its land area, but all three islands also gained extensive areas of new land resulting from the emplacement on to the pre-existing solid rock bases of great thicknesses of pyroclastic deposits. Estimates of ash depths are of the order of 60–80 m and to this day the vast majority of the area of the three islands remains mantled in these unconsolidated ashes, with relatively little solid geology exposed at the surface. No evidence for any surviving plant or animal life was found by the scientific team led by Verbeek later that year, and in May 1884 the only life spotted by visiting scientists was a spider. The first signs of plant life, a 'few blades of grass', were detected in September 1884.

The question of how complete was the destruction has been hotly debated (Backer 1929; Docters van Leeuwen 1936) and it is conceivable that some viable plant propagules might somehow have survived to be uncovered by later erosion of the ash mantle (Whittaker *et al.* 1995). However, there is no evidence for survival, and indeed the densest populations of early plant colonists were located on *terra nova*. The islands can be taken to have been as near completely sterilized as makes no practical difference. The fauna and flora have thus colonized since 1883 from an array of potential source areas, the closest of which, the island of Sebesi, is 12 km distant. All the nearest land areas, including Sebesi, were also badly impacted by the 1883 eruptions. Here then was a group of three islands, each mantled in sterile ashes and in time receiving plant and animal colonists. The potential of the islands as a natural experiment in dispersal efficacy and ecosystem recovery was appreciated, although, unfortunately, only botanists seized the opportunity at an early stage, with surveys in 1886 and 1897. More recently, a team of zoologists led by Ian Thornton has worked hard to remedy the deficiency, producing an important series of papers in the 1980s and 1990s, ably synthesized in Thornton's book (1996). During the late 1980s and early 1990s several points of dispute arose between the Thornton group and mine, over points of detail and interpretation of the island ecology of Krakatau. However, what strikes me now is the similarity of the positions that have emerged, albeit with some remaining differences in emphasis. To employ a metaphor for this, Thornton and his colleagues take the view that the cup of the ETIB is half-full, Whittaker and colleagues that it is half-empty. Each group has concluded that overall equilibrium has not been reached and that a more complex, successional model of island assembly is required to describe the Krakatau data.

It is important to appreciate the dynamism of the platform upon which ecosystem assembly has taken place. The early pace of erosion of the ash mantle must have been dramatic. It created a deeply dissected 'badlands' topography which, even after 110 years, is geomorphologically highly dynamic. The extensive new territories around the coasts were also subject to extreme rates of attrition, and steep cliffs formed rapidly around much of the shoreline. Shallow shelving beaches, which provide the most favourable points for the colonization of many plant and some animal species, are restricted in their extent. In 1927, after nearly 46 years of inactivity, a new island began to form in the centre of the 1883 caldera, finally establishing a permanent presence, Anak Krakatau, in 1930. Through intermittent activity it grew by the mid-1990s to an island of over 280 m in height and 2 km in diameter (Thornton *et al.* 1994). Over this period it has caused significant and widespread damage to the developing forests of two of the three older islands, Panjang and Sertung, but has only indirectly impacted on Rakata (Whittaker *et al.* 1992*b*; Schmitt and Whittaker 1998) (Fig. 8.6).

Fig. 8.6. A towering eruption column from Anak Krakatau in July 1994. The island was about 280 m in altitude at this time. The foreground woodland is dominated by the sea- and wind-dispersed pioneering tree *Casuarina equisetifolia*. Further inland, the barren cone is testimony to the harsh and disruptive environment. Disturbance from the volcano has 're-set' the colonization on Anak Krakatau and has also impacted on succession on the nearby islands of Sertung and Panjang (see text). (Photo: RJW 1994.)

8.5.2. Community succession

Although the first food chains to establish may well have been of detritivores, and micro-organisms, consideration of the real business of community assembly begins with the higher plants. The system must be understood both in terms of successional processes and in relation to constraints on arrival and colonization. The broad patterns of succession will be described first. Partial survey data and descriptions of the plant communities are available from often brief excursions in 1886, 1896/97, 1905–08, most years between 1919 and 1934, 1951, 1979, 1982/83, and 1989–95 (Whittaker *et al.* 1989, 1992*a*; unpublished data). The coastal communities established swiftly, most of the flora being typical sea-dispersed (thalassochorous) members of the Indo-Pacific strand flora. In 1886, 10 of the 24 species of higher plants were strand-line species. By 1897, it was possible to recognize a *pes-caprae* formation (named after *Ipomoea pes-caprae*) of strand-line creepers, backed in places by establishing patches of coastal woodlands, of two characteristic types. First, patches which were to become representative of the so-called *Barringtonia* association (more generally typified by the tree *Terminalia catappa*), another vegetation type typical for the region, and secondly, stands of *Casuarina equisetifolia*. The latter is a sea- and wind-dispersed pioneering tree species, which also occupies some precipitous locations further from the sea. In the coastal areas it typically lasts for just one generation, as it fails to establish under a closed forest. It is notable that as early as 1897 the coastal vegetation types were already recognizable as similar to those of many other sites in the region (Ernst 1908). The coastal communities continued to accrue species over the following two decades, but have since exhibited relatively little directional compositional turnover, apart from the loss in many areas of *Casuarina*. They may thus be described as almost an 'auto-succession' (Schmitt and Whittaker 1998).

In the interiors, a much more complex sequence of communities has unfolded. To varying degrees this can be understood in relation to the considerable differences in habitats between and within islands, to differential landfall of plant species within the group, and to the dynamics of the physical environment, especially the disruption originating from the new volcanic island, Anak Krakatau. However, from the partial survey data available, it appears that the early stages were broadly similar in the three older islands. In 1886, the majority of cover in the interior was supplied by 10 species of ferns, the balance of the species being made up of wind-dispersed grasses (two species) and Asteraceae (2–4 species). By 1897, the interiors had become clothed in a dense grassland, dominated by *Saccharum spontaneum* and *Imperata cylindrica*, interspersed with small clusters of young pioneer trees. Ferns dominated only the higher regions of Rakata and the balance of species had also shifted in favour of the flowering plants. By 1906, the woodland species had increased considerably in the interiors, although remaining patchy, and the fern communities were gradually receding upwards. The grasslands were tall and dense and so difficult to penetrate that proper exploration of the interiors was greatly hampered prior to 1919. The woodlands of the lowlands continued their rapid development, with *Ficus* spp., *Macaranga tanarius*, and other animal-dispersed (zoochorous) trees to the fore.

Forest closure took place over most of the interior of each island during the 1920s, such that by 1930 very little open habitat remained. This key phase of system development (1919–32) was fortunately the subject of detailed investigations by Docters van Leeuwen (1936). As the forests developed, habitat space for forest-dependent ferns, orchids, and other epiphytic plants became available, and their numbers increased rapidly in response. Conversely, the pioneering and grassland habitats were reduced, species populations shrank, and some species may have been lost. Rakata is a high island, of *c.* 735 m elevation, and altitudinal differentiation of forest composition was evident as early as 1921. The highest altitudes thereafter followed a differing successional pathway, in which the shrub *Cyrtandra sulcata* was for many years a key component. While most vegetation changes appear to have been faster in the lowlands, spreading up the mountain of Rakata, one key species, the wind-dispersed pioneering tree *Neonauclea calycina*, first established a stronghold in the upper reaches, before spreading downwards. By 1951 it had become the principal canopy tree of Rakata from just below the summit down to the near-coastal lowlands. It remains so as of the 1990s, although the forests are actually rather patchy in their composition, with a number of other important canopy and subcanopy species varying in importance from place to place within the interiors (Fig. 8.7).

The patterns of development on the much lower islands of Panjang and Sertung were broadly similar up to *c.* 1930, although differences in the presence and abundance of particular forest species were noted. Since 1930, both islands have received substantial quantities (typically in excess of 1 m depth) of volcanic ashes over the whole of their land areas. Historical records and studies of ash stratigraphies demonstrate that some falls of ash have been very light, but *c.* 1932/35 and 1952/53, and possibly on other occasions, the impact has been highly destructive (Whittaker *et al.* 1992*b*). For instance, Docters van Leeuwen (1936) described how in March 1931 the most disturbed forests of

Fig. 8.7. Rakata Island (summit 735 m above sea level), Krakatau. After sterilization in 1883, a successional sequence unfolded, producing a more or less continuous forest cover on the island by around 1930. Over 60 years later the forests continue to change as earlier tree species begin to decline and others increase in abundance. While the early pioneer vegetation inland featured many wind-dispersed species, most inland forest trees and shrubs are animal-dispersed—their colonization has been effected by frugivorous birds and flying foxes (fruit bats). (Photo: RJW 1994.)

Sertung resembled 'a European wood in winter', and how grasses re-invaded (possibly resprouted) within the stricken woodlands. Since then, forests dominated to a considerable extent by the zoochorous trees *Timonius compressicaulis* and *Dysoxylum gaudichaudianum* have become characteristic of large areas of both islands. Whittaker *et al.* (1989) argued that the eruptions had, in effect, deflected the successional pathways followed on these islands and quite possibly slowed the pace of species accumulation through the loss of bridgehead populations and the re-setting of succession to earlier stages. The relative significance of volcanism, in relation to differential land-fall and other environmental differences within the islands, in influencing the patterns of forest community development, remains the subject of debate (Bush *et al.* 1992; Whittaker *et al.* 1992*b*; Thornton 1996). While less damaging eruptions in the 1990s have been found to have impacts on forest development and turnover (in the sense of canopy turnover rather than island ecology), and apparently to differentiate between different species, the lack of data from Panjang and Sertung between 1934 and 1982 provides something of an obstacle to resolving this debate (Schmitt and Whittaker 1998). However, this need not be of great concern in the present context, as it is not a question of whether, but to what relative extent, each factor has influenced community development. Recent studies of forest dynamism and gap fill processes have demonstrated more generally that the forests of each of the islands are responsive to the vagaries of a highly variable physical environment. Geomorphological activity, storm damage, and drought (ENSO-related) may each be added to volcanism as causal agents for the mortality of plants at all stages of their life cycle (Schmitt and Whittaker 1998).

Although different forest types have been recognized on the Krakatau islands, these 'communities' are not discrete and forest successional pathways are in practice more complex than simple summary successional schema (e.g. Whittaker *et al.* 1989, Fig. 15) might be taken to imply. It is notable that of all the vegetation types of the Krakatau islands, only the *pes-caprae* formation and *Barringtonia* association were assigned with any confidence to phytosociological types by the earlier plant scientists. As Docters van Leeuwen (1936, pp. 262–3) put it, 'All other associations in the Krakatau islands are of a temporary nature: they change or are crowded out.' While the coastal systems are similar to those of other locations in the region, those currently recognized from the interior by Whittaker *et al.* (1989) lack documented regional analogues of which we are aware. The interior forests of the Krakatau islands continue to accrue new species of higher plants, and the balance of species in the canopy is undoubtedly in a state of flux, with strong directional shifts in the importance of particular species being evident over the period since 1979 (Bush *et al.* 1992; Whittaker *et al.* 1998). The business of forest succession is ongoing.

8.5.3. A dispersal-structured model of island recolonization

The community-level changes I have just sketched out have been underpinned by trends within the colonization data (Whittaker *et al.* 1989, 1992*a*). Figure 8.8 simplifies these trends into three time-slices, restricting the treatment to the island of Rakata, for which the best data (least confounded by volcanic disturbance) are available. The figures are for all species recorded from the island, whether still present or not (Whittaker and Jones 1994*a*). Phase 1 represents the pioneering stage during which wind-dispersed (anemochorous) pioneers, first ferns, and then grasses and composites (Asteraceae), initiated the colonization of the interior. As emphasized above, colonization of the strandlines was rapid during this phase, gradually diminishing thereafter.

Phase 2 represents the period during which the extensive grasslands waxed and waned, as, increasingly, animal-dispersed (zoochorous) trees and shrubs spread out from their initial clumps to form woodlands. The interiors have been filled almost exclusively by species which are primarily either wind- or animal-dispersed. The relative balance between these two groups shifts dramatically between phase 1 and 2. Few ferns

Phase 1

Sea
Rapid colonization of strand-line
Simple succession
1.64

Animal
Delayed colonization of interior
0.14

1.0 Ferns
1.0 Other
Wind
Rapid colonization of interior by pioneers (ferns, grasses, composites)

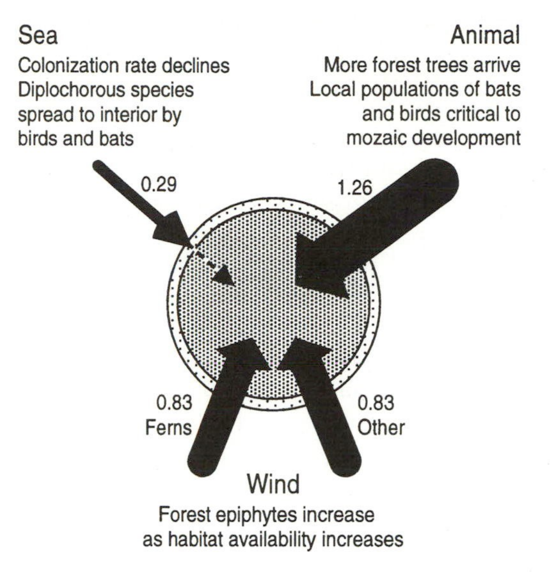

Phase 2

Sea
Colonization continues at a reduced pace
1.18

Animal
Colonization rate increases as attraction to frugivores increases
1.32

0.18 Ferns
0.86 Other
Wind
Pioneers lose ground

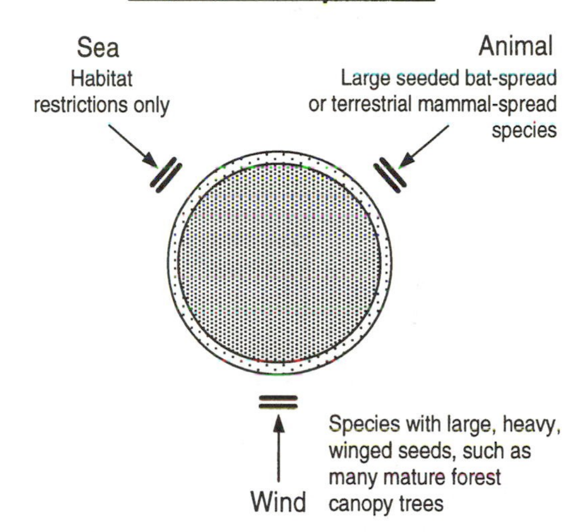

Fig. 8.8. Plant recolonization of Rakata Island (Krakatau group) since sterilization in 1883. The three phases correspond with survey periods and represent convenient subdivisions of the successional process. Phase 1, 1883–1897; phase 2, 1898–1919; phase 3, 1920–1989. Arrow widths are proportional to the increase in cumulative species number in species/year (these values are also given by each arrow). The flora is subdivided into the primary dispersal categories, i.e. the means by which each species is considered most likely to have colonized: animal-dispersed (zoochorous), wind-dispersed (anemochorous), and sea-dispersed (thalassochorous). The model distinguishes between strand-line (outer circle) and interior habitats and, in the fourth figure, identifies constraints on further colonization. (From Whittaker and Jones 1994*a*.)

colonized during phase 2 despite the efficiency of their dispersal system (microscopic spores). The interpretation offered for this is that there are actually relatively few ferns in the regional species pool which typify such extreme, seasonally droughted pioneer sites, and they arrived very quickly during phase 1. Wind-dispersed early successional flowering plants also continued to accrue during phase 2, including Asteraceae and terrestrial orchids, but very few trees and shrubs. The vast majority of arboreal species are zoochorous, and after a very slow start, in which only two species of zoochores of any sort were found by 1897, their rate of arrival took off in the early 1900s.

Phase 3 runs from the point when forest closure became extensive, to the 1989 datum. During this phase, the colonization rate of sea-dispersed (thalassochorous) species slackens, which reflects the relatively limited source pool, but also the limited array of coastal habitats available. In the interiors, it was only when forest habitats became available that the second wave of ferns, those of forest interiors, could build in numbers. Hence their rate of discovery increased rapidly in the period of forest formation, peaking around 1920. The ease of dispersal of ferns, with their microscopic propagules, can be demonstrated by comparison of the size of the Krakatau flora (all islands 1883–1994) with the native flora of West Java. The ratio for spermatophytes is 1:10.1 and for ferns 1:4.2, indicating Krakatau to have a remarkably species-rich fern flora (Whittaker *et al.* 1997). Wind-dispersed flowering plants colonizing the newly available forests of phase 3 were predominantly orchids, many of which are epiphytic, together with a mix of other epiphytic and climbing herbaceous species. The single largest group recorded for phase 3, however, is the zoochorous set, mostly being trees and shrubs.

More detailed analyses of dispersal mechanisms suggest further structural features in the data. For instance, anemochorous species can be split into those with dust-like propagules, plumed seeds, and winged seeds or fruits. The larger, winged propagules have limited dispersive ability and, although significant in the regional pool of forest canopy trees, only one or two species of this category have colonized Krakatau (Whittaker *et al.* 1997): they are thus noted in Fig. 8.8 as improbable colonists. Colonization of the zoochorous species of Krakatau, with the exception of a few human-introduced cases, can be attributed to the actions of frugivorous birds and bats. The features of many zoochorous fruit types do not limit dispersal to one or other group of frugivores and thus a definitive assessment of the roles of each in the colonization of Krakatau is not yet possible. Whittaker and Jones's (1994*b*) best estimates are given in Table 8.3. Fruit bats swallow only the smallest seeds (<1 or 2 mm) and it was assumed that they would be unlikely to carry larger seeds in their mouths or claws over the many kilometres of sea separating Krakatau from source areas. Thus, small-seeded zoochores might be introduced either by birds or bats, but larger-seeded species can be considered strictly bird-assisted colonists. Beyond a certain threshold seed size it is likely that only the largest fruit-pigeons can effect an introduction. However, once a species has established on an island, a wider range of dispersal agents may be involved in local dissemination.

Where plant species have two rather contrasting dispersal vectors, they are termed diplochores. Twenty-four sea-colonist, but potentially animal-spread species occur on Krakatau, of which 14 are bat-spread, four may be bat- or bird-spread, and six are bird-spread. This mixed group of diplochores includes several of the earliest colonists of the Krakatau strand lines. It is plausible that they may have had an important role kick-starting the succession of zoochores. As shown, zoochores were laggardly colonists. There cannot be much advantage in visiting a barren island if you happen to be a frugivorous bird or bat. Once the fringing plant communities of sea-dispersed species established and began to fruit, however, the islands would have provided a suitable food supply in the form of the fruit of the diplochorous species. From the early accounts it appears that many of the first true zoochorous plants to be found were concentrated around the coastal fringes. Admittedly the interiors were less well explored, and certainly pockets of trees quickly became

Table 8.3

*Estimates of the numbers of plant species found on Krakatau between 1886 and 1992 for which birds and bats have a dispersal role. Under the heading of zoochory (animal dispersal), two modes of transport may be distinguished: those seeds which are eaten, swallowed and which pass through the gut or which are eventually spat out (endochorous) and those transported by external attachment to the animal (exochorous). The data set includes all four Krakatau islands and all records (i.e. including some species which have not maintained a presence) (source: Whittaker and Jones 1994*b*)

Dispersal mode	Number of species
Endochorous introduction (bird and/or bat)	124
Exochorous introduction (bird)	10
Human introduction, endochorous spread (bird and/or bat)	15
Sea-colonist, endochorous-spread	24
Total zoochorous introduction and/or spread	173
Of the 124 species introduced endogenously:	
Either bird or bat-colonist	43
Bird-colonist, bat-spread	31
Bird-dispersal	50

dotted throughout the interiors, but the first points of landfall, and the first roosting points for frugivores and, it seems likely, the first food supply, must have been provided by the coastal vegetation.

The first frugivores observed on the islands were birds, six species being observed in 1908 (the first bird survey), whereas it was not until the next zoological survey in 1919 that bats were noted. Birds may thus have been the more important group in terms both of simply the number of zoochorous colonists and being earlier in arrival (although survey data are too poor to be conclusive). However, bats have had a differing role to birds. First, they have spread some species that birds do not transport. For instance, the important coastal and near-coastal tree *Terminalia catappa* only invaded the island interiors after the arrival of the bats. Secondly, Whittaker and Jones (1994*b*) suggested that they might be more important to the early seeding of open habitats because of differences in their foraging behaviour (in part mediated by interactions with predators).

The point of detailing these patterns and processes here is to illustrate that on real islands (in the sea and some kilometres in area) succession is complex and demonstrates hierarchical interdependency across trophic levels. In order to understand how compositional patterns develop in the vegetation and flora, it is necessary to consider the ecological attributes of the plant species, their habitat relationships, and their hierarchical ecological links with animals (Bush and Whittaker 1991). Animals effect their dispersal, in cases their pollination, and of course also act as seed predators. In turn the vegetation provides the habitat and food resources of the animals; no fruit supply means no resident frugivores.

8.5.4. Colonization and turnover—the dynamics of species lists

Given the basic elements of vegetation succession and colonization patterns, broken down into plant dispersal types, it is possible to interpret much of the pattern in the rates of immigration and turnover of both plants and animals of the Krakatau islands. What follows here flows on from the consideration of turnover in Chapter 7, but it is placed here because of the structural features evident within it. An important caution applies to these studies. The Krakatau islands are large, topographically complex, and sufficiently difficult to explore that some parts have never, or only rarely, been penetrated by scientists. While considerable efforts have been expended in exploration in recent years, we continue to discover additional plant species which, from their stature or abundance, we know must have colonized many years before. For both plants and animals, surveys over the 1883–1995 period have thus been partial, have involved considerable turnover in personnel (and thus expertise and methods), differing areas of search, and provide no means of assessing quantitatively the efficiency of their collecting endeavours. They have also taken place at irregular intervals. For these reasons it is, in my view, necessary to treat estimates of biogeographic rates for the Krakatau islands with a great deal of caution. Although I present some extinction and immigration rates here and have done elsewhere, on the whole I prefer to present the data in different terms, and eschew the use of inferential statistics as unwarranted. While this limits the power of the analyses, I contend that this is appropriate given the limits of the data. The immigration and extinction rates calculated for Krakatau are, of course, not as defined in the ETIB, rather they represent 'arrival in the lists' and 'departure from the lists'. Arguably the best data for resident status are for birds (Thornton *et al.* 1990, 1993), but even here significant uncertainties remain. Yet in my view the patterns depicted provide useful insights into the forces structuring island biotas, they are also drawn from the longest running natural experiment that is available. They are the best we have.

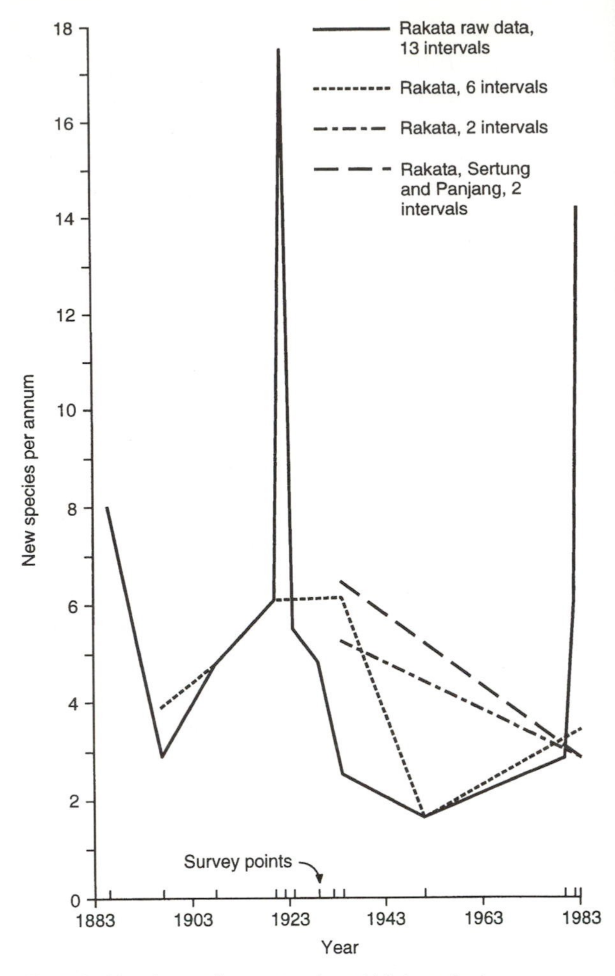

Fig. 8.9. Numbers of new species of higher plants recorded on Krakatau over a variety of intervals, expressed as an annual rate. The figures exclude re-invasions. (Redrawn and corrected from Whittaker *et al.* 1989, Fig. 16.)

Whittaker *et al.* (1989) presented an analysis of higher plant colonization for 1883–1983, which illustrates the smoothing effect of calculating immigration rates over greater intervals of time, ignoring the intervening survey data (Fig. 8.9). The repeat surveys of Docters van Leeuwen (1936) in 1919 and 1922 produce a pronounced spike in the immigration data (another occurs for the same reasons at the end of the series), but the peak for the 1920s remains even when this survey frequency effect is removed. The low point, *c.* 1951, reflects survey deficiency. Bush and Whittaker (1991) therefore calculated 'immigration' and 'extinction' rates for plants using survey data grouped into adjacent surveys, and ignoring the 1951 datum. Reasoning that for most plants, especially large forest trees, temporary disappearance from the lists was more likely to reflect survey inefficiency than genuine turnover, they provided upper and lower estimates: first, the recorded turnover and, secondly, assuming the minimum turnover allowable from the data (Fig. 8.10) (see also Whittaker *et al.* 1989, 1992*a*). The overall trend in species richness is also shown in this figure, and while the lack of adequate surveys between 1934 and 1979 makes the precise shape of the curve unknowable, additional surveys since 1989 continue to turn up additional species records.

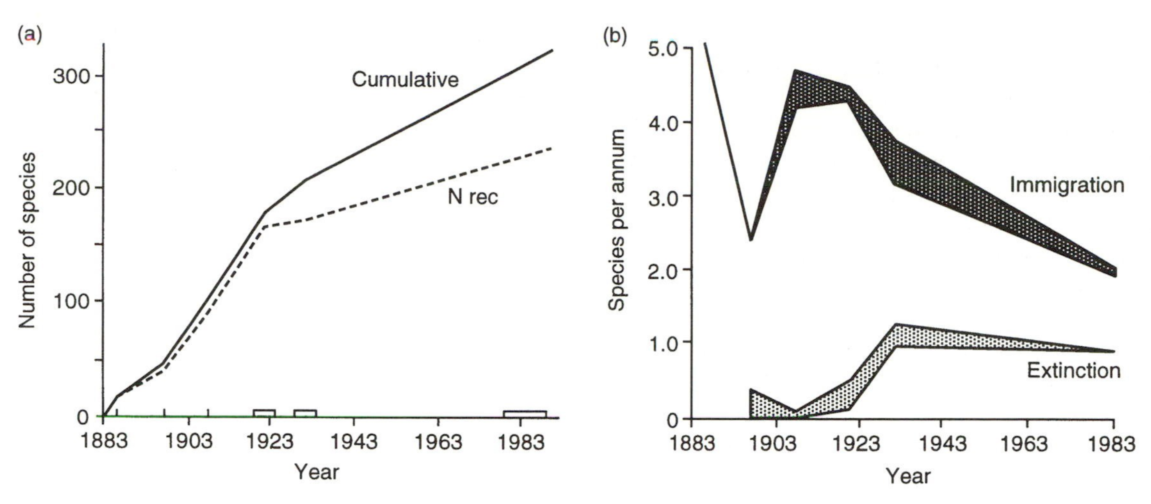

Fig. 8.10. Plant recolonization data for Rakata Island (Krakatau) 1883–1989 (redrawn from Bush and Whittaker 1991). Surveys are grouped into survey periods, and the mean year of each period is used to estimate rates. Shaded lines represent the difference in rate according to whether minimum turnover is assumed or if 'recorded' immigration/extinction is assumed. (a) Cumulative species totals and number of species recorded at each survey (Nrec), curves for higher plants. Nrec figures are on the basis of the assumption of minimum turnover, i.e. species are only counted as going extinct if they fail to be found in all subsequent surveys. This assumes that temporary absences are artefacts of sampling rather than the true extinction and subsequent re-invasion of the species. (b) Immigration/ extinction curves for spermatophyte species. Rates are calculated as follows: (i) assuming recorded turnover, i.e. that all extinction and immigration records are real (immigration rate = number of species not recorded at time 1 but recorded at time 2/time 2 − time 1; extinction rate = number of species recorded at time 1 not found at time 2/time 2 − time 1); (ii) assuming minimum turnover (immigration rate = number of species not previously recorded, but found at time 2/time 2 − time 1; extinction rate = number of species recorded at time 1 not found at or after time 2/time 2 − time 1).

Providing regional source pools are not destroyed by human action, it seems probable from our analyses of forest dynamics (e.g. Bush *et al.* 1992) that species number can continue to rise for some time to come.

What is evident in the analyses of 'immigration' and 'extinction' rates is that discovery of new species on Rakata peaked during the period of forest formation and closure, and the rate of loss of species from the lists also has a peak, lagging behind the 'immigration' peak, and reflecting the loss of early successional habitats. Underlying these trends is a marked species-stability of the earliest assemblages of species, particularly those of the coastal fringes. Of those species of coastal plants which established early and on each of the islands, scarcely any have disappeared. By comparison of the 1883–1934 grouped data with the 1979–1983 grouped data, Whittaker *et al.* (1989) showed that 42 of 47 persisting sea-dispersed species were originally found on all three islands, whereas 28 of 41 which were missing from the 1979–1983 lists were originally known from just one of the islands. Several of these missing species were recorded as juveniles or 'ephemerals', and are known not to have established breeding populations. The disappearance of some others can be attributed to the loss of particular habitats, such as the brackish lagoon which formerly featured on the spit of Sertung Island. The loss of these ephemeral non-colonists and habitat specialists contributed to the increased 'extinction' rate of the 1920s.

The full analyses of survival (in the lists) as a function of island combination and dispersal type performed by Whittaker *et al.* (1989, their Table 9) were consistent with a view that most species lost from the lists were in one or more of the following categories: not properly established initially; loss attributable to habitat destruction; loss of pioneers attributable to successional loss of habitat; and species initially restricted in distribution. We have recently repeated some of these calculations, using additional survey data from 1989 to 1994, and while the total number of 'extinctions' of spermatophytes across the island group was reduced from approximately 113 using the 1979–1983 data to 78 using the 1979–1994 data, the broad patterns described in our earlier paper were very little changed (Whittaker, Field, and Partomihardjo, in preparation). The trends of arrival and persistence in the lists as described at the level of dispersal guild thus appear to be robust. Rates derived for 'immigration', 'extinction', and 'turnover' are much less so.

Similar problems of data quality and frequency afflict surveys of other organisms. The butterfly data (Fig. 8.11) and land-bird data (Fig. 8.12) show similar trends coincident with forest closure, but contrasting trends towards the end of the data series, with butterflies seemingly climbing more steeply and birds approaching an asymptote. In the first two decades, the poverty of the vegetation arguably presented a limited array of opportunities for butterfly species, many of which require the presence of fairly specific food plants. The rate of arrival appears to have peaked during the period of most rapid floral and habitat diversification. Extinction also peaked following forest closure. Bush and Whittaker (1991) attribute four of the losses to succession, as open habitats were lost, other losses being possibly due to sampling deficiencies and to the inclusion of likely migratory species in the calculations. Thornton *et al.* (1993) took this view in replotting the data, thus finding a lower peak in the extinction rate, although the general trends in rates were the same.

Figure 8.12 shows the trends in the bird data as calculated by Thornton *et al.* (1993). These data are similar to the butterfly data but have an additional survey point, for 1951. This survey, by Hoogerwerf (1953), was missed by MacArthur and Wilson (1967) who suggested from the three survey collations for Rakata and Sertung up to 1934 that equilibrium might already have been reached. In fact, species number has increased slightly since the 1930s (Thornton *et al.* 1993). There is evidence in these data of successional effects involving different trophic levels, a point which was recognized by MacArthur and Wilson (1963, 1967). The data also indicate a degree of turnover, the precise figure involved being highly sensitive to assumptions made as to breeding status of birds (Thornton *et al.* 1993). While the colonization curve is now fairly flat, it is perhaps too early to judge whether the birds of Rakata have achieved a dynamic equilibrium. Both Thornton's group

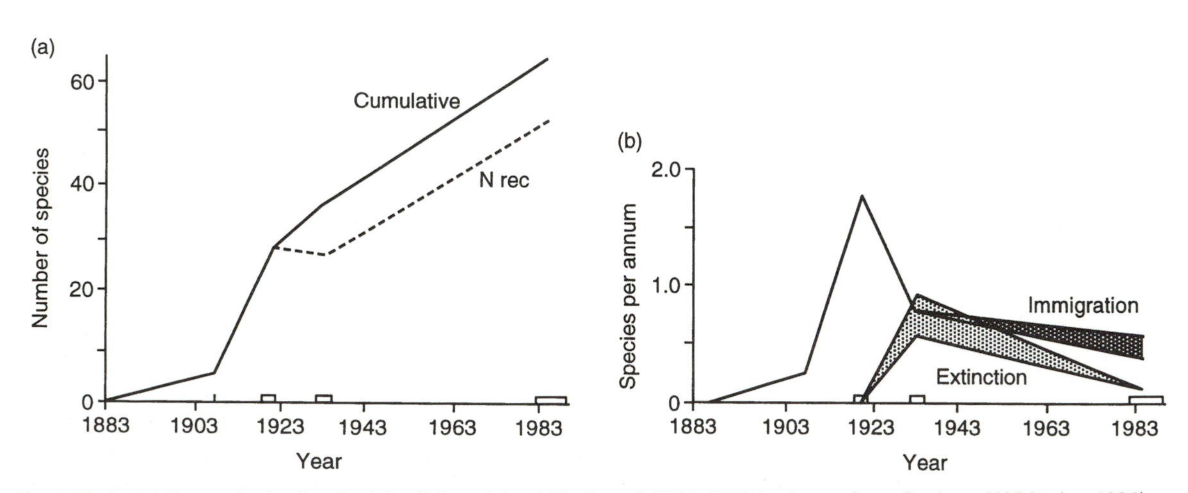

Fig. 8.11. Butterfly recolonization data for Rakata Island (Krakatau) 1883–1989 (redrawn from Bush and Whittaker 1991). The peaks in immigration and extinction appear to correlate with habitat succession as grassland gave way to forest habitat on the islands. (a) Cumulative total and number of species recorded at each survey (Nrec), as Fig. 8.10. (b) Immigration and extinction curves, as Fig. 8.10.

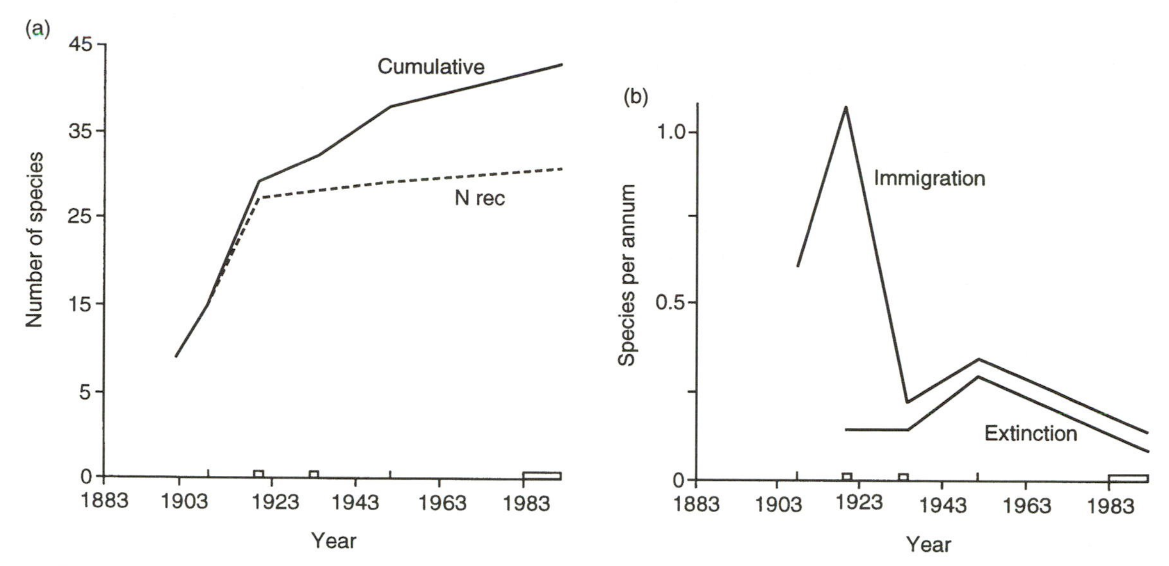

Fig. 8.12. 'Resident' land bird recolonization data for Krakatau (redrawn from Thornton *et al.* 1993). (a) Cumulative total and number of species recorded at each survey (Nrec), assuming minimal turnover. (b) Immigration and extinction curves, as species per year for inter-survey periods, as Fig. 8.10.

and Bush and Whittaker (1993) concluded that for the three taxa examined there is no evidence as yet of an equilibrium number of species having been met, with much of the real turnover recorded being successional in nature. In so far as it can be judged, 'extinction' rates appear to have declined rather than (as the ETIB expects) having risen over the most recent inter-survey period.

Another of the Krakatau data sets, that for reptiles, also indicates both an important role for abiotic factors and a low turnover rate. Rawlinson *et al.* (1992) show that there have been only two extinctions of stabilized reptile populations, one due to habitat changes (canopy closure and coastal erosion) and the other to catastrophic alteration of habitat by eruptions of Anak Krakatau. No evidence was found of an approach to a dynamic equilibrium. Thus, for Krakatau, the fit of the data to the expectations of the ETIB appears to be poor, the best relationship being for birds, although even here the fit is imperfect.

Some differences remain in the viewpoints reached by the Thornton and Whittaker/Bush groups, although there is broad agreement on the basic position. One point stressed by Bush and Whittaker (1991) was that the changes in direction of rates of 'immigration' and 'extinction', i.e. their non-monotonicity, was contrary to the ETIB. MacArthur and Wilson's (1967) stated position was that the precise shapes of the I and E curves could be modified within the confines of the ETIB, providing that they remained monotonic, 'When a new set of curves must be derived for a new situation, the model loses much of its virtue…' (p. 64). Departure from monotonicity suggests that the intersects may be unstable or perhaps that there may be alternative stable states. Taken at face value, the rate changes shown in Figs 8.10–8.12 are not consistent with the ETIB, although it is fair to point out, as Thornton *et al.* (1993) have done, that such deviations were recognized within the broader view of *The theory of island biogeography*. Furthermore, the rates shown in the figures are not I and E in the strict sense of the ETIB. In this sense, the Krakatau data can hardly be used as a test, but in so far as the originators of the theory and others have chosen to do just that, the fit to the model is demonstrably poor.

The Krakatau data are flawed, yet, as argued above, the basic patterns of species assembly appear to be relatively robust. Much of the turnover is successional, and quite a lot involves species which may well not have established breeding populations. If the turnover attributable to the loss and gain of habitats and to succession is set to one side, undoubtedly some turnover remains. Precisely how much is real, in the sense of extinction of established populations, is very hard to quantify. Overall, turnover can be judged to be heterogeneous rather than homogeneous *sensu* Rey (1985).

8.5.5. The balance between stochasticism and determinism

Is the assembly of the Krakatau system governed by chance, or is it deterministic and predictable in its behaviour? It is difficult to provide definitive answers to this question because of the deficiencies of the historical data and because there are not enough other similar 'experiments' to compare it with. However, some interesting insights have emerged from studies of the new island, Anak Krakatau.

Anak Krakatau has been termed an 'analogue' or a 'replicate' for the early stages of recovery on the older islands (e.g. Zann *et al.* 1990), but it is probably best to view the island in Thornton's (1996; Thornton *et al.* 1992) terms as a 'model within a model'. The creation and development of the new island has modified the conditions for arrival, mortality, and survival within the island group as a whole (Whittaker and Bush 1993). It has impinged upon developments on the other islands, not least through its volcanic activity. Its own history of environmental development has been distinctive, and its colonists can be assumed to have been drawn predominantly from within the archipelago, rather than from the more distant source pools of the post-1883 period. Indeed, some populations of animals move between the islands in foraging. As an experiment in island colonization, it thus cannot be viewed as a replicate in the strict sense. Nevertheless, if a more liberal view is taken, then it can be shown statistically that there has been a bias in the early patterns of species assembly on Anak Krakatau towards those species which colonized the archipelago in the early post-1883 phase (Thornton *et al.* 1992). This applies to several groups of plants and animals (e.g. birds, reptiles, and bats). The appropriateness of inferential statistics is acknowledged by Thornton (1996) to be debatable, but the patterns are clear enough to be convincing of a strong degree of successional structure as a general feature of the data; 'It is unlikely that this combination of similarities is due to chance. Rather, it strongly suggests the existence of a substantial deterministic component of the colonization process up to the stage of the beginning of forest diversification' (Thornton 1996, p. 269). The exception to the pattern was provided by the butterfly data set. This was probably due to the dependence of butterfly species on the availability of particular habitats and food plants, which differed sufficiently between old and new runs of the experiment to buck the trends.

The recolonization of Anak Krakatau has not been an uninterrupted process. Partomihardjo *et al.* (1992) have analysed the succession of floras following, first, the island's appearance in the centre of the Krakatau caldera in 1930, and, subsequently, following wipe-outs of the vegetation in 1932/3, 1939, 1953, and damaging, but not entirely destructive eruptions around 1972 (Thornton and Walsh 1992; Whittaker and Bush 1993). Surveys have been carried out only intermittently, but we have in essence three or four runs of the assembly experiment on Anak Krakatau, for each of which a species list is available. This rather small-scale natural experiment has demonstrated a strong degree of repetition, a temporal nestedness, with a core of constant species, which make it back each time and are added to (Whittaker *et al.* 1989; Partomihardjo *et al.* 1992). The first assemblage was really just a seedling flora, a few members of which were not found in the second assemblage. Yet, of 32 species identified from these two surveys, 30 recolonized subsequently. The eruptions in 1972 are presumed to have severely reduced the 'third' flora, yet all but two of its 43 species were recorded between 1979 and 1991.

If the Anak Krakatau flora is broken down into arbitrary functional guilds, there are basically three sets of species:

1. *Strand-line species.* These are the largest set, representing 58 of 125 spermatophytes

found between 1989 and 1991. Propagules of these species are sea-dispersed, and are produced in large numbers locally on the other islands. This is a very consistent set. Forty-two of these species have been found on all four Krakatau islands.

2. *Pioneers of interior habitats.* These species are wind-dispersed ferns and grasses, with a few composites. Again they are locally available, are very dispersive, and produce large numbers of propagules. They are adapted to a highly dispersive life style and are very capable of 'finding out' such open environments, on which their survival depends (they are presumed to be not good competitors). As with the strand species, they are in essence a subset of those found on the other islands post-1883.

3. The third set of species, the *second phase colonists of interior habitats*, only managed to become established in the most recent flora. It will be interesting to establish the extent to which it may repeat following any further wipe-out. Comparisons with the older islands suggest that in relation to the sequence on the older islands this group of species is likely to be less predictable in colonization sequence because of the more variable patterns of seed production and dispersal (mostly by birds or bats) within the archipelago. For instance, some of the contemporarily most abundant tree species on the other islands (e.g. *Arthrophyllum javanicum*, *Timonius compressicaulis*, and *Dysoxylum gaudichaudianum*) were recorded on Anak Krakatau at arguably an earlier stage of successional development than in the post-1883 sequence.

Whittaker and Jones (1994*a*) have formulated these and other observations from Krakatau into an informal 'rule table' in which the dispersal, successional, turnover, and compositional characteristics of five such functional plant guilds were tentatively put forward should the opportunity arise to evaluate them elsewhere. Within plant successions there may be both autogenic 'facilitation' effects and diffuse competitive effects between guilds, culminating in later successional communities shading out earlier ones. Yet much of the structure may be attributable to a form of relay floristics, mediated through dispersal attributes of the plants and dependent, for some guilds, on hierarchical links between plant and animal communities (Bush and Whittaker 1991; Whittaker 1992; Thornton 1996). Just as in other island systems, as well as there being structure as to which species assemble and in what sequence, there may be particular sets of species which fail to colonize. Whittaker and Jones (1994*a*) highlight those types for which they regard Krakatau as probably undersampling the regional species pool (Fig. 8.8). More formal comparison of the Krakatau floras with other sites in the region has provided support for these observations (Whittaker *et al.* 1997). Later successional species with poor dispersal adaptations, those which have large, winged, wind-scattered propagules, those dispersed by terrestrial mammals, and large-seeded bat-fruits (lacking diplochory) remain deficient on Krakatau, whereas highly dispersive forms such as ferns and orchids have been 'over-sampled'.

The Krakatau plant and animal recolonization data thus suggest a general trend in the degree of predictability through time. While the draw of early pioneers is fairly predictable as a function in large measure of their superior dispersability, precisely which later successional species happen to establish breeding populations is much less predictable. Thornton (1996) has made a further, interesting point regarding the elements of chance and determinism. He draws the analogy between community assembly and the construction of a jigsaw, arguing that the further on in the process a species joins the system the less influence it can have over subsequent events, as it operates within successively narrow bands set by what has gone before. Thus a late-joining species may be unpredictable in its identity, yet may be predicted to have relatively little impact on community trajectories. The parallel with the assembly rules of Diamond is clear. At early stages of the recovery process, a number of species combinations or community pathways are possible, but as the system becomes more complex, and the jigsaw more complete, the number of pieces that will fit in to a given gap will decline, eventually to just one. This form of analogy has some intuitive value for groups such as birds, but is harder to sustain for plants. Successional pathways are not as discrete and limited as such a picture implies. The islands are continually subject to the vagaries of a varied environment, ranging from volcanic eruptions, through storms and landslides, to droughts. In such a disturbed setting competitive replacement within natural plant communities simply cannot be detected on a one-on-one basis (if it ever can), and returning to the analogy, the jigsaw is continually being disturbed and re-formulated but never quite completed.

To conclude, the island ecological rates for Krakatau contain a lot of sampling noise, through which clear structural features are evident, revealing island recolonization to be in essence a special case of succession. Rates of turnover have, in addition, been affected by the physical environmental dynamics of the system. Hierarchical features across trophic levels are evident and it has also been argued that, even within a single taxon, not all species additions have equivalent effects. As mentioned in Chapter 7, the bird data from Anak Krakatau indicate that while the addition of another insectivore or frugivore may not perturb the existing community structure significantly, the addition of the first predatory birds caused a major community perturbation (Thornton 1996). Similarly, the appearance of the first zoochorous fruits, and the first birds, and the first bats, post-1883 must also have constituted important events. A more recent example is the crossing of a threshold population size of fig trees on Anak Krakatau, in providing for resident populations of fig wasps and of frugivores (Compton *et al.* 1994). Different taxa reveal a variety of trajectories of their colonization curves, and greater and lesser degrees of turnover. The appropriateness of placing these data into equilibrial frameworks will be considered in the next section.

8.6. Disturbed island ecology revisited

Immigration and extinction curves are theorized in the ETIB to have smooth concave forms, the former falling and the latter rising to a point of intersection, the dynamic equilibrium. The patterns of arrival and disappearance from the Krakatau lists do not correspond with this expectation, in part because of hierarchical, successional features evident in the data which are simply absent from the model. The big early kinks in the rates relate to the key switch from open to closed habitats. It is conceivable, however, that this switch marks the end of significant autogenically derived (i.e. biotically driven) switches in trends. In which case, once beyond the early phases of succession, the ETIB may become more realistic, i.e. at some stage, an island must fill up with species and some form of equilibrium be established. However, Bush and Whittaker (1993) argue that the time-scale of succession in such complex ecosystems as lowland tropical moist forests is too lengthy for population processes within particular (but not necessarily all) taxa to regulate in the manner envisaged in equilibrium theory. The argument rests on the premise that the dependency of many animal groups on plants for habitat and food resources is such that their patterns of colonization and turnover will be tied to the dynamics of the plant communities. Forest succession is a slow process, the life span of individual canopy trees can exceed 300 years, and even early successional species may live for decades. If a true biotic equilibrium in the plant system were to be established it must be after a period of several generations of biotically mediated interactions, i.e. hundreds of years. However, such conditions are unlikely to be reached in very small patches because of varying forms of environmental variability and disturbance. Events such as hurricanes, volcanic eruptions, flooding, landslides, and other high-magnitude phenomena occur within continental and island landscapes, and typically have periodicities of less than several hundred years (Chapter 2). This line of reasoning led Bush and Whittaker (1991) rather boldly to question the applicability of equilibrium island thinking not only to islands like Krakatau, but more generally to forested systems in the lowland tropics. Thornton (1996) rejects this form of extrapolation, perhaps rightly, but at the least Krakatau shows that in the specific case, a century has not been enough, and I would argue that applications of island models to large, complex, forested isolates should not simply assume that equilibrium conditions apply in the absence of evidence (see Weins 1984).

Figure 8.13 is Bush and Whittaker's (1993) attempt to generalize the Krakatau island ecological trends and to set out alternative trajectories, allowing for taxa of differing ecological roles and response times. It encapsulates the temporally bumpy rates associated with successional processes and provides for two possible scenarios whereby an equilibrium might be reached. The data for (list) rate changes indicate that equilibrium might be approached by a declining immigration rate, combined with two alternative extinction trends, each in opposition to the ETIB. In the first, extinction rate is low and turnover rather heterogeneous (interactive). In the second, it is essentially squeezed out of the system altogether (non-interactive). Given the problems of collecting adequate survey data,

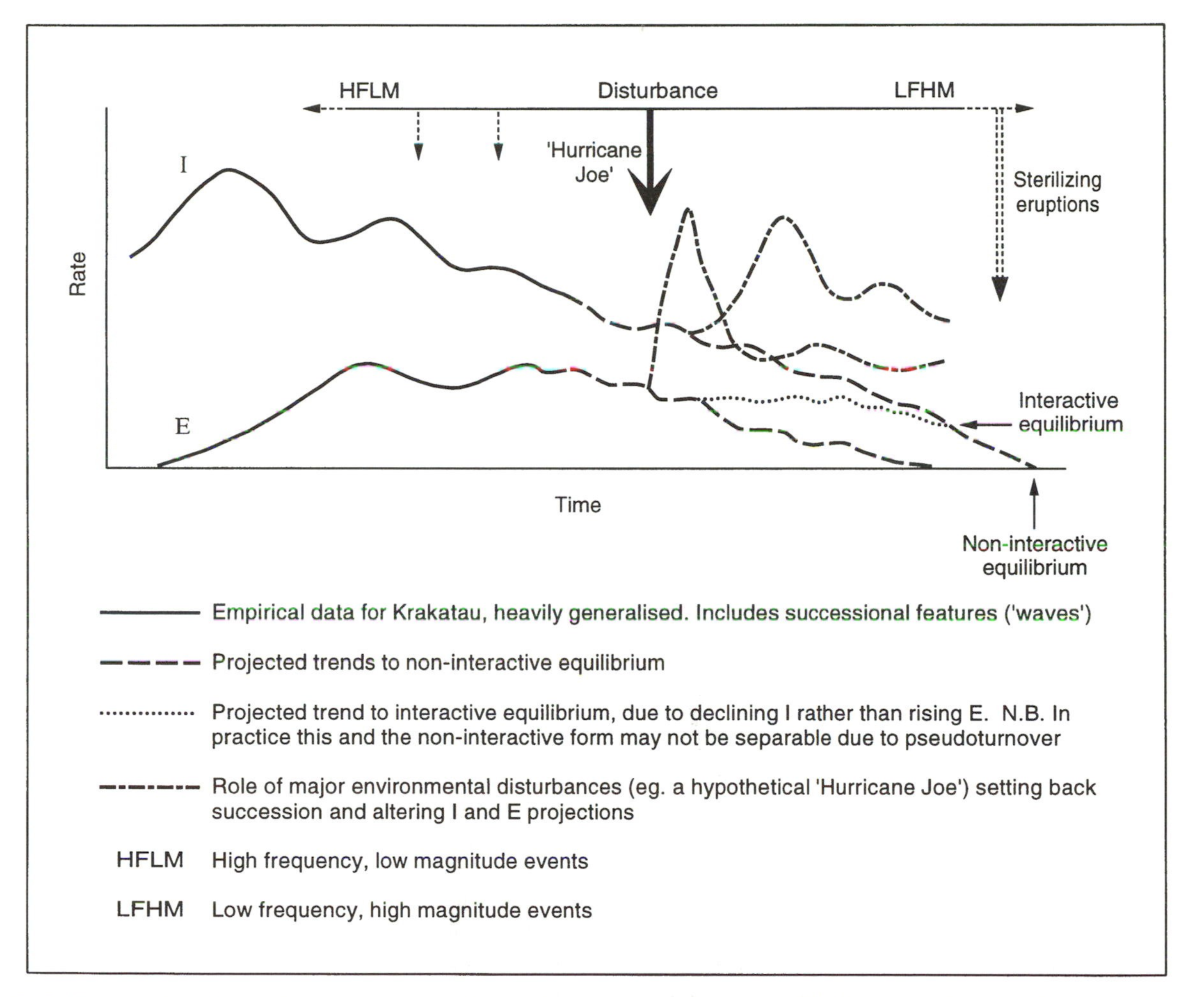

Fig. 8.13. A model of island immigration and extinction incorporating disturbance. The figure provides a highly generalized representation of data for the recolonization of Krakatau and three hypothetical projections. (From Bush and Whittaker 1993.)

and the high degree of pseudoturnover which is a feature of empirical data from such large and complex systems, the two conditions may not be readily distinguishable in practice. The third form of projection recognizes that the islands experience environmental change, which may be dramatic and destructive. The longer the period prior to attaining equilibrium, the greater the likelihood that an intermediate- or high-magnitude event will cast the system away from equilibrium and into the perpetual or dynamic non-equilibrium condition of Fig. 7.9 (section 7.4.7). In this projection, the biota chases perpetually moving environmental 'goal-posts'.

The relevance of the three projections in Fig. 8.13 to particular Krakatau taxa will be dependent on features of their ecology such as generation times and dispersal attributes, and their position within the biological and successional hierarchy (cf. Schoener 1986). Within the conceptual space of Fig. 7.9, different taxa or guilds (or the same taxa in different island contexts) may thus occupy different positions: the reptiles arguably towards the static non-equilibrium; the plants of Anak Krakatau towards the dynamic non-equilibrium; the coastal flora at the archipelago level towards the static equilibrial position; and the birds of Rakata, at least at times, relatively close to a dynamic equilibrium but with less homogeneous patterns of turnover than indicated for a pure ETIB position. The two diagrams are thus complementary, in showing the conceptual space (Fig. 7.9) and temporal features characteristic of different points within it (Fig. 8.13).

It might be argued that islands devastated by events such as volcanoes and hurricanes are extremes; Krakatau is just a special case. My response to this is that less dramatic environmental perturbates may also perpetuate non-equilibrium conditions. Several of the arguments I have given have also been developed by other authors working on a variety of systems. For example, Lynch and Johnson (1974), who may have coined the term 'successional turnover' in its island ecological sense, argued that where turnover is successional in nature, such as produced after a fire, then the turnover is determined not primarily by area–distance effects, but by the timing, extent, and nature of changes occurring in the habitats. They regard use of an equilibrium model in such situations as inappropriate. They also argue that interpretation of faunal change is especially difficult if both equilibrium and non-equilibrium (e.g. successional) turnover are involved, as applies above. I know of relatively few studies of turnover in relation to recurrent disturbance. Perhaps this is because highly disturbed islands do not fit within the paradigm and are rejected as objects for studies of island ecological turnover? A number of studies alluding to the significance of environmental fluctuation and disturbance were, however, given in section 7.4. The following provide a few more examples pointing to the possible wider relevance of disturbance.

During post-hurricane succession in the Caribbean, it has been found that the animal community may continue to change long after the initial 'greening-up' of the vegetation, through invasions and local extinctions as the vegetation varies through successional time (Waide 1991). Initially, nectarivorous and frugivorous birds suffer more after hurricanes than do populations of insectivorous or omnivorous species, a consequence of their differing degrees of dependence on re-establishment of normal physiognomic behaviour in the vegetation. In turn, animals can have key roles in plant succession, for example, as dispersers of seeds (Elmqvist *et al.* 1994).

Abbott and Grant (1976) found evidence of non-equilibrium in land-bird faunas on islands around Australia and New Zealand, a result that did not appear to be attributable to sampling error or to humans. They suggested in explanation that islands at high latitudes (such as these) are generally subject to irregular climatic fluctuations and may not therefore have fixed faunal equilibria (cf. Russell *et al.* 1995). Abbott and Grant (1976) argued that the extent to which climate fluctuates about a long-term average value determines the extent to which species number does the same. 'In this sense, species number "chases" and perhaps never reaches a periodically moving equilibrium value, hindered or helped by stochastic processes' (Abbott and Grant 1976, p. 525). Abbott and

Black (1980) studied turnover in vascular plant species on 76 aeolianite limestone islets. They noted a variation in turnover pattern between 1975 and 1977, and between 1977 and 1978, which they suggested may have been related to an increase in rainfall or to a cyclone in 1978. They also discuss studies of turnover from sand cays in the Caribbean and Great Barrier Reef region, pointing out that the dynamism of their floras may relate to the variation through time in island shape and area, and to disturbance: 'as most cays occur in tropical regions, they are subject to frequent cyclonic or hurricane disturbances which often cause waves temporarily to inundate cays and wash vegetation away' (p. 405) (see also Sauer 1969). In contrast, in their studies of more physically stable islands, in a temperate zone less prone to freak weather events and storm surges, turnover was of a very low order over the course of their short data series. Another small-island illustration of disturbance was provided by Buckley (1981, 1982) who studied plants on sand cays and shingle islands on the northern Great Barrier Reef. He drew similar conclusions to those of Abbott and Black (1980) and Bush and Whittaker (1993). Depending on the relative time-scales of changes in the island environment and response by the biota, the biotas might be: always in equilibrium; in equilibrium most of the time but occasionally disturbed from equilibrium; or never in equilibrium, but always lagging behind changes in environment.

Extreme climatic events, such as freezes, have been recognized as important determinants of island avifaunas. An illustration is provided by Paine's (1985) observations of the extinction of the winter wren (*Troglodytes troglodytes*) from Tatoosh Island, a 5–6 ha island, 0.7 km from the north-western tip of the Olympic Peninsula (Washington, USA). The loss of the species followed an extremely cold period in December 1978, and it took 6 years to re-establish a breeding population, which was presumed to be because of the difficulty of a relatively sedentary species invading across a water gap.

If environmental fluctuations can be important to some small to moderate-sized near-continental islands, why do we not hear more of the adverse effects of natural disturbance in remote oceanic islands? Might they not be anticipated to be worst affected, being unable to replenish species-stocks swiftly after such events as cyclones? Arguably, they are affected, but the effects of variability are largely 'built in' to their ecologies. Miskelly (1990) reports widespread reproductive failure and delayed breeding by Snares Island snipe (*Coenocorphya aucklandica*) and black tit (*Petroica macrocephala dannefaerdi*), two endemic land birds, on the Snares Islands, in the New Zealand subantarctic. Snipe also suffered unusually high adult mortality. He attributed these phenomena to the pronounced El Niño (ENSO) event of 1982–83, which produced heavy rainfall and significantly lowered temperatures on the Snares Islands, thereby reducing the invertebrate food supply. Neither bird species invested much energy in reproduction during the ENSO event compared with subsequent years. The poor breeding seasons in this study contrast dramatically with the greatly enhanced reproductive success of land birds on the Galápagos, which benefited from improved plant growth and seed production enabled by the high rainfall (Gibbs and Grant 1987). Very heavy rainfall stimulated the production of plants and therefore the caterpillars feeding on them, important to finches in the breeding season. A significant increase took place in the population size of Darwin's ground finch on Daphne Major Island. Gibbs and Grant (1987) concluded that rare events can have a major influence on key population processes, including processes of significance to evolutionary changes, in long-lived birds living in temporally varying environments (see also Grant 1986). These events did not involve species gains and losses, but warranted attention from ecologists due to the importance of the endemic species involved.

Whittaker (1995) argued that island endemics are unlikely to be lost very often in response to even large-scale disturbances, such as hurricanes. First, because species and ecosystems have evolved within the context of the disturbance regimes. Secondly, and linked to this, they tend to occur on larger oceanic islands, which may well provide refuges unaffected by the perturbation somewhere within their land area. By this line of thinking, small oceanic islands, such as cays and atolls, should support relatively impoverished systems with few endemics restricted to them.

Support for this general line of reasoning is provided by a study of the Laysan finch (*Telespiza cantans*), one of the endangered Hawaiian honeycreepers. It occurs on the small coral island of Laysan in the Hawaiian archipelago, although in prehistorical times it also occurred on Oahu and Molokai, and a small population has recently been introduced to Pearl and Hermes reef. Laysan Island is only 397 ha in area and of that only 47% is vegetated. Morin (1992) found that weather conditions played a crucial role in regulating reproductive success. In 1986, a severe storm caused almost total mortality of eggs and chicks regardless of clutch size. Later that year, the number of fledglings per nest increased as clutch size increased. The species is omnivorous and eats some part of almost every plant on the island, as well as invertebrates, carrion, and eggs. It also has life history parameters conditioned by long experience of the fluctuating conditions on the island. Reproduction can be spread over a long breeding season, with the potential for multiple broods, and the species has a long reproductive life span. Notwithstanding this flexibility of response, it was concluded that stochastic weather events and predation are the two key factors currently limiting the Laysan finch population, in contrast to most of the other honeycreepers of Hawaii, whose populations are limited primarily by past and present human activities.

The South Pacific islands of Samoa have two flying fox (fruit bat) species, *Pteropus samoensis* (endemic to Samoa and Fiji) and *Pteropus tonganus*. Pierson *et al.* (1996) discuss the response of each of them to two severe cyclonic storms, Ofa, which struck in 1990, and Val, in 1991. Flying foxes are important elements of the ecosystem, possibly warranting keystone status, because of their roles as seed dispersers and plant pollinators. Activity patterns and foraging behaviour were disrupted for both species. Prior to the hurricanes *P. tonganus* was much the commoner, with a population on the small island of Tutuila of about 12 000, contrasting with fewer than 700 *P. samoensis* individuals, and it was thought that the more restricted endemic might be the more vulnerable to storm effects. However, the endemic species was found to have survival strategies not observed in the more common and widely distributed *P. tonganus*, and the latter experienced the more severe declines, of the order of 66–99% per colony. The numbers of *P. samoensis* seemed to be relatively unaffected. Its survival strategies included making greater use of leaves in the aftermath of the storm and also feeding on the fleshy bracts of a storm-resistant native liane. *Pteropus tonganus*, on the other hand, seems to have adopted the unfortunate behavioural response of entering villages to feed on fallen fruit, where they were killed by domestic animals or people. Given that flying foxes can live for 20 years or so, individuals may experience several severe storms in a lifetime. Pierson *et al.* (1996) make the point that island species of the tropical cyclone belt have evolved with recurrent cyclones. It appears in this case that the species of narrowest geographical range actually has evolved more effective responses to cyclone events. Such events, of themselves, are unlikely to lead to extinction of the species, providing other threats, such as habitat loss and hunting, are constrained (where the two are combined, however, extinction becomes more likely (Christian *et al.* 1996)). Although wind damage is typically patchy, it seems reasonable to suggest that areas of high topographic complexity, e.g. volcanic cones and deep valleys, are the most likely areas to retain patches with some foliage, and so should be given priority in reserve design.

It is difficult from the brief selection of examples given here and in section 7.4 to draw conclusions as to the relative importance of environmental variability (inclusive of severe disturbance) in the ecologies of different categories of island. Conceivably, some island systems (large and remote) might respond to a severe El Niño event, for example, by exhibiting considerable flux in population sizes, biomass, and productivity, without much impact on species turnover rates. In other cases (small and near-shore islands), a severe cold snap in the winter might have little effect on biomass and productivity but might greatly impact on populations of small birds, leading to measurable alteration of species turnover rates. I am not arguing that all islands are subject to environmental variation and therefore all are out of equilibrium. Rather, I am suggesting that some islands will be out of equilibrium some of the time, and a number perhaps all of the time. When human impact is considered, it becomes even more difficult to find undisturbed islands, 'in practice it is next to impossible to find a set of islands without environmental change or evidence of the severe impact of man's activities' (Case and Cody 1987, p. 408). It is not enough to assume that a system is in equilibrium, any more than to assume that it is not. This should be a matter of empirical investigation, particularly if other theories are to be constructed on the basis of equilibrial assumptions (cf. Weins 1984). As shown by the studies of Long, Karkar, and Krakatau, where disturbances are severe, or environmental change is very considerable, there may be long-term ecological changes which unfold over the course of decades or centuries. In such cases, a successional model of island ecology can provide a more appropriate representation of turnover and compositional patterns than do simpler stochastic models based only on parameters such as area and isolation.

8.7. A general critique of island ecological theory

...the equilibrium model and its derivatives suffer from extreme oversimplification by treating islands as functional units with no attention to internal habitat diversity and by treating species as functional units with no allowance for genetic or geographical diversity. This is not even good as a first approximation, because it filters out the interpretable signal instead of the random noise. The authors are in such a hurry to abandon the particulars of natural history for universal generalization that they lose the grand theme of natural history, the shaping of organic diversity by environmental selection... (Sauer 1969, p. 590, on MacArthur and Wilson 1967).

Throughout this chapter and the previous one there have been certain continuing strands of debate: to what extent do equilibrium models apply; to what degree are species numbers or identities predictable; and what forces structure island biotas? Many of the problems with the essentially stochastic core model of MacArthur and Wilson have been discussed already. Table 8.4 summarizes some of the objections that have been raised. Others might be added, for instance, the predictive capacity of the ETIB is damaged by the indeterminate nature of the relationships between immigration and extinction rates, on the one hand, and physical factors such as area and isolation.

Yet there remain those who defend the theory and argue it 'holds up well'. One of these is Rosenzweig (1995). He argues that it is inaccessibility that defines 'islandness' in a biological sense, and that it follows from this that a biological definition of an island is preferable to physical definitions such as being surrounded by water. His definition is: 'An island is a self-contained region whose species originate entirely by immigration from outside the region' (p. 211). This is an odd definition as it could apply to a deglaciated landscape rather better than to some real islands which mix relictual lineages with those evolved *in situ*. The subtext of the definition clarifies its meaning, however. By self-contained, it is meant that each species is a **source species**, i.e. it has an average net rate of reproduction sufficient to maintain positive population. The alternative condition is that species are **sink species**, which are dependent on influx from outside to maintain their populations. In Rosenzweig's islands, there are no sink species. This is a form of slight of hand, a means of ring-fencing the ETIB, as the definition effectively removes isolates dominated by metapopulation phenomena (cf. the rescue effect) and evolution from the domain of ETIB. This is to define islandness by how well the

Table 8.4
Some of the limitations identified in MacArthur and Wilson's equilibrium theory (ETIB)

Author	Limitations of ETIB
Sauer (1969)	Theory ignores autoecology—but species are not interchangeable units (e.g. see Armstrong 1982 on the effects of rabbits introduced onto islands)
Lynch and Johnson (1974)	The data are rarely adequate for testing turnover
Hunt and Hunt (1974)	Turnover can be confounded by trophic-level effects
Simberloff (1976)	Re: mangrove data set, most turnover involves transients, i.e. is pseudoturnover
Gilbert (1980)	Turnover at equilibrium has not been demonstrated.
Williamson (1981, 1989*a*,*b*)	Immigration, extinction, and species pool are each poorly defined. ETIB is imprecise on reasons for extinctions, and most turnover is ecologically trivial (cf. Simberloff 1976)
Pregill and Olsen (1981)	Ignores historical data and role of environmental change
Haila (1990)	ETIB has a narrower domain than originally thought (may be operational on the population scale)
Bush and Whittaker (1991)	Ignores successional effects and pace, and the hierarchical links between taxa

ETIB applies to an island. I consider that this argument is valid, but prefer to state it the other way around: the equilibrium model only holds relevance to a particular subset of islands. This position is similar to that taken by Haila (1990) (and below).

Brown (1981) observed that the equilibrium theory has three characteristics which he claimed any successful general theory of diversity must possess. First, it is an equilibrium model, thus historical factors, climatic change, successional processes, and the like are acknowledged, but side-stepped. The theory explains rather the ultimate limits, the theoretical patterns of diversity. Secondly, it confronts the problem of diversity directly, number of species being the primary currency of the model. It also takes account both of biological processes and some, at least, of the characteristics of the environment. Thirdly, it is empirically operational, making robust, qualitative predictions that can be tested. He notes that the fact that it has been repeatedly and unequivocally falsified does not diminish the contribution it has made to advancing our understanding of diversity. He therefore characterizes it as a useful beginning:

> The model implies that the determination of diversity is a very stochastic process. Species continually colonize and go extinct, the biota at equilibrium is constantly changing species composition, and all of these processes occur essentially at random. MacArthur and Wilson knew that this was not so, but '... it was a useful simplifying assumption ...' (p. 883).

He then called for new thinking, to dig deeper and to seek more deterministic theories of species diversity and community organization: the subject matter of the present chapter.

There are two lines of defence for the equilibrium theory. The first is that the various criticisms of the theory are not fatal to it as a predictive tool, all that is required is modification. The second is that it has heuristic value even if not predictive value (in practice, of course, it turns out to be very difficult to test the predictions). By this second line of reasoning, critics of the theory are seen as carping over detail and are criticized for failing to see the tremendous contribution to ecology made by MacArthur and Wilson (1963, 1967). This may seem a perfectly reasonable stance to take, although relatively few critics of the model would deny this anyway, but what seems to have followed from the 'heuristic value' defence is that the theory has continued to be used in its narrower form as an assumption on which other ecological arguments are based, for instance in conservation biology. Simberloff (1976) pointed this out some 20 years ago when he described the theory as having achieved paradigm status. As argued above, it is not sufficient to assume ecological systems to be in equilibrium, this must be demonstrated (Weins 1984). In systems in which abiotic forcing dominates, equilibrial models do not provide adequate fits with the data. In those in which long-term ecological processes operate, again more complex models may be required. The mantra that only equilibrium models will bring salvation should finally be discarded.

One of the features imposed by the equilibrium paradigm has been the restriction of focus to species that are resident on islands. While this has not always been adhered to (e.g. Krakatau plant data lack such detail), it does result, for example, in seabirds and (often) migrant birds being excluded from consideration in island ecological studies. Such species may be of considerable significance to the ecology of an island and may act to introduce other fauna and flora. There is nothing new in this insight. In the nineteenth century, Charles Darwin undertook experiments that showed how land shells might reach an island carried on the feet of wildfowl (Burkhardt 1996).

An illustration of the problems concerning notions of residency comes from the fruit bats of Krakatau. The largest flying fox species is *Pteropus vampyrus*. It has been present intermittently over the past two decades as a colony of several hundred individuals. *Pteropus* have been observed to gather in large roosts on islands but fly to mainland sites in order to feed (and vice versa), and their nightly range can be as much as 70 km (Dammerman 1948; Tidemann *et al.* 1990; Whittaker and Jones 1994*b*). Within mainland areas *Pteropus* is also known to be highly mobile, exhibiting movements that may represent non-seasonal nomadism in response to variation in food availability, rather than regular seasonal migration. Krakatau may thus provide just one roost site of a number used by the same colony within a larger geographic area. Their intermittent use of it may reflect on variation in food supply in the region, human interference at other roost sites, and avoidance of periodic ash falls on Krakatau. These animals may act as important seed vectors for the introduction and/or spread of Krakatau plant species.

The large fruit-pigeon, *Ducula bicolor*, is another important species of frugivore, which for island ecological purposes is considered as a Krakatau resident (Thornton *et al.* 1993). It has been described as roaming in large flocks between offshore islands in the region (Dammerman, 1948). The two species of Krakatau *Ducula* (the other is *D. aena*) are believed to have played crucial roles in introducing zoochorous plants—particularly those with larger seeds—to Krakatau (Whittaker and Jones 1994*b*). The particular identities and timing of the trees introduced by such means may have enormous ramifications for the species of animals which may subsequently be supported within the islands. These species of frugivores have been highly significant to the structure inherent in the assembly process for Krakatau. Arguably, the behaviours of these species exhibit characteristics of connectedness to other locations which are more akin to metapopulation models (Chapter 9) than to the notions of residency demanded by equilibrium models. In short, for some animal taxa and some guilds, there are potentially complex hierarchical links between their turnover patterns and floral development and turnover. Other forms of hierarchical links may also be found, such as those linking predator–prey groups in an area. How can these different forms of island ecology be resolved within a single framework?

8.7.1. Scale and the dynamics of island biotas—Haila's critique

There are, as noted above, two ways of interpreting the equilibrium theory, a strict explanatory model that predicts the number of species in a given taxon on islands, or as a deductive scheme that points out various potentially important factors affecting insular communities. For those who recognize such distinctions, Haila (1990) regards this wider interpretation as making the theory a *research programme* rather than a *paradigm*. His conclusion is that as an explanatory model, it has a narrower 'domain of validity' than originally supposed, i.e. that in many situations the hypothesis is simply inappropriate, but that when the scale of island matches implicit assumptions of the hypothesis it retains value. This critique is a helpful one, in that it indicates the territory in which the theory might reasonably be tested, and the circumstances where other lines of theorization are needed.

Island ecological analyses have been undertaken for a wide variety of forms of insular systems. While the situations are typically described by similar variables and terms ('immigration' and 'extinction'), they are actually rather different, requiring differing frameworks of analysis. The solution to this problem, Haila (1990) argues, is to clarify what is mean by an 'island', relative to the ecological processes at work. The processes themselves form a continuum, with characteristic coupled space–time scales, as shown in Fig. 1.1. He distinguishes: (1) the individual scale; (2) the population scale 1—dynamics; (3) the population scale 2—differentiation; and (4) the evolutionary scale. Haila illustrates this framework with examples taken from northern European archipelagos, although his data provide only indirect evidence of some of the processes at work.

On the **individual scale**, a patch of land is an island if some crucial phase of the life cycle of individual organisms obligatorily takes places within its boundaries. Consider those birds which breed on an island but overwinter elsewhere, such that their life cycle is influenced by factors external to the island. Birds of prey, on the other hand, may include several small islands of an archipelago within their territory. In such circumstances individuals may be viewed as an integral part of the regional population, such that the insularity of the environment on this individual scale has few population- or community-level consequences, other than through being inferior or superior places for reproduction.

When islands are larger and more isolated, they may support populations that are dynamically independent of those on other land areas. At this point the relevance of the ETIB model may be apparent. To be an island on the **population dynamic scale**, two criteria are identified. First, the whole life cycle of the organisms concerned must be confined within the island and, secondly, the island population must be able to demonstrate independence from the mainland dynamics for several generations. For instance, there is no evidence that bird populations on single islands from the Åland archipelago are dynamically independent. Although about 40 km from the Swedish mainland and 70 km from mainland Finland, and up to some tens of square kilometres in area, island population dynamics commonly occur in parallel with those of the mainland. Only for a few sedentary bird species on the largest and most isolated Baltic island is there evidence of dynamic independence.

Should the criteria for independence expressed above be fulfilled for long periods, the island populations become increasingly independent as genetic units, and this allows for the second scale of population processes, **differentiation**. Exemplification of this comes from several studies, two cases being the various subspecies of wrens (*Troglodytes troglodytes*) on different North Atlantic archipelagos (Williamson 1981), and the studies of Berry (e.g. 1986) on the founder effect in small mammals on offshore islands around Britain. The relevance of the equilibrium theory would seem to be fairly slight in such circumstances. The natural progression in this scheme is to the **evolutionary scale**, by which Haila (1990) means processes leading to the divergence of species, i.e. taxonomic differentiation to a greater degree than indicated by the term 'subspecies' or 'variety' or by slight niche shifts. Empirical evidence to support the claim that speciation and extinction on islands constitute an equilibrium is flimsy (Chapter 7).

Of these four scales, the ETIB is restricted mainly to the *population dynamics* scale, in that the other scales do not appear to satisfy the criteria necessary for its operation. The scales of Haila's framework grade one to another, such that particular studies may be at the interface, and the applicability of the theory uncertain. Even within a single taxon, such as birds, or bats on the Krakatau islands, some members may effectively be island bound while others may have wider territories inclusive of the islands under study (above). It is not, then, that the theory holds for all or most islands within a certain space–time configuration, but that the effects of the processes it represents may be prominent within this conceptual territory, while elsewhere (in the remaining subject space) they are subsumed by other dominant processes. Haila's (1990) scale framework is predominantly a biological one. He makes reference to environmental dynamics, but not prominently. It will, however, be apparent that particular forms of environmental change and disturbance will also have characteristic scale patterns, which determine their relevance to interpretation of the differing scales of island ecological process.

8.7.2. Concluding remarks

The bulk of this chapter has been concerned with the identification and causes of structural features of island biotas, either as they vary in space or in time. Notwithstanding the debates concerning the statistical tests of such patterns, the evidence for non-random patterns from a variety of empirical studies is overwhelming (Schoener and Schoener 1983; Weiher and Keddy 1995). Competition is merely one of the forces involved, but a role it does have (e.g. Diamond and Gilpin 1982; Grant 1986), as also does predation (Schoener 1986). Progress in this area has been made. The tools of analysis of island assembly studies, such as Diamond's incidence functions, have been refined and added to, for example, by new indices for calculating nestedness. Yet, over the years, many more researchers have been involved in the species numbers games than have focused on species compositional structure. It appears to me that theories of island assembly hold more promise, are at least as interesting, and may be of more practical value to conservationists (Chapter 9; Worthen 1996) than, for example, are attempts to refine geometric models of the relationship between coastline configuration and immigration rate.

The rationale of studying islands as natural laboratories has often been cited. In the present book, the consideration of islands began with the more distant (evolutionary) islands, and has progressed through less distant, but still mostly real (ecological) islands. As we have seen in earlier chapters, scale factors are crucial to the answers to questions in island biogeography, particular effects finding expression on particular types of island but being subsumed by other processes on different types of island (more, or less isolated; larger, or smaller, etc.). The next stage is to examine how theories derived from these natural laboratories have been applied to some of the most pressing concerns of our time—the implications and management of the process of fragmentation and attrition of natural ecosystems within land masses. In doing so, the scale of isolation is often reduced further; there is therefore one further theoretical domain of island ecology that has been reserved for the following chapter, that of metapopulation theory.

8.8. Summary

Island biotas are not simply random draws from regional species pools. Instead, they typically exhibit compositional structure, that is to say that some species, species combinations, or species types, are found more frequently, and some less frequently, than might be expected by chance. A comprehensive island assembly theory was formulated by Jared Diamond, in his studies of birds on islands in the Bismarck and other groups around New Guinea. His analyses were founded on several forms of distributional patterns. Incidence functions describe the frequency of a species as a function of island species richness (or sometimes island size). Those species found only on the most species-rich islands, which were also the larger, land-bridge islands, were termed high-S species. Supertramps were, in contrast, found only on small, remote, species-poor islands, including islands recovering from past disturbance. Between these two extremes are the 'lesser tramps', found on varying numbers of the islands. Checkerboard distributions, were also observed, whereby taxonomically and ecologically related species have mutually exclusive but interdigitating distributions across a series of islands. Within particular guilds of birds, Diamond found evidence for compatibility rules, i.e. that a certain degree of niche difference is necessary to enable coexistence. After these various effects were considered, it was apparent that some combinations of species were more frequent and others less frequent than expected, suggesting combination rules. In total, seven distinct assembly rules were derived empirically on the basis primarily of the distributional data. The interpretations of these patterns rested on the assumption that through competition the component species are selected and co-adjusted in their niches and abundances, thereby 'fitting' with each other to form relatively species-stable communities.

The approach drew pointed criticism from Simberloff and colleagues, who in a twin-pronged attack questioned both the primacy given to competition and the extent to which the distributional patterns departed from a random expectation in the first place. On the first count, Diamond and Gilpin responded by emphasizing the extent to which factors other than competition (e.g. predation, dispersal, habitat controls, and chance) were in fact integral to the theory. On the second count, they contested the basis for the null models, demonstrating that the findings depend heavily on the biological assumptions that particular authors build in to their null models. The debate has continued and re-appeared in other guises. Competition is undoubtedly a force in shaping island biotas but it is extremely difficult to distinguish the precise extent of its effects. Subsequent studies using similar types of distributional 'tools' have demonstrated several common features, often finding a role for island area, but also involving a range of habitat determinants.

Nested distributions are where smaller faunas (or floras) constitute subsets of the species found in all richer systems. Statistical tests of the degree of nestedness have been developed and applied to many island data sets. Nestedness is a common feature of insular biotas, varying in importance depending on the nature of the islands and the taxon being considered. The extent to which nestedness may be attributable to selective immigration or selective extinction remains uncertain, but hopefully further research can clarify this in time.

Island recolonization constitutes in essence just a special case of ecological succession. This is illustrated by reference to the recolonization of Krakatau. In this natural experiment, dispersal attributes, hierarchical relationships across trophic levels, and habitat changes are each seen as significant in structuring island species assembly. Turnover from the species lists can be attributed largely to a combination of succession, habitat loss, in a few cases predation, and to the comings and goings of 'ephemerals', i.e. it is heterogeneous in nature. The fit with equilibrium theory is poor, and while some taxa or 'guilds' may have reached an asymptote, overall the system has yet to equilibrate, if indeed it is destined ever to do so.

In a reconsideration of disturbance in island theory it is argued that as ecological (especially successional) processes operate over such lengthy periods, and in the context of natural fluctuations in environment, many islands may, in practice, be out of equilibrium, thus exhibiting compositional characteristics that follow on from their particular pattern of directional dynamic. The chapter closes with a reassessment of island ecological theories, in which it is again emphasized that different theories find expression on different types of island. Haila's distinction between individual scale, population dynamics scale, population differentiation scale, and evolutionary scale being one useful approximation. It follows from these arguments that no single model or interpretation of island ecology holds primacy. The evidence for deterministic structure, and the interpretations offered for it, point to the relevance of a range of ecological forces which may be relevant to understand island ecologies. It cannot be assumed as an act of faith that a particular insular system is behaving according to equilibrium models. The relevance of this to conservation biology will be one of the themes explored in the next chapter.

9

Island theory and conservation

...the abundance of publications on the theory of island biogeography and applications [of it] has misled many scientists and conservationists into believing we now have a ready-to-wear guide to natural area preservation...

(Shafer 1990, p. 102.)

The classical paradigm in ecology, with its emphasis on the stable state, its suggestion of natural systems as closed and self-regulating, and its resonance with the nonscientific idea of the balance of nature, can no longer serve as an adequate foundation for conservation.

(Pickett *et al.* 1992, p. 84.)

The fragmented world of biological communities in the future will be so different from that of the past, that we must reformulate preservation strategies to forms that go beyond thinking only of preserving microcosms of the original community types.

(Kellman 1996, p. 115.)

9.1. Habitats as islands

Over the course of the Quaternary there have been repeated cycles of climatic changes. These changes have driven massive alterations in the distribution of species across the Earth's surface. Individual species have experienced alternating episodes of expansion and of contraction and fragmentation of ranges. Species have continually re-sorted themselves into differing combinations, thus assemblages of the past frequently differed radically from those found today (e.g. Bush 1994). There is nothing new about range alterations. Now, however, it is humans who are driving range shifts and extinctions. We have done this for a long period of our history, and we are doing so in an accelerating fashion and on a global scale (Saunders *et al.* 1991; Bush 1996). Humans have brought about extinctions through various means: hunting, mixing up species from different regions, and through habitat alteration and loss (Groombridge 1992).

The loss of species is something which I imagine that all natural scientists view with dismay—many with alarm—but there are a lot of uncertainties concerning the big picture. We are currently unable to say to within an order of magnitude how many species we share the planet with (Gaston 1991; Groombridge 1992). Most land plants and animals are believed to be rain-forest species, but we do not know how great an extent of tropical forests exists, nor with much precision how fast it is being destroyed on a global scale (Whitmore and Sayer 1992; Grainger 1993). We know there to be a crisis, we don't know its magnitude. Natural scientists have responded to this crisis by increased attention to conservation biology.

A full consideration of the theoretical and operational principles of 'nature' conservation and the long history of the conservation movement are beyond the scope of this book. The aim here is an apparently simple one, to examine the application of island ecological ideas to habitat islands. We are thus concerned with the short- and long-term implications of habitat destruction and fragmentation.

It is generally accepted amongst conservation biologists, that the ongoing fragmentation and reduction in area of natural habitats is causing species extinctions at local, regional, and global levels (Whitmore and Sayer 1992).

The remaining areas of more-or-less natural habitats are increasingly becoming mere pockets within a sea of altered habitat; in short, they are **habitat islands**. It was therefore to island biogeography that many scientists turned in the search for predictive models of the implications of fragmentation and for guidance as to how best to safeguard diversity. The implications of increasing insularity for conservation have long been recognized, and indeed Preston drew attention to this issue in his seminal paper (Preston 1962). He observed that in the long run species would be lost from wildlife preserves, for the reason that they constitute reduced areas in isolation.

The most basic question is probably that of how many individuals are enough to ensure the survival of the last population of a species, i.e. what is the minimum viable population? We will consider this first. Many species of conservation concern are not, however, restricted to a single locality, or at least, not yet. Rather, their survival in a region might be dependent on a network of habitat islands. Most of this book has been concerned with real islands in the sea. Is it realistic to expect habitat islands within continents to behave according to the same principles as real islands? Instead of the barrier of salt water, a forest habitat island might be separated from another patch of forest by a landscape of meadows, hedgerows, and arable land. The implications for species movements between patches might be radically different. As we have seen, however, this is probably too simple a distinction and too simple a question. Scale of isolation is crucial as well as the nature of the intervening landscape filter. Mountain habitat islands within continents, but surrounded by extensive arid areas, may be much more effectively isolated than real islands that happen to be located within a few hundred metres of a mainland. Many isolates are actually sufficiently close to one another that their populations are in effect linked: they form metapopulations, the second theoretical element that we will consider. When we have added these tools to our armoury, we can then consider the implications of fragmentation and how 'island' approaches have contributed to their understanding.

While few, if any, reliable conservation principles have been derived uniquely from island theories, it is, I believe, important to understand their role in the development of conservation thinking. Over the past three decades there has been a shift in ecology from what Pickett *et al.* (1992) term the classical paradigm (**the balance of nature**) which, as the above quote summarizes, sees the world as naturally tending towards balance, a tendency driven by the actions of the biota. The framework of ideas which Pickett *et al.* (1992) believe provides a more realistic basis for conservation planning and management may be labelled the **flux of nature** paradigm. This framework does not deny the possibility of equilibrium, but recognizes equilibrium as merely one special state. It stresses process and context, the role of episodic events, and the openness of ecological systems. As we will discover, these themes are relevant to habitat island biogeography and form important features of the emerging applied picture. Intriguingly, MacArthur and Wilson's ETIB has, in this broader theoretical context, been variously seen as an equilibrium and a non-equilibrium theory, the latter because of its stress on islands as open systems exchanging propagules. Traditionally, however, the 'island theory' input to conservation debates has been based on the general assumption that a 'natural' area is in a state of balance, or equilibrium. Humans then intervene to remove a large extent of the area, fragmenting the remaining parts, which are thus cast out of biotic equilibrium. Subsequently, their newly acquired insular properties dictate that species will be lost from the fragments as a new biotic equilibrium is sought. This means of conceptualizing the problem is, in my view, entirely consonant with the classical paradigm. The thesis offered in this book is that much island thinking has underestimated the role of physical environmental factors, and that ecological responses to environmental forcing factors are often played out over too long a time frame for equilibrium to be reached (Chapter 8). Consequently, some of the theoretical island ecological effects are, in practice, fairly weak. These points are addressed in the present chapter in the applied, conservation setting. Hence we will consider the physical effects of fragmentation as well as the biotic, and we will examine what sorts of factors in practice cause species to be lost from fragments.

9.2. Minimum viable populations and minimum viable areas

9.2.1. How many individuals are needed?

What is the **minimum viable population** (MVP)? By this we mean the minimum size that will ensure the survival of that population unit, not just in the short term, but in the long term. It is often defined more formally in terms such as the population size that provides 95% probability of persistence for 100 or for 1000 years.

Population sizes may be viewed as undergoing stochastic fluctuations, such that very small populations may disappear altogether from an area 'by chance'. But, there is more to it than this of course. Small populations isolated within a nature reserve may lose genetic variability as they pass through 'bottlenecks' (Chapter 5). They may then lack the genetic flexibility to cope with the normal fluctuations of environment, in short they have then lost **fitness**. It is generally held as axiomatic that an increase in **inbreeding** in small populations reduces fitness in animals, although unambiguous demonstrations of the effect in natural populations are relatively scarce (Madsen *et al.* 1996). The basic rule of conservation genetics is that the maximum tolerable rate of inbreeding is 1% per generation. This translates into an effective population size of 50 to ensure short-term fitness according to the calculations of I. A. Franklin in 1980 (reported in Shafer 1990). However, there will still be a loss of genetic variation at such a population size, and Franklin recommended 500 individuals in order to balance the loss of variation by gains through mutation. As the figure of 50 came from animal breeders, rather than studies of endangered species in the wild, and that of 500 from a study of bristle number in *Drosophila*, it can be appreciated that these recommendations, despite being widely cited, were really rather approximate rules of thumb (Fiedler and Jain 1992). They provide simply a rough guide to the sort of numbers of individuals that might be needed to maintain a population in the long term. Why is there such a degree of uncertainty as to the numbers required for survival?

The loss of genetic variation carries the danger that the remaining individuals, even if they do not accumulate deleterious genes, may lack the flexibility to respond successfully in altered circumstances, such as might be brought about by climatic change, or interaction with a new pathogen or competitor. In such circumstances, the population may crash to extinction. Island evolution from tiny founder populations illustrates that this is not inevitable if numbers expand after a bottleneck. But intuitively we expect that the loss of a broad genetic base is likely to increase the probability of extinction. We learnt in section 5.1 that putting a population through a bottleneck of just a few individuals may not be as damaging to the genetic base as one might anticipate, providing the population is allowed to expand again fairly quickly. In illustration, the great Indian rhino was once very common, but within the Chitwan National Park in Nepal (one of its two main havens) the species was reduced to an effective population of only 21–28 animals in the 1960s. In this case the bottleneck was short. Rhinos have a long generation time, and it has since recovered to about 400 individuals. Genetic variation within the population is still remarkably high (Tudge 1991). However, the *Drosophila* studies of Hampton Carson and colleagues showed us that sustaining the bottleneck, or repeating it at short intervals, may have the effect of reducing the base, i.e. increasing homozygosity within the genome. The implications are that you may be able to rescue some, perhaps many, species from a bottleneck of short duration, but that you cannot assume that this will apply to a sustained bottleneck. Just because a suite of threatened species are hanging on in their reserves now, it doesn't mean that the same level of biodiversity can be sustained for a long time in a fragmented landscape.

A further complication in assessing genetic effects is that where a species is split into numerous separate populations in fragmented habitats, there may be multiple bottlenecks involved. This may actually result in increased genetic differentiation *between* populations, even though it may bring reduced variation *within* populations (see Leberg 1991). Such effects may be of significance to lengthening persistence in metapopulation scenarios (below).

In many species, only a proportion of the *adult* population participates in breeding; these animals form the **effective population size** (Shafer 1990). A study of grizzly bears in the Yellowstone National Park showed that the 1% inbreeding rule was actually broken not at an overall population size of 50 but at 220 individuals and below (Shafer 1990). Lacy (1992) has observed that crudely estimated MVPs for mammals based simply on data for inbreeding effects from captive populations actually vary over several orders of magnitude, and that, as yet, it is too early to generalize on MVPs in the wild. Instead of having a universal specific number of individuals, the MVP should, ideally, be worked out by study of each particular threatened species. Unfortunately, the resources are rarely available.

Pimm *et al.* (1988) have undertaken an analysis of island turnover which illustrates how some other autecological characteristics may be important to determining minimum population sizes. They analysed 355 populations belonging to 100 species of British land birds on 16 islands. They found that, as theory demands, the risk of extinction decreases sharply with increasing average population size. They examined extinction risk for large-bodied species, which tend to have long lifetimes but low rates of increase, and for small-bodied species. They found that at population sizes of seven pairs or below, smaller-bodied species are more liable to extinction than larger-bodied species, but that at larger population sizes, the reverse is true. This is an apparently odd result, but one which they reasoned should be expected on theoretical grounds. Why should this be so? Imagine that both a large-bodied and a small-bodied species are represented on an island by a single individual. Both are doomed to die, but the larger-bodied species is liable to live longer and so, on average, such species will have lower extinction rates per unit time. On the other hand, if the starting population is large, but it is then subject to heavy losses, the small-bodied species may climb more rapidly back to higher numbers, whereas the larger-bodied species might remain longer at low population size, and thus be vulnerable to a follow-up event. They also found that migrant species are at slightly greater risk of extinction than resident species. Numerous factors might be involved in migrant losses, such as events taking place during their migration or in their alternative seasons' range (e.g. Russell *et al.* 1994). Further studies will be necessary to establish the generality of this particular finding (see criticisms of Pimm *et al.* (1988) by Tracy and George (1992) and by Haila and Hanski (1993), refutation of criticisms by Diamond and Pimm (1993), and further comment by Rosenzweig and Clark (1994)).

9.2.2. Minimum viable populations and disturbance

Isolates subject to significant environmental change or disturbance may need to have much larger populations than is otherwise the case to ensure survival. A number of studies have recognized the *potential* of environmental catastrophes in this context but, typically, they have not attempted to evaluate the general significance of such catastrophes (see Pimm *et al.* 1988; Williamson 1989*b*; Menges 1992; Korn 1994). Mangel and Tier (1994) argue that environmental catastrophes may often be more important in determining persistence times of small populations than any other factor usually considered, and therefore should be considered explicitly when formulating conservation measures. They go on to provide computational methods for modelling persistence times which do take account of catastrophes and which allow for quite complicated population dynamics.

Perhaps their most important conclusion is that minimum viable populations are larger than those derived from variants of the MacArthur and Wilson model, or indeed other analyses of population viability which ignore catastrophes (Fig. 9.1) (see also Ludwig 1996).

In 1989, just such a catastrophe hit Puerto Rico, in the form of Hurricane Hugo, which caused extensive damage to the forests of the Luquillo mountains (Fig. 9.2). The single remaining wild population of the Puerto Rican parrot (*Amazona vittata*) lives in the Luquillo forest. In 1975 the parrot population had reached a low of 13 wild birds, due to a combination of habitat destruction, capture for pets, and slaughter (for food and to protect crops). Given protection, the population built up from this low point, to pre-hurricane levels of 45–47, from which it was reduced by the hurricane to about 25 birds (Wilson *et al.* 1994). Breeding

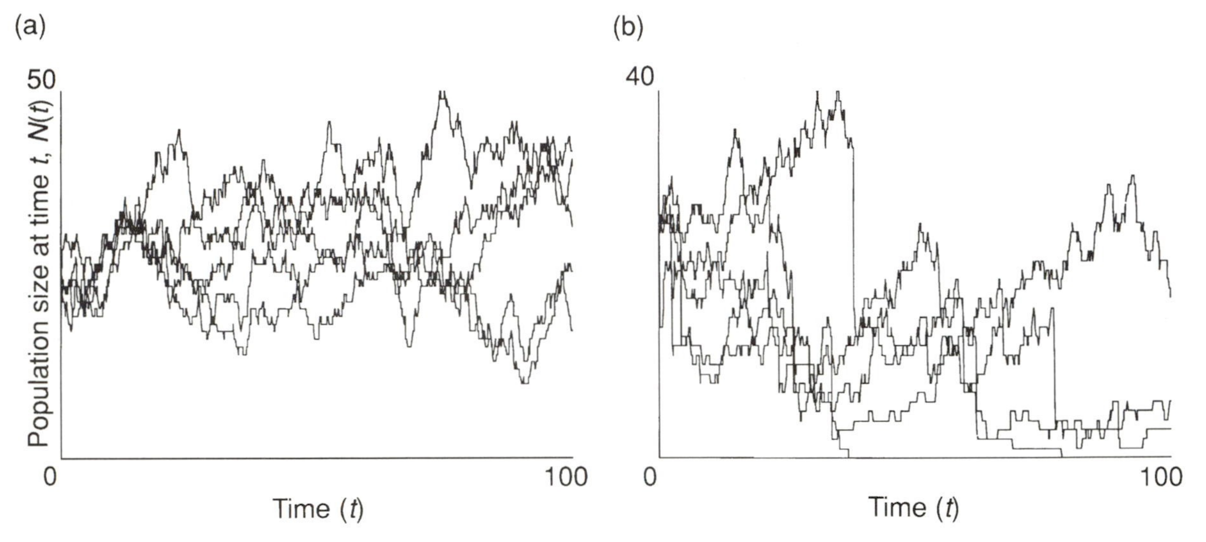

Fig. 9.1. Simulation runs of population trajectories. (a) In the absence of catastrophes all five trajectories persist for the 100 time units shown. (b) In the presence of catastrophes two populations become extinct and two are at low population sizes after 100 time units. (From Mangel and Tier 1994, Fig. 5; original figure supplied by M. Mangel.)

Fig. 9.2. The forests of the Luquillo Mountains in Puerto Rico were seriously disturbed by Hurricane Hugo in September 1989. Although regrowth is rapid—this photograph was taken in June 1993—the impact on the forest composition and dynamics will be evident for decades to come. The single wild population of the endemic Puerto Rican parrot, *Amazona vittata*, was reduced from about 45 to 25 individuals by the hurricane. (Photo: RJW 1993.)

resumed in 1990, with some success in fact. Six pairs bred in 1991, the highest total since the 1950s and 11 young fledged the following year. The post-breeding population went up to between 34 and 37 parrots. While this indicates that the risk attributable to a major storm could be overestimated, the number of birds is presently far too small for comfort and with this reduction to a single, localized population, a natural disturbance phenomenon of this sort has the potential of placing it beyond the brink. As endemic species tend not to occur on the smallest of islands, extinctions in the global sense caused primarily by natural disturbances are likely to be rare events on ecological timescales. One possible example of a species finished off in this way, by hurricanes in 1899, is the St Kitts subspecies of bullfinch *Loxigilla portoricensis grandis*, although the last record of this bird was actually several years earlier, in 1880 (Williamson 1989*b*).

The relevance of environmental disturbance or stochasticity to survival in habitat islands is not amenable to easy generalization, but in particular cases, e.g. in fire-prone regions, it may be crucial. Attempts to calculate the viability of single populations, i.e. whether the population exceeds the minimum requirement, are termed **population viability analyses**. Formal models have a role in such analysis, but as Bibby (1995) remarks, 'how do you estimate the probability of a cat being landed on a particular small island in the next five years?' Management regimes thus have to take account of ecological and life-history data, and managers must monitor the performance of a population in the wild, remaining adaptive to changing knowledge and circumstances.

9.2.3. How big an area?

How does a figure for a minimum viable population translate to a **minimum viable area** (MVA)? With some species and communities, e.g. butterfly populations, a fairly small area may suffice to maintain the requisite number of individuals. In general, the higher up the trophic chain, the larger the area needed (for an exception to this generalization see Mawdsley *et al.* 1998). Simply considering the home range of a few vertebrates is illustrative. It has been calculated that a single pair of ivory-billed woodpeckers (*Campephilus principalis*), may require 6.5–7.6 km^2 of appropriate forest habitat; that the European goshawk (*Accipiter gentilis*) has a home range of about 30–50 km^2; and that male mountain lions (*Felis concolor*) in the western United States may have home ranges in excess of 400 km^2 (Wilcove *et al.* 1986). For some species, reserves must be really rather large if their purpose is to maintain a MVP entirely within their bounds.

There is of course, one very large proviso attached to the calculation of a minimum critical area. The approach assumes that the area concerned acts rather like a large enclosed paddock in a zoo, with freedom of association within it, but no exchange with any other paddocks or zoos. Unless the reserve does indeed contain the only population of the species, or they are completely immobile creatures, there is always the possibility of immigration of other individuals from outside the reserve. This leads us back to earlier island ecological themes. The compensatory effects discussed in section 8.2.7 (Lomolino 1986) suggest that even for certain seemingly very isolated mountain ranges, some migration of the mammalian species involved might occur between isolates from time to time. Where a significant degree of exchange takes place, the concept of a fixed critical minimum area loses its meaning. In practice, this must be assessed on a taxon by taxon, and probably species by species, basis.

9.2.4. Applications of incidence functions

Area requirements of particular species can be examined by means of incidence functions (section 8.2.2). In this applied biogeographical setting, incidence functions are often calculated as a function of area, or even of isolation, rather than of species richness (Wilcove *et al.* 1986). The information returned from area-based incidence functions is not equivalent to estimating minimum viable areas. This is because the incidence functions constitute the product of population exchange between the network of suitable habitats within a region. They therefore tell you the properties of 'islands' on which a target species currently occurs, not those on which it may persist in the long term, or in an altered biogeographical context.

Simberloff and Levin (1985) examined the incidence of indigenous forest-dwelling birds of a series of New Zealand islands, and of passerines of the Cyclades Archipelago (Aegean Sea). In both systems they found that most species are found remarkably predictably, with each species occupying all those and only those islands larger than some species-specific minimum area. A minority of species in each avifauna did not conform to this pattern, possibly because of habitat differences among islands and because of anthropogeneous extinctions. While stochastic turnover might be occurring, the structure in the data indicates that the habitat and human influences dominate. These findings are consistent with the position expressed by Lack (1976) that most absences of birds from islands within their geographic range can be explained by unfulfilled habitat requirements.

Hinsley *et al.* (1994) quantified the incidence functions of 31 bird species in 151 woods of 0.02–30 ha in a lowland arable landscape in eastern England over three consecutive years (Fig. 9.3). For many woodland species the

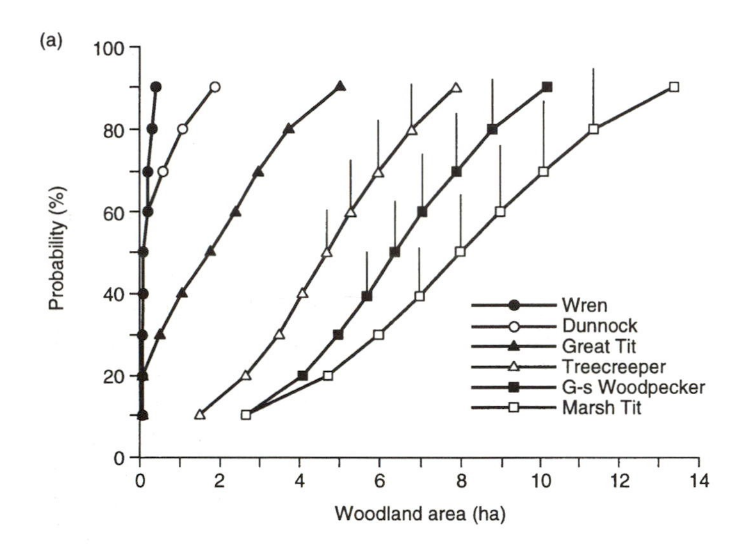

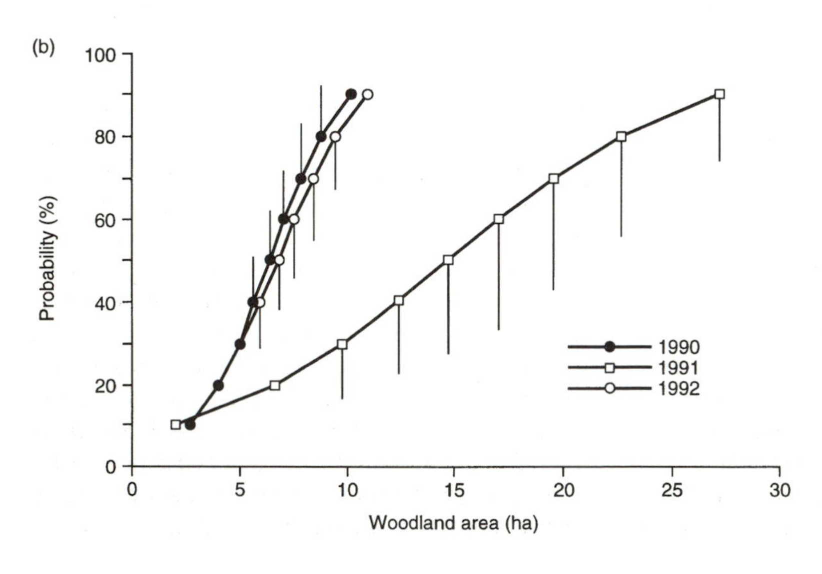

Fig. 9.3. Incidence functions for the probability of breeding as a function of woodland area, based on 151 woods of 0.02–30 ha area, in a lowland arable landscape in eastern England. (a) Data representative of widespread and common woodland species (wren, dunnock, great tit) compared with that for more specialist woodland species (treecreeper, great spotted (G-s) woodpecker, marsh tit). All relationships are for 1990, except that for the marsh tit, which is for 1991. (b) Interannual variation in the incidence functions for the great spotted woodpecker, reflecting a period of severe weather in February 1991. Small woods were reoccupied in 1992. Vertical bars represent 1 SE; those below 10% are not shown. (Redrawn from Hinsley *et al.* 1994.)

probability of breeding was positively related to wood-land area, although some breeding did occur even in the smallest of the woods. Only the marsh tit (*Parus palustris*), nightingale (*Lusciniea megarhynchos*), and chiffchaff (*Phylloscopus collybita*) failed to breed in woods of area less than 0.5 ha in any year of the study, and each was uncommon in the area. Some species were less likely to breed in smallish woods than others. Small woods appear to be poor habitat for specialist woodland species, but are preferred by others. Specialist woodland species were more likely to disappear from small woods after severe winter weather than were generalists, and could take in excess of a year to recolonize. Variables describing the landscape around the woods were important in relation to woodland use by both woodland and open country species. For instance, long-tailed tits (*Aegithalos caudatus*) were found to favour sites with lots of hedgerows around them, while yellowhammers (*Emberiza citrinella*) preferred more open habitats, small woods, and scrub.

In conclusion, incidence functions can be useful tools, particularly if repeat surveys are available to establish temporal variability in their form, but they do not of themselves provide sufficient data on which to base conservation policies. Notions of threshold population sizes and area requirements, as ever in island ecology, must be offset against dispersal efficacy and habitat requirements.

9.3. Metapopulation dynamics

In practice, most species are patchily distributed and are best regarded as a population of subpopulations. If such geographically separated groups are interconnected by patterns of gene flow, extinction, and recolonization, this constitutes what is termed a **metapopulation**. The first metapopulation models were constructed by Richard Levins in papers published in 1969 and 1970 (Gotelli 1991). The basic idea can be understood as follows. Imagine that you have a collection of populations, each existing on patches of suitable habitat. Each patch is separated from other nearby habitat patches by unsuitable terrain. While these separate populations each have their own fairly independent dynamics, as soon as one crashes to a low level, or indeed disappears, that patch will provide relatively uncontested space for individuals from one of the nearby patches, which will soon colonize. Thus, according to Harrison *et al.* (1988), within a metapopulation, member populations may change in size independently but their probabilities of existing at a given time are not independent of one another, being linked by mutual recolonization following periodic extinctions, on time scales of the order of 10–100 generations.

Studies of the checkerspot butterfly (*Euphydryas editha bayensis*) in the Jasper Ridge Preserve (USA) provide one of the better empirical illustrations of metapopulation dynamics (Harrison *et al.* 1988). The butterfly is dependent on food plants of serpentinite grasslands. The study area is of 15 × 30 km and includes one large patch (2000 ha) and 60 small patches of suitable habitat. The large 'mainland' patch supports hundreds of thousands of adults and is effectively a permanent population. The smaller populations are subject to extinctions, principally due to fluctuations in weather, but patch occupancy may also be influenced by habitat quality. A severe drought in 1975–1977 is known to have caused extinctions from three of the patches, including the second largest patch, and so was assumed to have eliminated all but the largest population. Unfortunately, only partial survey data were available from the period and this assumption cannot be tested. By 1987, eight patches had been recolonized. Small patches over 4.5 km from the 'mainland' patch were found to be unoccupied. By consideration of an index of habitat quality, Harrison *et al.* (1988) showed that the distribution of populations described an apparent 'threshold' relationship to both habitat and distance. Patches had to be both good enough and near enough to the 'mainland' in order to be inhabited at that time. A similar relation, from a separate study, is shown in Fig. 9.4.

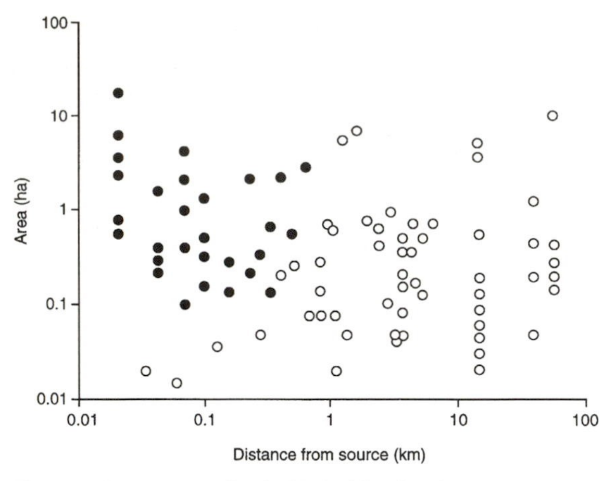

Fig. 9.4. Occupancy of suitable habitat by the silver-studded blue butterfly (*Plebejus argus*) in North Wales in 1990. Most patches larger than 0.1 ha were occupied (filled circle), provided that they were within about 600 m of another occupied patch. Beyond this distance, no patches were occupied (open circle), regardless of patch size. (Redrawn from Thomas and Harrison 1992.)

From models of the dispersal behaviour of the populations, assuming a post-1977 recolonization of the patches, it was predicted that more patches would become occupied in time. How many depends on the assumptions of the model. They offered two differing scenarios. The first assumed that colonization continues until reversed by the next severe drought, an event with an approximately 50-year periodicity in the study area. The second assumed that some extinction events occur between such severe events, thereby requiring a continuous extinction model and producing an equilibrium number of populations, with some turnover. The authors concluded that there was evidence for both modes of extinction and also for variations in dispersal rates over time in the checkerspot butterfly data. We may note that the fit with equilibrium theory is thus uncertain. On the other hand, their data show a degree of independence of population dynamics between the isolates, combined with a demonstration of recolonization (in at least three cases), thereby fulfilling requirements of metapopulation theory.

How does this view of extinction and recolonization differ from that of the ETIB? In the latter, the emphasis is on species number as a dynamic function of area and isolation. Whereas in the checkerspot butterfly story the focus is on the status of populations of a single species, with extinction seen principally as a function of alterations in the carrying capacity of the system. It is consistent with the metapopulation approach—although not necessarily with all metapopulation *models*—that rates of movement and of population gains and crashes may vary greatly over time. As Haila (1990) suggested, within a metapopulation system, the dynamics of the islands are to some extent interdependent. If the patches were much more isolated, their population fluctuations would be entirely independent, and one of the alternative theoretical models would be required. If, on the other hand, a more dispersive species was being considered, the entire system might constitute a single functional population, albeit one spread across a fragmented habitat. Thus, metapopulation models form a bridge between the study of population ecology and island theories of the ETIB form (Gotelli 1991).

Relaxing the single population assumption of the MVP model to allow for such ebb and flow between local populations will lengthen projected persistence times. Hanski *et al.* (1996) coin the term **minimum viable metapopulation** (MVM) for the notion that long-term persistence may require a minimum number of interacting local populations—and that this may depend on there being a **minimum amount of suitable habitat** (MASH) available.

Conservationists have suggested the creation of metapopulations for endangered species as a means of maintaining populations across areas of increasingly fragmented habitat. The most celebrated example of this strategy is that of the Interagency Spotted Owl Scientific Committee. They proposed the creation and maintenance of large areas of suitable forest habitat in proximity, so that losses from forest remnants will be infrequent enough and dispersal between them likely enough that northern spotted owl (*Strix occidentalis caurina*) numbers will be stable on the entire archipelago of forest fragments. Much of the work on metapopulations to date has been theoretical (Hess 1996) and Doak and

Mills (1994) have expressed some concern that despite intensive study of the owl, the metapopulation models have to assume a great deal about owl biology. The safety of the strategy cannot be demonstrated simply by running the models and so, in effect, the experiment is being conducted on the endangered species. Criticisms of the strategy led to a US district judge ruling against it on the grounds that the plan carried unacknowledged risks to the owl (Harrison 1994).

Similar concerns were expressed by Wilson *et al.* (1994) concerning plans to subdivide the small captive flock of the Puerto Rican parrot into three groups following the principle that metapopulations ensure better prospects of survival. The captive flock, of about 58 birds, is maintained within the same forest area as the wild population. In this case, the suggestion was that on subdividing the captive flock, one of these populations should be transferred to a multi-species facility in the continental United States, with individuals subsequently exchanged between the populations. Wilson and colleagues pointed out the danger of the accidental spread of disease, picked up in this multi-species context, back to Puerto Rico. They also pointed out that mate selection is idiosyncratic in this, as in some other '*K*-selected' species, and that subdividing the population, while still at a small size, might be counter-productive for this reason also. The recent resilience shown by the wild population to the devastation of Hurricane Hugo (above), notwithstanding the losses incurred, suggests that *in situ* conservation may be the best bet.

As with the other models considered in this chapter, it is important to understand that metapopulation models represent simplified abstractions, allowing 'what-if' scenarios to be explored (Harrison 1994). It is therefore dangerous to assign primacy to any single model in determining conservation policy.

9.3.1. The core–sink model variant

In a recent review, Dawson (1994) has noted two features which are not dealt with adequately by the majority of metapopulation models, although on first principles one might assume them to be rather common situations. First, in practice, if one patch is big, and the others small, then they may have a **core–satellite** or **source–sink relation**, in which the core is essentially immortal (Fig. 9.5). This behaviour is not consistent with the original models of metapopulations, which mostly assume equivalence in size and quality of patches. It will be

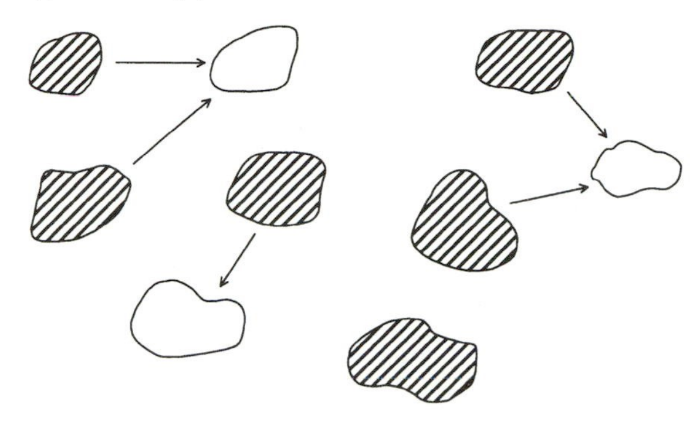

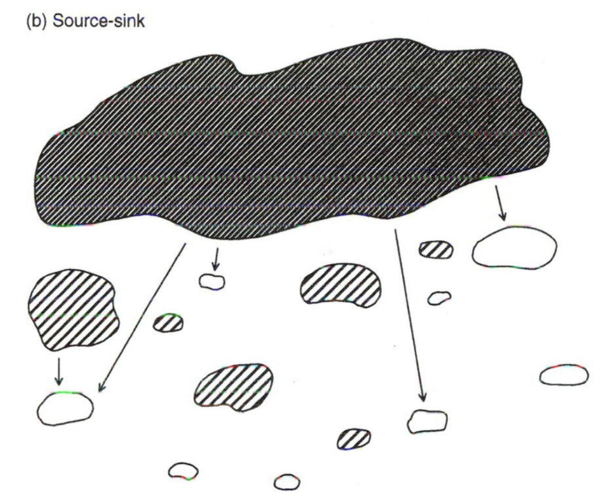

Fig. 9.5. Metapopulation scenarios. (a) In classic models, habitat patches are viewed as similar in size and nature: the occupied patches (hatched) will re-supply patches which lose their populations (unshaded), as indicated by the arrows, and at a later point in time will, in turn, lose their population and be re-supplied. (b) Much more commonly, large, effectively immortal patches (source or 'mainland' habitat islands) (dense shading) re-supply smaller satellite or sink-habitat patches after density-independent population crashes (e.g. related to adverse weather): close and larger patches may be anticipated to be swiftly re-occupied (hatched) but more distant patches take longer to be re-occupied, especially if small (unshaded).

noted that the checkerspot butterfly example from Jasper Ridge fits this core–satellite description. So, too, does Schoener and Spiller's (1987) study of orb spiders on Bahamian islands. Larger populations of spiders were found to persist, whereas small populations repeatedly re-immigrate from the larger areas and then become extinct again. If smaller fragments around a large one act not as sources of mutual re-supply, but effectively as population sinks for excess individuals from the large (core) island, then they may be of little relevance to calculations of persistence of the whole metapopulation (Table 9.1; Kindvall and Ahlén 1992; Harrison 1994). Dawson (1994) concluded that it is usually survival on the large habitat fragments that ensures species survival, not an equilibrium turnover on many smaller patches (but, see Thomas *et al.* 1996 for an exception).

Secondly, even seemingly isolated populations may be maintained in effect by immigration. The populations may indeed be interacting, but not by local extinction followed by recolonization so much as by continued supplementation. In other words, the 'rescue effect' may operate. Gotelli (1991) has shown that building this effect into metapopulation scenarios has a highly significant impact on the outcome, suggesting that the original models may commonly be misleading.

Most metapopulation models assume no distance effects (Dawson 1994; Fahrig and Merriam 1994). In practice, dispersal abilities vary from species to species, so that in turn the applicability of a metapopulation model to a particular system of isolates may also be species specific. The dispersal capabilities of species are often of crucial concern. They are crucial to classical island ecological theory, crucial to metapopulation theory, and crucial even if you discount each of these approaches. It is easy to underestimate the power of natural dispersal to re-supply isolated sites because the process is so difficult to observe. The regular supply of migrant moth species to the remote Norfolk Island, 676 km from the nearest land mass, illustrates the power of long-distance dispersal (Holloway, 1996). A survey over a 12-year period suggested that the resident Macrolepidoptera total 56 species, and migrants as many as 38 species. In addition, some of the resident species are also reinforced by the occasional arrival of extrinsic individuals. Quite remarkable. The recolonization of the Krakatau islands so swiftly, and by so many species of plants and animals provides another form of illustration. There are many such demonstrations to be found in the biogeographical literature, but the Norfolk Island data are, in my view, particularly valuable, because quantifying the extreme, the tail of the distribution, is generally difficult. Despite my enthusiasm for dispersal, it is important to recognize that some species are poor dispersers. For them to survive in fragmented habitats may require active intervention by people, to enable dispersal of viable propagules between isolates (Primack and Miao 1992; Whittaker and Jones 1994*a*,*b*).

Table 9.1

Problems that have been identified with classic metapopulation models

1.	Occurrence of core–satellite relations
2.	The rescue effect is not dealt with adequately
3.	Distance effects and varying dispersal abilities
4.	The occurrence of 'sink' habitats
5.	Metapopulations are very difficult to replicate, tests are problematic
6.	Much extinction is deterministic

The importance of continual immigration into so-called 'sink habitats' is such that in some populations the majority of individuals occur in sink habitats (Pulliam 1996). Furthermore, a habitat may contain a fairly high density of individuals but be incapable of sustaining the population in the absence of immigration. Density can, therefore, be a poor indicator of habitat quality. Thus, while some suitable habitats will be unoccupied because of their isolation from source populations, other 'unsuitable' habitats will be occupied because of their proximity to source populations. As pointed out earlier, if these effects operate, snap-shot incidence functions may be misleading.

As Gotelli (1991) concludes, the empirical support for metapopulation models is fairly thin on the ground and tests are liable to be difficult.

Metapopulations are difficult to replicate and the time-scale of their dynamics (many generations of the organisms involved) may be of the order of decades. In addition, it is readily appreciated that populations are subdivided on many scales and so the delimitation of the local population is often subjective. If the metapopulation concept is to be of value as a theoretical framework it must be extended from the original, simplistic models, to allow for the differing degrees of population connectivity, and differing forms of inter-patch relationship, to be found in real systems (Harrison 1994; Hanski 1996). This is not to dismiss the results of such modelling exercises, merely to note that they must be placed in context, in particular, the context of the spatial structure of the landscape. Fahrig and Merriam (1994) identify the following factors as potentially significant:

1. differences among the patch populations in terms of habitat area and quality;
2. spatial relationships among landscape elements;
3. dispersal characteristics of the organism of interest; and
4. temporal changes in the landscape structure.

Some of these points will be considered in the following sections. The various forms and derivatives of metapopulation dynamics are usefully summarized by Hanski (1996).

9.3.2. Deterministic explanations of extinctions within metapopulations

Extinction models tend to be based on stochastic variation, in some cases emphasizing demographic stochasticity, sometimes taking genetic stochasticity and feedback into account, and more recently incorporating environmental stochasticity. However, Thomas (1994) has argued that, in practice, most extinctions are deterministic, and that they can be attributed directly to hunting by humans, introductions of species, and loss of habitat (Chapter 10). In many cases documented in the literature, virtually the entire habitat was lost or modified, causing 100% mortality. Stochastic extinction from surviving habitat fragments is minor by comparison. According to this view, stochastic events are superimposed as decoration on an underlying deterministic trend. To put this in a metapopulation context, if the reason for patches 'winking out' is due to habitat changes rather than—for sake of argument—a period of unfavourable weather, then the local habitat is likely to remain unsuitable after extinction and so will be unavailable for recolonization. In cases, fairly subtle changes in habitat may produce a species extinction which might be mistakenly classified as 'stochastic'. Thomas cites examples of his own work on British butterflies, whereby even changes from short grass to slightly longer grass can result in changes in the butterfly species found in the habitat (cf. Harrison 1994).

Thomas (1994) usefully reminds us that there are many reasons for species colonization of a patch, and they are not all island effects. He suggests six circumstances in which natural colonizations by butterflies are observed:

1. succession following disturbance;
2. where new 'permanent' habitat is created close to existing populations;
3. introductions of species outside former range margins;
4. regional increases in range;
5. turnover-prone peripheral patches (which act in essence as 'sinks' and are therefore unimportant to overall persistence); and
6. seasonal spread.

Thomas regards butterflies as persisting in regions where they are able to track the environment, and becoming extinct if they fail to keep up with the shifting habitat mosaic, or if the shift in habitat regionally is against them. Existing metapopulation models therefore need to be superimposed on an environmental mosaic, which in many cases will itself be spatially dynamic. 'An environmental mosaic perspective shifts the emphasis on to transient dynamics and away from the equilibrium (balance) concept of metapopulation dynamics, for which there is little evidence in nature' (Thomas 1994 p. 376). Thomas' comments could have been written about any one of the 'bandwagon' theories which appear to have characterized the conservation science literature. In short, this is saying, consider first the real world rather than the fancy models. The possibility of incorporating patchy disturbance phenomena into metapopulation models has since been developed by Hess (1996).

9.4. Reserve configuration—the 'single large or several small' debate

Given a finite total area which can be set aside for conservation as a natural landscape is being converted to other uses, what configuration of reserves should conservationists advocate? At one extreme is the creation of a single large reserve, the alternative is to opt for several smaller reserves amounting to the same area, but scattered across the landscape. This question, reduced to the initials **SLOSS**, was cast in terms of the assumptions and predictions of the ETIB (e.g. Diamond and May 1981). As one of the main aims of conservation is to maximize diversity within a fragmented landscape, might not this body of theory, which focuses on species numbers, provide the answer as to the optimal configuration of fragments?

Following the island analogy, the first factors to be included are isolation and area. Increased isolation of reserves reduces migration into them from other reserves. If a reserve is created by clearance of surrounding habitat, then it follows that on initial isolation the immigration curve should be depressed. The contiguous area of habitat is also reduced and thus extinction rate should increase. At the point of creation, therefore, the habitat island contains too many species, it may even gain fugitive displaced populations, and the result is that it becomes **supersaturated**. It follows that it should in time undergo 'relaxation' to a lower species number, a new equilibrium point (Fig. 9.6). Given knowledge of isolation and area of patches, it should be possible to estimate the number maintained at equilibrium in a variety of configurations of habitat patches. On these grounds, Diamond and May (1981) favoured larger rather than smaller reserves, short rather than long inter-reserve distances, circular rather than elongated reserves (minimizing edge effects), and the use of corridors connecting larger reserves where possible (Fig. 9.7). These widely cited suggestions spawned a largely theoretical debate in which much appeared to hang on the validity and interpretation of the ETIB. Having considered the ETIB, MVP, and metapopulation ideas ahead of it, some of the limitations of the SLOSS approach will be immediately apparent and so can be dealt with quickly.

The simplest criticism of this approach follows the line that the ETIB has been refuted. It is a flawed model. Hence it provides no firm foundation for the development of conservation policy. If policy makers adopt such theories as though providing formal rules, as some did in this case, this criticism is justified (Shafer 1990). The position adopted in this book, however, is that the ETIB has a place in a larger framework of island ecological theory. It is not that the theory is either true or false, but that the effects it models may be either strong or weak. If the effects are very weak for the habitat island system being considered, as often is the case, then the answer to the SLOSS question will not be supplied by the equilibrium theory. Furthermore, deployment of the ETIB often comes down to the use of species–area equations. Such an approach may give a rough idea of numbers of species on habitat islands, but not which habitats contribute most to richness, nor which species are most likely to be lost from the remnant (Saunders *et al.* 1991; Simberloff 1992; Worthen 1996).

To revisit the basis of this thinking, interpretations of species–area relations based on equilibrium theory have it that the regression line provides the equilibrium number of species. Boecklen (1986) cautions against such assumptions. The statistical interpretation of the scatter is simply that the regression line represents the average number of species for a given area and a point above it represents a positive error.

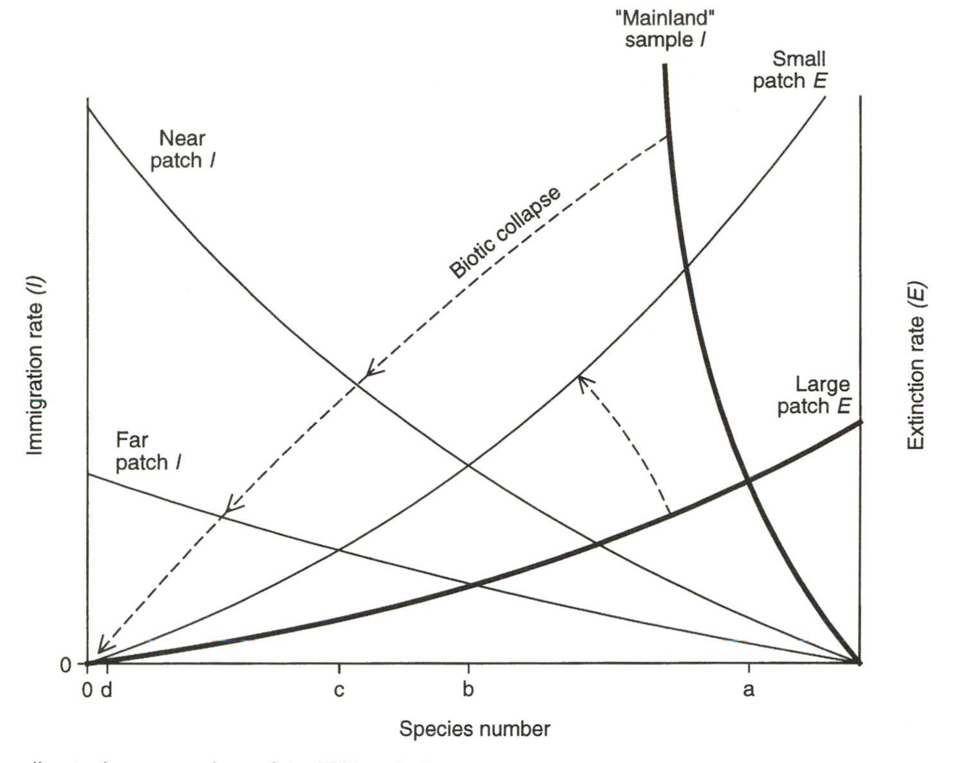

Fig. 9.6. According to the assumptions of the ETIB, reducing area causes supersaturation as immigration rate declines and extinction rate rises. This causes loss of species, i.e. 'relaxation' to a lower equilibrium species number. In the extreme scenario of 'biotic collapse', the immigration rate is so low that the equilibrium species number is zero. (Redrawn from Dawson 1994, Fig. 4.) In practice this abstract theory may lack predictive power as to either numbers lost or rate of loss, being founded on assumptions of the system passing from equilibrium (pre-fragmentation), to non-equilibrium (at fragmentation), to a new equilibrium condition (when *I* and *E* balance again): circumstances which may not apply in complex real-world situations. This is not to deny species losses, they do occur, but very often they are not a random selection, but instead are drawn from predictable subsets of the fauna or flora, structured in relation to: particular ecological characteristics of the species in question; habitat or successional changes in the isolate; and the nature and dynamics of the matrix around it. Equally, isolates may acquire increased numbers of some species, or gain new species, sometimes of an 'undesirable' nature judged in terms of their impact on the original species.

If a newly created habitat island lies above the line it doesn't necessarily follow that it is supersaturated and destined to lose species. It could just be that habitat quality or heterogeneity allows it to maintain more than the average number of species (see also Boecklen and Gotelli 1984). In his analyses of a large US bird census data set, Boecklen (1986) found that habitat heterogeneity is a significant predictor of species number even after area has been factored out. The significance of habitat effects in the data set points to the possibility that in particular cases several small reserves can incorporate a wider (or better) array of habitat types and thus support more species than the single large option.

Most recent commentators regard these theoretical debates as having contributed little of direct practical value to conservation (e.g. Saunders *et al.* 1991). However, the *implications* of different reserve configurations still require resolution, irrespective of the fate of the theoretical frameworks that spawned them.

9.4.1. Dealing with the leftovers

In most cases, natural scientists are unable to exert great influence on the basic ground-plan

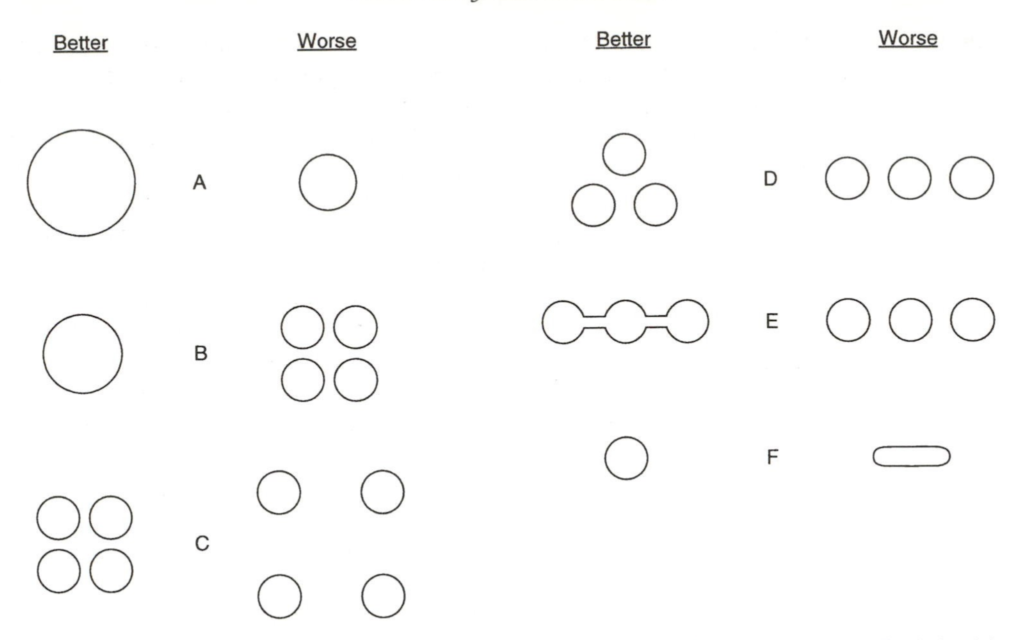

Fig. 9.7. The suggested geometric principles for the design of nature reserves which were supposedly derived from island biogeographic studies, and which were at the centre of the so-called SLOSS, or single large or several small debate (redrawn after Diamond 1975b). These 'principles' have been challenged on both theoretical and practical criteria (see text).

of fragmentation, and even where they do have a role to play, it is never a free hand. There is, moreover, a range of biogeographical, economic and political considerations which can be crucial to the strategy adopted. This has been recognized even by the advocates of the SLOSS 'principles'. For instance, the proposed reserve system for Irian Jaya, on which Jared Diamond acted as an adviser, rightly paid more attention to pre-existing reserve designations, distributions of major habitat types, and centres of endemism, and to economic and land-ownership issues, than it did to the SLOSS principles (see Diamond 1986). It makes sense that if areas can be identified which harbour large numbers of endemic species, they warrant designation.

Most often, conservationists today are in the position of dealing with and managing a system of habitat fragments, and perhaps modifying its configuration slightly. Fragmentation may have been carried out in a very selective fashion. Thus, on the whole, farmers take the best land and leave or abandon the least useful. In much of southern England, this has meant that most remaining native woodland is on very heavy, clay-rich soils, with the lighter soils in agricultural use. The message is that the species mix in your fragments may not be representative of what was in the original landscape. To give just one example, the short-leaved lime (*Tilia cordata*) does not favour the heavier soils, and it is now known as a species of hedgerows. In the past it was probably an important forest tree (Godwin 1975). Equally, the juxtaposition of landscape units may be critical. For instance, some Amazonian forest frogs require both terrestrial and aquatic environments to be present within a reserve in order to provide for all stages of the life cycle (Zimmerman and Bierregaard 1986).

It must also be recognized that humans may interfere directly in events within reserve systems. Bodmer *et al.* (1997) point out that although much work has focused on extinction rates caused by deforestation, many of the recorded extinctions in the past few hundred years have been the result of overhunting. They collected data on the relative abundance of large-bodied mammals in the north-eastern Peruvian Amazon in areas with, and without, persistent hunting pressure. They found that species with long-lived individuals, low rates of increase, and long generation times are most vulnerable to extinction. In one reserve in north-eastern Peru, it was found that nearly half the meat harvested by hunters in the buffer zones of the reserve originated from mammals that are categorized as vulnerable to overhunting. Such results can be used to help formulate appropriate policy towards such activities.

9.4.2. The answer depends, in part, on the type of organism

The optimal configuration of areas is liable to vary depending on the type of organism being considered. If forced to generalize, larger reserves are more appropriate for large animal species needing a large area per individual, pair, or breeding group, and/or requiring 'undisturbed' conditions, e.g. the Javan rhino (*Rhinoceros sondaicus*) (Hommel 1990). Species requiring large areas are often the ones most threatened by us and in need of protection, e.g. top carnivores, primates, and rhinoceros. Over the past decade none of the primates on the endangered species list have moved off it, they have simply been joined by others. Top carnivores, such as big cats, may require large territories, whereas it might be possible to keep the highest diversity of butterfly species by means of a number of small reserves, each targeted to provide particular key habitats. They should not, however, be too small, lest they consist principally of 'sink' populations. However, trophic level and body size are not the only considerations; for instance, birds of prey may have large territories incorporating both good and poor habitat. Conservation of large birds of prey in Scotland, such as the golden eagle and the osprey, has required a range of measures taken to discourage and prevent killing of the birds within a mixed-use landscape. It has also required intensive but highly localized efforts to protect the nesting sites from poachers. The 'reserves' required are tiny compared to the ranges of the individuals.

While, in general, most studies advocate relatively large reserves at the expense of larger numbers, in cases it may be better to have a set of several small reserves. First, they may incorporate more different habitats. Secondly, competition may, in theory, lead to the exclusion of species of similar niches from any given reserve and so it may be good to have several reserves so that different sets of species may 'win' in different reserves (but see section 9.8). Thirdly, there is an epidemiological risk inherent in having 'all your eggs in one basket' (section 9.10.2), and, lastly, particular species may rely upon small islands. For instance, small estuarine islands in the Florida Keys area appear to provide important breeding sites for several species of waterbirds (Erwin *et al.* 1995).

On both practical and theoretical grounds, the answer to the SLOSS debate is necessarily equivocal (e.g. Boecklen and Gotelli 1984; Shafer 1990; Worthen 1996). The answer depends on the ecology of the species (or species assemblage) being considered, but also on a range of system properties, which we will now explore. The discussion will mostly be limited to biological and physical parameters. However, it should be recognized that that there are often overriding practical considerations. For instance, the costs involved in establishing and maintaining conservation and game parks in developing countries depend on questions such as the following (Ayers *et al.* 1991). Is perimeter fencing required? Is it necessary to patrol to prevent poaching or encroachment into the reserve? How many wardens do you need? Do you have to equip and manage park headquarters? In many situations, fewer larger reserves make more sense from these practical, financial, and inevitably political perspectives.

9.5. Physical changes consequent upon fragmentation

A useful review of the physical environmental consequences of fragmentation is provided by Saunders *et al.* (1991). They illustrate how differences in fluxes of radiation, wind, water, and nutrients across the landscape can be significant to the composition and survival of remnant biota. Large environmental changes can be involved, particularly in conversion of forest to non-woodland cover (as Lovejoy *et al.* 1986).

- *Radiation fluxes.* In south-western Australia, elevated temperatures in fragmented landscapes reduced the foraging time available to adult Carnaby's cockatoos (*Calyptorhyncus funereus latirostrus*) and contributed to their local extinction.

- *Wind.* When air flows from one vegetation type to another it has to equilibriate to the height and roughness characteristics of the new system, requiring roughly 100–200 times the height of the vegetation to do so, i.e. 2–4 km for a 20 m high wood. Relatively little work has been done on such phenomena. We can say, however, that at the edge of a newly fragmented woodland patch, increased desiccation, wind-damage, and tree-throw can each be significant (e.g. Kapos 1989). Increased wind-turbulence can affect the breeding success of birds by creating difficulties in landing due to wind shear and vigorous canopy movement. Wind-throw of dominant trees can result in changes in the vegetation structure, and allow recruitment of earlier successional species.

- *Water* and *nutrient fluxes.* Removal of native vegetation changes the rates of interception and evapotranspiration, and hence changes soil moisture levels (Kapos 1989). In parts of the wheatbelt of Western Australia, the new agricultural systems cause rises in the water table, which can bring stored salts to the surface, and this secondary salinity has caused problems both to agriculture and the remnant patches. In the fenlands of eastern England, drainage for agriculture has led to peat shrinkage and a drop of 4 m in land level in 130 years. Remnant areas of 'natural' wetland now require pumping systems to maintain adequate water levels (see also Runhaar *et al.* 1996).

A slightly longer list of changes, related to edge effects upon fragmentation, is given in Table 9.2, in which some of the knock-on biological effects are noted. The newly fragmented system is likely to adjust to the new physical

Table 9.2

Classes of edge-related changes triggered by the process of forest fragmentation, as informed by the Minimum Critical Size of Ecosystem project. The first-order effects may lead to second-order and, in turn, third-order knock-on effects (after Lovejoy et al. 1986)

Class	Description of change
Abiotic	Temperature
	Relative humidity
	Penetration of light
	Exposure to wind
Biological	
First-order	Elevated tree mortality (standing dead trees)
	Treefalls on windward margin
	Leaf-fall
	Increased plant growth near margins
	Depressed bird populations near margins
	Crowding effects on refugee birds
Second-order	Increased insect populations (e.g. light-loving butterflies)
Third-order	Disturbance of forest interior butterflies, but increased population of light-loving species
	Enhanced survival of insectivorous species at increased densities (e.g. tamarins). NB Does not apply to birds (Stouffer and Bierregaard 1995)

conditions. For instance, in time the structure of a woodland edge fills and becomes more stable. Yet, just as there is a wide range of disturbance regimes across real islands, so must there be for continental habitats, and as we create new habitat islands, we inevitably alter disturbance regimes (Kapos 1989; Turner 1996). Furthermore, land-use of the matrix in which the habitat islands are embedded typically continues to change. It might therefore be anticipated that many habitat islands will be characteristic of the dynamic non-equilibrial position identified in Fig. 7.9 (cf. Hobbs and Huenneke 1992). Indeed, it has been suggested that smaller habitat islands may be subject to greater disturbance impacts than larger islands, thereby making such sites less suitable for, or removing altogether, particular species (McGuinness 1984; Simberloff and Levin 1985; Dunstan and Fox 1996).

9.6. Relaxation and turnover—the evidence

If habitat islands behave according to the expectations of the ETIB, following fragmentation they should be supersaturated with species. Subsequently, their numbers should 'relax' to a lower level than originally occurred in the area (Fig. 9.6). An equally important assumption is that immigration and extinction will continue to occur, both during the relaxation period and subsequently, when the island has equilibriated. Thus, if the theoretical effects are strong, even high-priority species—the very species that you wish to conserve—may in time disappear from a reserve. As we have noted, however, the empirical evidence from a wide range of real and habitat islands is that turnover tends to be heterogeneous (i.e. structured). Some species tend to be highly stable in their distribution across habitat islands. Most turnover involves 'ephemerals', species marginal to the habitat, or successional change (Chapters 7 and 8). The metapopulation studies examined above provide further insights into turnover. They suggest a tendency for large populations to persist, while smaller satellite populations may come and go, without jeopardizing the metapopulation. None the less, species do disappear from particular habitat islands, and from entire landscapes, and in cases the loss is a global one. Earlier, we noted that many species losses are attributable to deterministic causes, meaning hunting, habitat changes, and the like. The relaxation effect, in contrast, is based on stochastic processes. How strong is this island relaxation effect?

One of the earliest and most commonly cited examples of relaxation on ecological time-scales is that of birds lost from Barro Colorado Island (BCI), in Panama. It is not an ideal study in the present context in that, unlike most habitat islands, it is actually now surrounded by water. The island was formerly a hilltop in an area of continuous terrestrial habitat, but it became a 15.7 km^2 island of lowland forest when the central section of the Panama canal zone was flooded to make Lake Gatun in 1914. Of about 208 bird species estimated to have been breeding on BCI immediately following isolation in the 1920s and 1930s, 45 had gone by 1970 (Wilson and Willis 1975). Of these, some 13 have been attributed to 'relaxation'. The other losses could be attributed to particular forms of ecological change. At the time of the analysis, much of the forest was less than a century old, following abandonment of farming activity prior to 1914. Many of the birds lost were typical of second growth or forest-edge, suggesting a probable successional mechanism as forest regeneration reduced the availability of these more open habitats. Some were ground nesters which may have been eliminated by terrestrial mammalian predators. The latter became abundant due to the disappearance of top carnivores (such as the puma) with large area requirements (Diamond 1984). This effect, of increasing numbers of smaller omnivores and predators in the absence of large ones, has been termed **mesopredator release** (Soulé *et al.* 1988). Some of the birds lost were members of the guild of ant followers, which have been found to be vulnerable to fragmentation elsewhere (below). In a simple sense, BCI provides evidence of relaxation, but the stochastic signal is seemingly much smaller than that produced by changing habitat and food-web relations, and it might therefore be asked 'which is the signal and which the noise in the system?' (cf. Sauer 1969; Lynch and Johnson 1974; Simberloff 1976).

That species numbers and composition may change as a consequence of fragmentation is not in dispute. However, it is remarkably difficult to find empirical evidence for the island relaxation effect operating in habitat islands (Simberloff and Levin 1985; and see review in Shafer 1990). Most studies, including that for BCI, lack precise information about the species composition prior to fragmentation (e.g. Soulé *et al.* 1988). The BCI study is also fairly typical in the losses recorded being mostly 'deterministic' in nature. The predictions of species loss through 'relaxation' derive from the ETIB, but theoretically relaxation may also affect metapopulations. Hanski *et al.* (1996) speculate that the pace of fragmentation in many regions has been so rapid that 'scores of rare and endangered species may already be "living dead", committed to extinction because extinction is the equilibrium toward which their metapopulations are moving in the present fragmented landscapes' (p. 527). However, their findings are heavily dependent on models of the data of a single species of butterfly, the Glanville fritillary (*Melitaea cinxa*). The effect identified appeared to be far from instantaneous, and might turn out in practice to be a relatively weak effect in relation to some of the other processes discussed in this chapter.

The assumption that fragmentation will necessarily lead to ongoing species losses through relaxation has been challenged by Kellman (1996). He argues that reference to past environmental change and species responses to it shows that many species are more flexible in their ecological requirements than generally recognized. Kellman contends that many plant communities are actually undersaturated and that systems of fragments may be able to sustain increased species densities for significant periods of time (although not necessarily indefinitely). This is likely to prove a highly controversial viewpoint. He has developed it based primarily on studies of gallery forests in the neotropics. These forests occupy river valleys within savanna areas and thus constitute narrow peninsulars of forest within a non-forest landscape. According to equilibrium theory, the tips of the peninsulars should be impoverished due to their increased isolation. This extension of island theory is termed the **peninsular effect**. It has been tested on the large scale for Florida, and for Baja California and, once latitudinal gradients have been accounted for, has generally been found to be a fairly weak effect (Brown 1987; Means and Simberloff 1987; Brown and Opler 1990). In the gallery forests, Kellman and his associates have found no evidence for floral impoverishment with distance from the peninsular base, possibly because the frugivores that disperse the seeds of the plants are capable of adjusting their ranging behaviour to accommodate the long, thin peninsular configuration, or indeed to move through the sparsely wooded savanna matrix outside the peninsulars. These peninsulars have a long history, and thus appear to provide evidence for adequate gene flow of plant populations to the peninsular tips over long periods. Where fragmentation proceeds rapidly, however, as is occurring in many tropical regions at the present time, there may not be sufficient opportunities for these ecological adjustments to take place. It is not therefore clear how good an analogue these peninsulars provide. Kellman's data concern only plants, and it would be interesting to know how other taxa behave within these peninsulars.

The Atlantic coast of Brazil supports a form of tropical moist forest known as the Atlantic forests, reduced over recent decades and centuries to only about 12% of its original area and greatly fragmented. According to species–area calculations, some 50% of the fauna should be lost, but thus far evidence for actual extinctions is thin (Brown and Brown 1992). Although one or two known species may be extinct, no known species of its old, largely endemic fauna can be regarded as *proven* to be extinct (Whitmore and Sayer 1992). It has been suggested that this may be due to the highly heterogeneous and naturally much-disturbed character of the Atlantic forest system. Species endemic to these forests may thus be pre-adapted to living in small populations, in limited, semi-isolated habitat fragments. It is, however, possible that species not known to science have been lost, and it is possible that species may be 'committed to extinction' (Brown and Brown 1992). Indeed, according to Brooks and Balmford (1996), independent estimates of the number of endemic Atlantic forest bird species classified as threatened actually compare fairly closely to the numbers of extinctions predicted by using species–area calculations. They therefore reject the notions that the endemic fauna is pre-adapted to fragmentation. They also doubt that there have been many cryptic losses. In their view, the data are consistent with species–area predictions, provided a time lag between deforestation and actual extinctions is allowed for. However, whilst such time-lagged effects may well occur, it is not entirely clear that the estimates referred to provide a particularly independent test, as the IUCN criteria are themselves heavily influenced by attention to area considerations and extent of fragmentation.

There is a long-term fragmentation project which sheds some light on these questions, and in which data were collected prior to fragmentation: the Minimum Critical Size of Ecosystem project. It is located near Manaus, in the Brazilian Amazon, and sites were isolated between 1980 and 1984. It should be noted that the forest patches in the study were isolated by only 70–650 m from intact forests. While some patches remained isolated, the land surrounding others was abandoned during the study and became in-filled by forest regrowth dominated by *Cecropia*. Stouffer and Bierregaard (1996) examined the use of forest fragments by understorey hummingbirds. The three species found prior to fragmentation persisted at similar or higher numbers in the fragments through the 9 years of the study. Use of the fragments did not differ between 1 and 10 ha fragments. Furthermore, it mattered little whether the surrounding matrix included cattle pasture, abandoned pasture, or *Cecropia*-dominated second growth. This particular guild of birds thus appears to be able to persist in a matrix of fragments, secondary growth, and large forest patches.

The response of the hummingbirds is very different, however, from that of the insectivorous birds that dominate the understorey bird communities. This guild was found to be much more responsive to fragmentation, and both abundance and richness declined dramatically (Stouffer and Bierregaard 1995). Three species of obligate army-ant followers disappeared within 2 years (it will be recalled that ant followers were also lost from BCI, above). Mixed-species flocks drawn from 13 species disintegrated over a similar time-scale, although three of the species persisted in the fragments. The 10 ha fragments were less affected than the 1 ha fragments. Over time, the species flocks reassembled in those 10 ha fragments that were surrounded by a rapidly infilling regrowth of *Cecropia*. Ordination of the compositional data showed that over this fairly short period, communities in those larger fragments connected by a landscape of regrowth converged on pre-isolation communities, while communities in the smaller, or more isolated fragments continued to diverge. Notwithstanding the apparent successful recovery of the larger, better-placed sites, particular terrestrial insectivores, such as *Schlerurus* leafscrapers and various antbirds, did not return to any fragments. It may be concluded from these studies that the impact of fragmentation on birds is strongly ecologically structured. Different 'guilds' of species respond in divergent fashion and the types of species most vulnerable to fragmentation are fairly predictable (Lovejoy *et al.* 1986; Stouffer and Bierregaard 1995, 1996; Christiansen and Pitter 1997).

Fragmentation may often be associated with compositional change and species losses, but it has yet to be satisfactorily demonstrated that the stochastic 'island' relaxation effect is generally a strong one. Given the significance of the assumptions of relaxation to the debate on species extinction rates, it is surprising that more attention has not been given to demonstrations of the effect (Simberloff and Levin 1985; Simberloff and Martin 1991; Whitmore and Sayer 1992).

9.7. Succession in fragmented landscapes

As we have noted in the preceding sections, the processes of land-use change which create fragmented systems typically initiate successional changes in the habitat island remnants. This was apparent in the BCI study. Another example is provided by Weaver and Kellman's (1981) study of newly created woodlots in southern Ontario. The species losses which occurred did not take the form of stochastic turnover, with different species 'winning' in different woodlots. Instead, the 'relaxation' involved the successional loss of a particular subset of species. With appropriate management, the observed losses could be avoided. If occurring at all in the absence of disturbance and succession, turnover appeared to be very slow-paced among the vascular flora. In this system, the ETIB effects were clearly weak, and subordinate to other 'normal' ecological processes.

In practice, nature-reserve management often pays considerable attention to ecological succession, particularly in small reserves. Failure to do so often leads to the loss of desirable habitats. This applies to many of the lowland heath reserves of southern England, which have long been anthropogenically maintained. Without appropriate management, most areas suffer woodland encroachment. The concepts of **minimum dynamic areas** and **patch dynamics** (Pickett and Thompson 1978) are thus relevant to management plans. Simplified representations of woodland dynamics view a stand of trees as going through phases of youth, building, maturity, and senescence, each of which may support differing suites of interacting species. Reserves should therefore be large enough, or managed, to ensure that they contain enough habitat patches at different stages of patch life-cycles to support a full array of niches. Another related, and often contentious issue, in regions in which fire is a prominent ecological feature, is how fire regimes are managed. There is often a cyclical pattern of post-burn succession and fuel accumulation, leading to an increased likelihood of fire, repeating the cycle. The maintenance of a particular fire regime and patch mosaic structure may be of crucial relevance to species diversity in a reserve system, but is often poorly understood (Short and Turner 1994; Milberg and Lamont 1995). Fire is also politically contentious because of the threat it poses to property.

Much active management of nature reserves is thus about keeping a mosaic of different successional stages, and not allowing the whole of the reserve to march through the same successional stage simultaneously. The relevance of such successional dynamics to species changes in reserve systems appears to have received much less attention in the theoretical conservation science literature than its significance warrants (but see Pickett *et al.* 1992). Moreover, continuing changes in the habitats of the matrix can also be extremely important to the fate of species populations in the reserves themselves (Stouffer and Bierregaard 1995, 1996).

9.8. The implications of nestedness

The SLOSS 'principles' were advocated most strongly for circumstances in which the biogeography of an area is poorly known. In cases, there may be enough distributional data to be able to target those areas which are species rich, or which possess one or more endemic species. Another form of non-random distribution relevant to the debate on reserve configuration is that of nested subsets, whereby species present in small, species-poor patches are also found in larger patches that support more species (section 8.3). The empirical discovery of a strong degree of nestedness implies that the larger reserves are crucial to maximizing diversity. A low degree of nestedness, on the other hand, would mean that particular habitat patches sample distinct species sets and that an array of reserves of differing size and internal richness may be required to maximize regional diversity (cf. Kellman 1996). As we established in Chapter 8, nestedness may be the product of selective extinction, differential colonization, or, in theory, nested habitat (Cook and Quinn 1995).

There are several different indices of nestedness and related compositional features of island systems (section 8.3). Cook (1995) undertook a comparative study of 38 insular systems using six different indices of nestedness, ordered both by area and richness, and the so-called saturation index (SI) of Quinn and Harrison (1988). This index provides an indication of the degree to which a small number of large sites contain more or fewer species than a larger number of smaller sites with the same total area. It is not an index of nestedness, but it is useful in comparison with nestedness indices as it more directly measures the effect of habitat subdivision. The analyses ordered by richness returned higher nestedness scores than those ordered by area in between 35 and 37 cases (depending on the metric used). This reveals, as Simberloff and Martin (1991) suggested, that the ordering strategy is important. None the less, the relationships between the richness- and area-ordered indices were strong. Nestedness is, therefore, a common and robust feature of both real and habitat island data sets (Table 8.2; Simberloff and Martin 1991). Cook found that the relationship between the SI index and the nestedness indices is variable. They return complementary information, together revealing the configuration of diversity and the extent to which rare species tend to be found within the largest sites, respectively.

Blake's (1991) study of bird communities in isolated woodlots in east-central Illinois shows how nestedness calculations can provide useful insights. He demonstrated a significant degree of nestedness, particularly among birds requiring forest interior habitat for breeding and among species wintering in the tropics. In contrast, species breeding in forest-edge habitat showed more variable distribution patterns. These results support previous conclusions that small habitat patches are insufficient for preservation of many species. That is to say, some species will be lacking from any number of small patches, but can be found in the larger, richer patches (cf. Newmark 1991). Knowledge of nested subset structure might thus be significant to predicting the ultimate community composition of a fragmented landscape (Worthen 1996). Blake's findings resonate with those reported for São Paulo (Patterson 1990; section 8.3), where it will be recalled that whereas the full bird data set was non-nested, when the transient species were removed and the analysis restricted to the sedentary species, they were found to be significantly nested. Tellería and Santos (1995) have demonstrated the apparent importance of nestedness of habitat. They studied the winter use of 31 forest patches (0.1–350 ha) in central Spain by the guild of pariforms (*Parus* and *Aegithalos, Regulus, Sitta*, and *Certhia*—tits, goldcrests, nuthatches, and treecreepers). They found that birds with similar habitat preferences tend to disappear simultaneously with reduction in forest size, thereby producing a nested pattern of species distribution.

Simberloff and Martin (1991) have argued that establishing nestedness across a whole system is no longer particularly exciting, but that some benefit can come from examining discrepancies from nestedness. Cutler (1991) developed an index, U, which was designed to account for both unexpected presences and unexpected absences. He used it to identify what he termed **holes** and **outliers**. Holes are where widespread species are absent from otherwise rich faunas, and outliers are where uncommon species occur in depauperate faunas. He applied his index to the data for boreal mammals and birds of the Great Basin (North America). He found that for mammals most of the departures from perfect nestedness were due to holes, whereas for birds most departures were due to outliers. Cutler speculates that this difference could be a function of the superior dispersability of birds, allowing them to generate greater numbers of outliers by recolonization events. 'Supertramp' species (Chapter 8) might also appear as outliers in analyses of this sort. While Cutler did not consider nestedness to be necessarily more probable in extinction-dominated than in immigration-dominated systems, he did note that the mammal data, which are believed to be extinction-dominated, provided striking evidence of deterministic structure. If it is assumed that the mountain-top species were once widespread, only to have gradually disappeared from particular mountain ranges over the course of the Holocene, then Cutler's analyses show the sequence of extinction to be far less variable than the rate of extinction.

This might appear to be an important and powerful message for conservation. However, Simberloff and Martin (1991) rightly caution that, to a large degree, 'relaxation' and extinction across these habitat islands have been inferred rather than demonstrated to have happened. The simple statistics can, moreover, mask more complex and important underlying details. In illustration, they show that the same nestedness score can be generated for species with widely differing underlying distributions. In the Maddalena archipelago, two species have a nestedness score of zero (by the Wilcoxon statistic). One, the rock dove (*Columbia livia*), is found across the entire range of island sizes, and is absent from only two of 16 islands. The second, the little ringed plover (*Charadrius dubius*), has been observed only once in the archipelago, breeding only in one year. As they point out, there may be different explanations for accordance or deviation from nestedness in particular species. The explanation of such patterns requires basic information on species' presence–absence, abundances, minimum area requirements, habitat use at different sites, and temporal regularity in occupying differing types of patch. In their view, given such data, it becomes possible to offer sensible input into conservation decision-making—input in which the contribution of the nestedness statistics is fairly limited.

Simberloff and Martin (1991) regard the rise of interest in nestedness in conservation with some concern, seeing it as another manifestation of the SLOSS debate, i.e. of largely theoretical interest, liable to be poorly understood by non-practitioners, and unlikely to lead to reliable insights into refuge design. They may be right to express such worries given past history. Species–area relationships did not and do not contain enough information to resolve the SLOSS issue. Formal considerations of compositional structure were, for too long, side-lined in this debate, and it is for this reason that I welcome the attention now being given to compositional structure and the forces that produce it. A nestedness index can provide one compositional descriptor and can perhaps aid identification of risk-prone species; but should not be given primacy in conservation planning. The discovery of nestedness at a particular point in time does not necessarily provide clear insights as to the probability of maintaining the same sets of species (or any particular species) over time. The isolates may be subject to turnover and/or species attrition in new ways dictated by the changing biogeographic circumstances of the landscape in which the fragments occur. As Worthen (1996, p. 419) put it, nestedness is not a 'magic bullet', '... no single index should be expected to distil the informational content of an entire community, let alone predict how it will react to habitat reduction or fragmentation'.

9.9. Edge effects

The boundary zone, or **ecotone**, between two habitats, being occupied by a mix of the two sets of species, is often richer in species per unit area than either of the abutting 'core' habitat types. In the reserve context, attention has focused on the reserve edge, typically where a woodland reserve is surrounded by a non-woodland matrix. For some species groups, possession of a large expanse of such edge habitat can increase the overall species number found on a reserve, an apparently beneficial effect (Kellman 1996). But, the argument goes, if the additional species supported by the edge habitat are those of the matrix, then they do not depend on the reserve and should be discounted in evaluating the merits of the reserve system.

In illustration, Humphreys and Kitchener (1982) classified vertebrate species into those dependent on reserves and those which persist in disturbed areas in the intervening agricultural landscape of the Western Australian wheatbelt (Fig. 9.8). They found that the small habitat islands, which were mostly 'edge' habitat, were disproportionately rich in ubiquitous species. Those species restricted to woodlots typically required a larger area (cf. Hinsley *et al.* 1994). This pattern was found separately for lizards, passerine birds, and mammals. By analysing all species within a taxon together, small reserves came out in a better light than they warranted.

In particular circumstances, ecotones may have negative implications for core woodland species (Wilcove *et al.* 1986), in that although they may support additional species or larger populations of non-woodland core species, these populations may have negative interactions

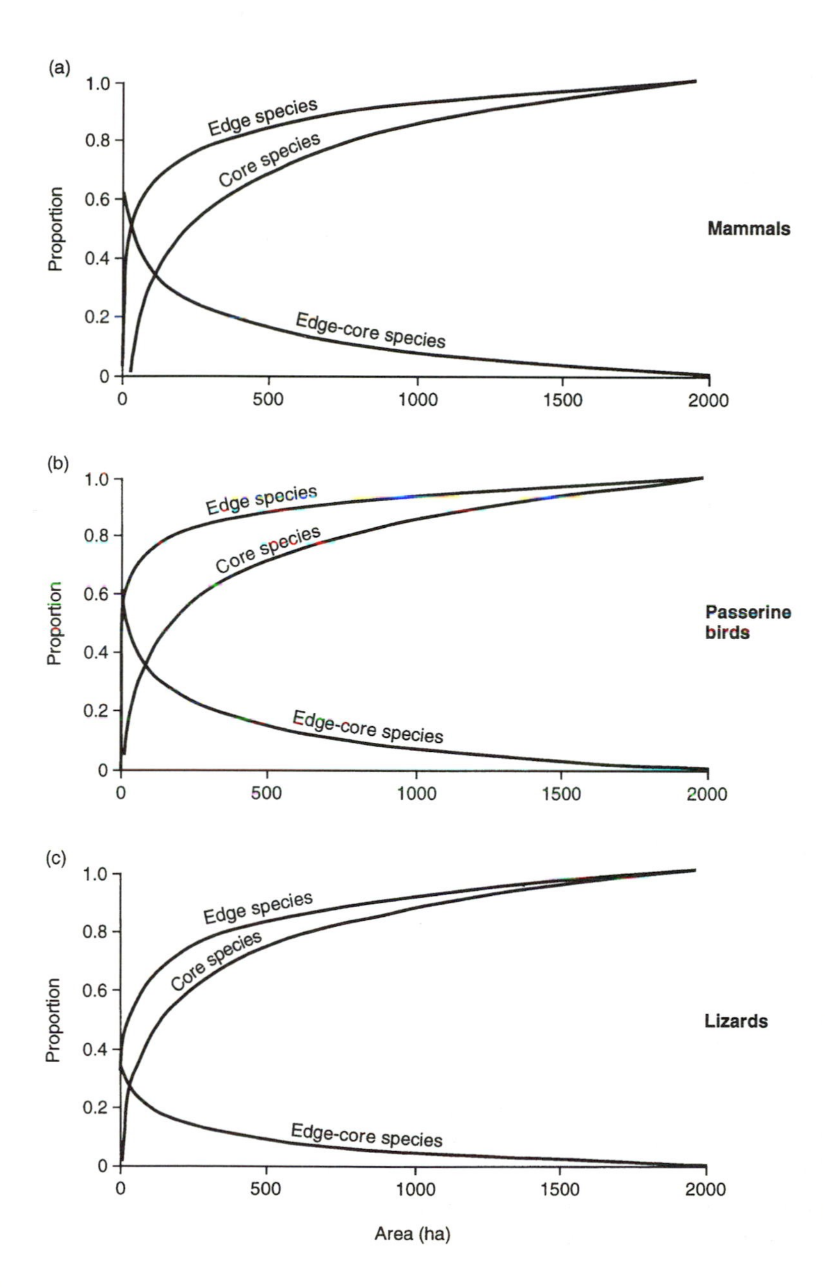

Fig. 9.8. The proportion of the core species (those dependent on the reserves) and the edge species (those which can also survive outside the reserves in disturbed habitat) which were present in a series of 21 nature reserves in the Western Australian wheatbelt in relation to the number of each present in a reserve of 2000 ha. The curves with negative slopes represent the difference between the curves for edge and core species, and represent the relative excess of edge species over core species compared with their distribution at 2000 ha. Data for (a) mammals, (b) passerine birds, and (c) lizards. (Redrawn from Humphreys and Kitchener 1982, Fig. 1.)

with deep-woodland species. Studies in North America have suggested that the nesting success of songbirds is lower near the forest edges than in the interior because of higher densities of nest predators (e.g. blue jay (*Cyanocitta cristata*), weasel (*Mustela erminea*), and racoon (*Procyon lotor*)) around forest edges. Wilcove *et al.* (1986) estimated that circular reserves of less than 100 ha will not support viable populations of forest songbirds simply because of this effect. Paton (1994) has undertaken a critical review of studies of nest predation in edge habitats. He concluded that, in general, there does appear to be evidence for a depression of breeding success, due either to enhanced predation or through parasitism. However, these forms of edge effect usually operate only within about 50 m of an edge, a narrower belt than some have claimed (e.g. Terborgh 1992). None the less, such an effect implies that the effective reserve area may be less than that described by the perimeter of the isolate. It has also been suggested that ecotones might enable the invasion of reserves by exotic species (Saunders *et al.* 1991). While this may be a problem if the fragment is disturbed, Brothers and Spingarn (1992) argue that as regards plants, the natural repair process, which produces well-developed woodland fringe vegetation, ensures reduced light levels within the woodland and may fairly swiftly discourage the invasion of light-demanding weedy species.

These effects can be amenable to experimentation. Burkey (1993) placed peanuts and hens' eggs at different distances from the edge of a patch of Belizian rain forest. Edge effects were evident although the 'edge' was only a minor road through otherwise continuous forest. His results demonstrate that edge effects may operate in very different fashions for different taxa. Egg predation rates were higher in a 100 m wide edge zone, but, conversely, seed predation rates were found to be higher 500 m into the forest than 30 m and 100 m from the edge.

The relationship between a reserve and its surrounding matrix is not subject to easy generalization. There are species that share both zones, and just as there are matrix species which may impact negatively upon 'core' reserve species, there may also be reserve species (dependent on it for breeding and cover) which exploit resources in the matrix. The concepts of source and sink populations may again be useful in this context, as the maintenance of maximal diversity across a landscape depends on sufficient source or core habitat for species of each major habitat. The identification of edge effects provides a part of such an analysis.

9.10. Landscape effects, isolation, and corridors

9.10.1. The benefits of wildlife corridors

The premise of much of the literature discussed in this chapter is that habitat connectivity is beneficial to long-term survival, as it enables gene flow within populations and metapopulations. Habitat connectivity might be achieved by having stepping stones, or **corridors**, of suitable habitat linking larger reserves together. In practice, habitat corridors act as **differential filters**, enabling the movement of some species but being of little value to others. As Spellerberg and Gaywood (1993) point out, we are not merely concerned with forest peninsulars or hedgerows. All sorts of linear landscape features, such as rivers, roads, and railways, may act as conduits for the movement of some species. Equally, they may represent barriers or hazards for others (Reijen *et al.* 1996). Harris (1984) points out that landscapes with great topographic relief channel their kinetic energy via dendritic tributaries, main channels, and distributaries, and that these in turn influence the landscape template in characteristic ways. The stream-order concept of the fluvial geomorphologist may be a means of understanding species distributions within the landscape (Fig. 9.9), and of understanding movements of species between disjunct habitat patches (Fig. 9.10). Studies of features such as hedgerows have not always found them to be as effective in connecting up woodlands as was hoped, with

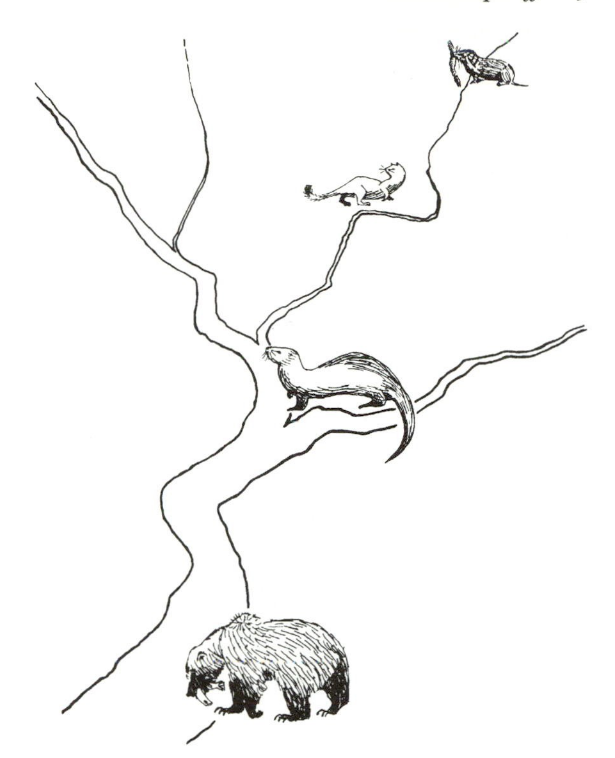

Fig. 9.9. Association of different-sized carnivorous mammal species with stream order and typical food particle size in accordance with the stream-continuum concept (redrawn from Harris 1984, Fig. 9.8).

some species moving as well through surrounding fields and others simply not moving well along the hedges. However, habitat corridors may be important in particular cases.

A useful illustration of how corridors can be beneficial comes from the study by Saunders and Hobbs (1989) of Carnaby's cockatoo (*Calyptorhyncus funereus latirostrus*) from the Western Australian wheatbelt. Studies from this area have already been discussed in this chapter and it may be helpful to add a little context. The wheatbelt is an area of 140 000 km^2 in the southwest of the state, 90% of which has been cleared for agriculture, over half of it (encouraged by the government) after 1945. Areas were set aside mostly for water catchments rather than as nature reserves, but there are now 639 designated conservation areas, mostly rather small and totalling only 6.7% of the area. Reportedly, 104 species of plants have become extinct in Western Australia, mostly from the wheatbelt; 132 more are rare or likely to become extinct. There could easily be more in each category. Of 46 native mammal species living in the wheatbelt before the Europeans arrived, 13 have gone from the region and, of these, nine are extinct from the Australian mainland altogether. Fewer than half of the 46 are still common. Of 148 bird species known from the area, only two have been lost so far, but many others are greatly reduced in numbers. Many Australian birds are poor colonizers and will not readily cross areas of unsuitable habitat, especially open areas. As an example, in a 10 year period, one reserve of 80 ha lost three out of seven of those passerine birds that are found only on remnant native vegetation.

Saunders and Hobbs (1989) report on the demise of the Carnaby's cockatoo, one of Australia's largest parrots and once the most widely distributed cockatoo in the region. It breeds in hollows in eucalypt trees and eats seeds and insect larvae from plants in the sandplain heath. It congregates in flocks, both to nest and forage. Individuals may live 17 years and breeding status is not attained until at least 4 years of age, each pair rearing a single offspring at a time. The widespread clearance of the land has removed extensive areas of the vegetation type in which they feed, replacing it with annual crops which are useless to the birds. In recent developments, wide verges of native vegetation have been left uncleared along the roads. These act to channel the cockatoos to other areas where food is available.

Cockatoos have not survived in areas of earlier clearances, carried out without these connecting strips, as once they have run out of a patch of acceptable habitat it takes a long time for the flock to find another patch of native vegetation. The more patchy the vegetation, the less successful the birds are at supplying adequate food to rear their nestlings. Furthermore, the narrow road verges of these early clearances result in a higher incidence of road deaths. The example illustrates that the degree and nature of connectivity of different landscape elements—in this case of the breeding and feeding habitats, and of the vehicular hazards—are critical to the survival of this species in these newly fragmented landscapes. Given the relatively

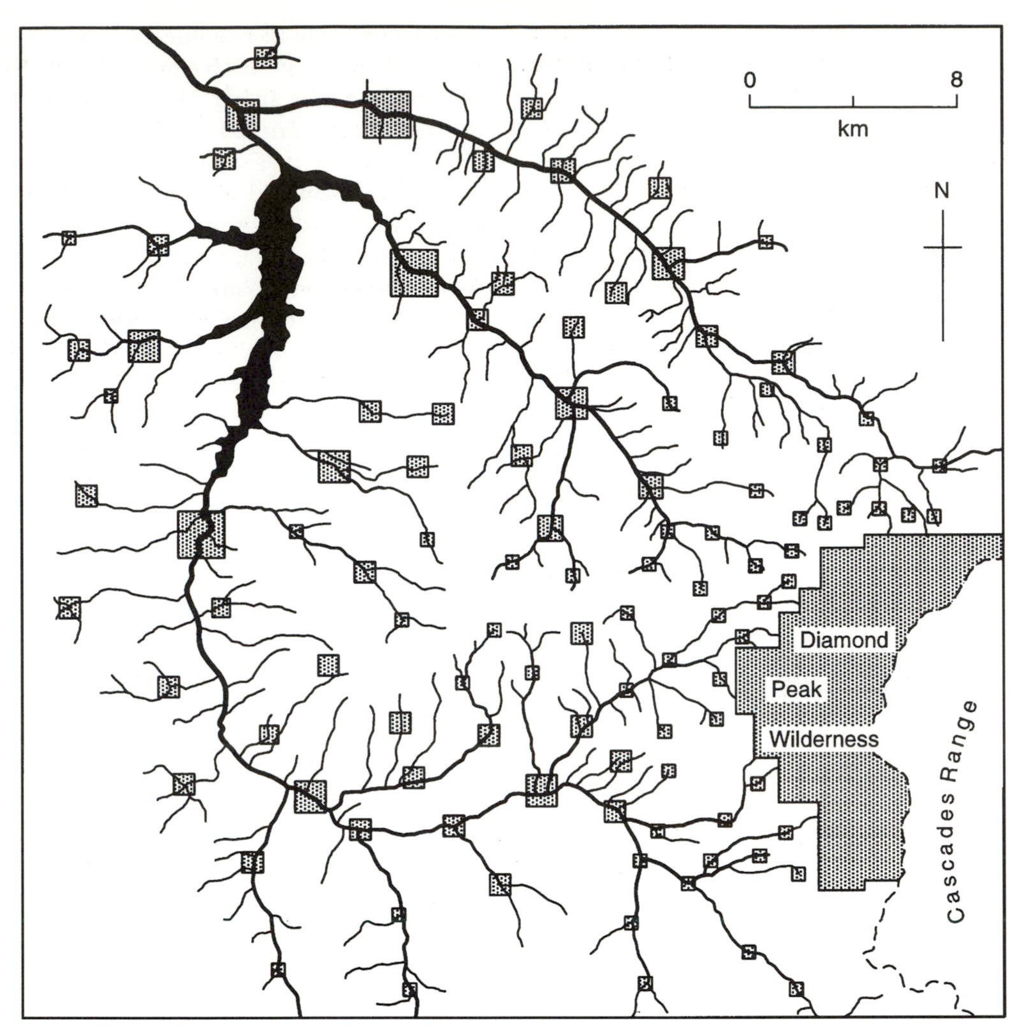

Fig. 9.10. A possible spatial and size-frequency distribution of old-growth habitat islands arranged along riparian strips at progressively greater distances from a current wilderness area in the Willamette National Forest, USA, designed to provide the optimal degree of connectivity between patches in a system structured by the dendritic pattern of drainage. In this hypothetical system, the most distant islands are generally larger in order to counter presumed lower dispersal. (Redrawn from Harris 1984, Fig. 10.2.)

long life cycle of these birds, and their flocking behaviour, failure can have a sudden and wholesale expression in the bird's disappearance. The big reduction in area is fairly recent, and the cockatoo is not yet in equilibrium with the new regime. It is still on a downward path and is viewed as a species in danger.

9.10.2. The benefits of isolation

It is important to retain balance in approach to all conservation issues. While much of the literature is concerned with maintaining population movements, there are also contexts in which isolation of populations of a target species might be desirable (Simberloff *et al.* 1992). For instance, it might be thought wise to have more than one separate population of an endangered species, lest a disease should devastate it in a single large reserve setting. There is evidence that epidemics can devastate wild populations. For instance, in eastern Australia *c.* 1900, a mysterious but poorly documented disease apparently caused the decimation or virtual extinction of many dasyurid marsupials. One of the species that succumbed in New South Wales was the native cat (*Dasyurus quoll*). A further problem here, and one that may well have been involved in these cases, is that alien species can carry diseases to which they are largely immune, but to which the native fauna have no defence. Anthrax introduced on domestic stock has recently been identified as the cause of 54% mortality among 10 species of large mammals in Etoshe national park in Namibia. Another example of a disease problem is the jarrah-dieback, a fungal disease of exotic origin, which kills a high proportion of native plant species in Western Australia. It has been spreading through the southern part of the state since its accidental introduction *c.* 1912. No adequate management or control technique has yet been devised. Exotic disease, transmitted moreover by exotic vectors, has also been implicated in the demise of some of the Hawaiian avifauna (Chapter 10).

Young (1994) evaluated 96 studies of natural die-offs of large mammal populations, defined as cases where numbers crashed by 25% or more. He noted that populations subject to large-scale phenomena such as drought and severe winters may not be protected from such die-offs by population subdivision. Where the crashes are produced by disease epidemics, subdivision may be beneficial, and the creation of linking corridors, and translocation efforts, may be harmful. The reported causes of the die-offs in the study were related to trophic level. Most herbivore die-offs were due to starvation, but carnivore die-offs were more often attributed to disease. An epidemic need not eliminate all the individuals in a population for it to have a crucial role in causing the eventual extinction of a population from a reserve. Combining these considerations with metapopulation modelling, Hess (1996) argues that there should theoretically be an optimal degree of movement within a metapopulation to enable the movement of propagules of the target species but not of disease. But much will depend on the nature of the intervening landscape matrix and the extent to which it filters target and 'pest' species.

It is not just the spread of exotic diseases which may be problematic, but also of exotic competitors and predators and of fire between or into reserves (Spellerberg and Gaywood 1993; Lockwood and Moulton 1994). It has been suggested that alien plant invasion can be reduced by isolating reserves and surrounding them with simplified cropland oceans (Brothers and Spingarn 1992). Short and Turner (1994) found that the decline and extinction of certain species of native Australian mammals were most parsimoniously explained by reference to the roles of introduced predators (foxes and cats) and herbivores (rabbits and stock). This explained the continued persistence of the native species on offshore islands from which the exotic species were absent. In these cases, isolation is beneficial (as Janzen 1983) and many reserves set in 'seas of altered habitat' within mainland Australia have proved to be insufficiently isolated to prevent their invasion by exotic 'pests' and 'pathogens'.

Is it possible to offer guidance about the situations when corridors may be beneficial and when not? First, as Dawson (1994) notes, all-purpose corridors do not exist, as each species has its own requirements for habitat, its own ability to move, and its own behaviour. Corridors therefore act as filters. He points out that many rare and threatened species are unlikely to benefit from corridors, because the corridor would have to contain their presumably rare habitat, i.e. rare species may require odd corridors. Suitable corridors may be necessary as links between preferred habitats for animals that undertake regular seasonal migrations, e.g. some fish, amphibians, reptiles, and for large mammals in both the seasonal tropics and the Arctic. On the other hand, for some populations, corridors may act as 'sinks', drawing out individuals from the main habitat area but not returning individuals to supplement it, in which case they may do more harm than good. In other cases, they may be fairly neutral in their ecological cost–benefit, but perhaps be quite expensive to purchase and set up if not already existing in a landscape. Simberloff *et al.* (1992) point out that numerous corridor projects have been planned in the

USA, potentially costing millions of dollars, despite the lack of data on which species might use the corridors and to what effect. In short, there is no firm theoretical basis for assuming that corridors are on balance a 'good thing', notwithstanding that they may be beneficial in particular cases. Each case must be judged on its merits.

9.10.3. Reserve systems in the landscape

As we have seen, the debate about habitat corridors has broadened from fairly simple theoretical beginnings to a consideration of how a whole range of differing natural, semi-natural, and artificial features are configured within a landscape. For instance, thousands of birds are killed each year in countries such as Britain and the Netherlands through collision with motor vehicles and with overhead power cables (Bevanger 1996; Reijnen *et al.* 1996). Research has shown that deaths through collisions with cables can be greatly reduced by consideration of important flight paths during construction, by design features of the gantries, and by attaching a variety of objects to the cables to enable birds to sight them (Alonso *et al.* 1994). While more research into the ways in which linear features function would be beneficial, what is required is that such information is integrated into improved management of whole landscapes (Spellerberg and Gaywood 1993). In short, a landscape ecological framework is needed.

Landscape components include both habitat patches and the matrix in which the patches are embedded. Modelling exercises typically favour clumped configurations of reserves, treating the matrix as a uniform 'sea'. In practice, the optimal configuration must take account of the details of the landscape involved (e.g. Fahrig and Merriam 1994). For instance, there may be greater opportunity for dispersal between two distant reserves linked by a river and its adjacent riparian corridor, than between two similar but closer reserves separated by a mountain barrier of differing habitat type (Fig. 9.10). An early attempt to place island theories into realistic landscape contexts was provided by Harris (1984). Recognizing that there were limits to the amount of land that would be put over to reserve use, Harris advocated a series of reserves placed within the landscape as dictated by geographical features such as river and mountains, to encourage population flows between reserves. Most reserves would remain in forestry management. Within such reserves, there should be an undisturbed core of forest that is never cut, providing habitat for species requiring undisturbed, old-growth conditions, around which commercial operations might continue, following a pattern of rotational partial felling. This would provide a mosaic of patches of differing successional stage, maximizing habitat diversity, while maintaining income (Fig. 9.11). His strategy aimed to satisfy the requirements of island theories, patch dynamic models, and economic realities. The implementation of such a policy requires concerted action from a wide variety of agencies and is thus easier to sketch out than to bring to a realization.

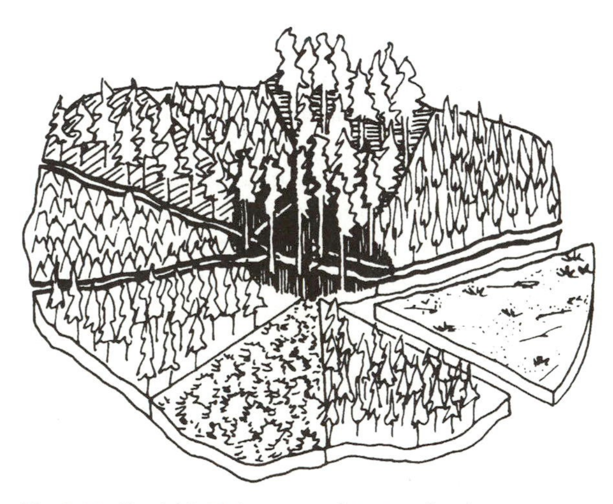

Fig. 9.11. Harris' (1984) proposed system for the management of forest habitat islands on a long-rotation system, such that different sectors are cut in a programmed sequence, but the core old-growth patch is left uncut. This recognizes the importance of patch dynamics to habitat diversity and how such management may serve both economic and conservation goals. (Redrawn from Harris 1984, Fig. 9.5.)

9.10.4. Species that don't stay put

In the Northern Territory of Australia, as in many other systems across the globe, mobility and massive population fluctuations in response to a highly variable climate are features of the wildlife. Woinarski *et al.* (1992) cite as a particular example the magpie goose (*Anseranas semipalmata*). They note that a conventional reserve network consisting of discrete National Parks (such as they have developed for plants in the region) will not cater for the need of the magpie geese to relocate in response to patchiness of rainfall. They identify four possible strategies for conservation in the Northern Territory:

1. The status quo, with improvements to ensure all vegetation types are included in the network. This would not solve the problem, as it is not just vegetation types but the year-to-year variation in carrying capacity across the region that matters.
2. Seek inclusion of known habitat of important species into the reserve system. For the magpie geese, this might mean inclusion of all wetlands. However, it would be difficult to decide which species to concentrate on and it would be impractical to gazette all the land in question because of competing claims for use.
3. Develop very large reserves which span extensive slices of the environmental gradient, just as does the present Kakadu National Park, a 20 000 km^2 reserve. However, this would require vast reserves in the semi-arid and arid region, in which, at the present time, the largest reserve is of 1325 km^2.
4. Supplement the representative reserve network with measures that protect wildlife habitat on large areas of unreserved land. They see this as the only viable answer in the long term, requiring a broad educational and political strategy to get there (Woinarski *et al.* 1992; Price *et al.* 1995).

So, wildlife conservation must in part be concerned with the practicalities of conserving outside reserves. The Northern Territory example may seem an extreme case when viewed from many parts of the world, yet the argument surely applies to some degree everywhere.

9.11. Does conservation biology need island theory?

Zimmerman and Bierregaard (1986) undertook an analysis of Amazonian frogs which demonstrated that a knowledge of the **autecological requirements** of the species would be critical to planning a reserve system (not that a system would necessarily be designed specifically for frogs). They undertook a prediction by a species–area approach and showed that it lacked relevance in the light of the available autecological data. Frog species differ in the habitat required for breeding. While some are strictly streamside breeders, others require landlocked permanently or periodically flooded terra firma pools. The authors concluded that if such data are available for enough species in the system being considered, then there need be no recourse to abstract principles. If they are not available, it is still preferable to conduct a rapid survey and collect biogeographic and ecological data. This is just one of many studies that have championed empirical over theoretical 'island' approaches. Does conservation biology need island theories?

9.11.1. A non-equilibrium world?

Worthen (1996) bemoans the fact that the bias towards species–area relationships over compositional–area relationships continues to be a feature of the literature, including some of the most widely used introductory ecology texts. This bias continues despite the demonstration, in numerous studies, of nestedness, community similarity of habitat fragments, the relevance of successional patterns, and a variety of other forms of compositional structure. These non-random patterns in how communities assemble, or alternatively how they change in response to fragmentation, demand that we reappraise the contribution of 'island ecology' within conservation biology.

Much of the earlier applied island literature seemed to assume that matrix contexts were relatively insignificant and that stochastic decay towards equilibrium would restore a new, depauperate balance. In the present chapter I have tried to show how the behaviour of habitat islands is more realistically captured within the notions of *flux of nature* described by Pickett *et al.* (1992). This framework stresses physical as well as biotic processes (e.g. succession), the role of episodic events (e.g. fires or hurricanes), and the differing degrees of connectivity of populations across 'filtering' landscapes. We have also seen that the nature of the matrix can be crucial to what goes on in the habitat islands within the landscape. How much of this is 'island biogeography' is a moot point, but insularity remains a key property and issue within these other topics.

The application of equilibrium (ETIB) assumptions to reserve systems has been neatly critiqued by Saunders *et al.* (1991, p. 23):

> It is commonly assumed that at some stage the remnant will re-equilibriate with the surrounding landscape. It is however, questionable whether a new stable equilibrium will be reached since the equilibration process is liable to be disrupted by changing fluxes from the surrounding matrix, disturbances, and influx of new invasive species. The final equilibrium can be likened to an idealised endpoint that is never likely to be reached, in much the same fashion as the climatic climax is now conceptualized in succession theory. Management of remnant areas will thus be an adaptive process directed at minimizing potential future species losses.

Another seemingly reasonable assumption generally made is that the alarming fragmentation of many 'natural' systems will continue and that if much wildlife is to survive, it will have to do so in reserve systems. Further, that we have very little time in which to produce guidelines and have any chance of influencing where and how much is retained (this particular line of argument is now at least a decade old; Diamond and May 1985). Increasingly, such lines of reasoning have been attacked on the grounds: first, that many valid generalizations are self-obvious platitudes; secondly, that models are often both too simplistic and effectively untestable; and, thirdly, that many theoretical generalizations have achieved the status of dogma (Doak and Mills 1994). I hope to have drawn attention in these pages to some of the flaws that afflict the underlying assumptions of the equilibrium or 'balance of nature' thinking. These assumptions have influenced conservation policy and they remain pervasive. Because of this, it is important to understand both the strengths and, particularly, the weakness of the 'island' theories discussed in this and the previous chapters. In my view, the stochastic consequences of reduced population sizes are not all-powerful effects. It follows that they may not provide reliable estimates of how many species really are 'committed to extinction', i.e. still around but heading for an inevitable nemesis. Some would disagree with me. For instance, Pimm and Askins (1995) argue that concerns about the precise causal mechanisms of extinctions, as expressed here, miss the point and that, in practice, simple predictions of species losses based on species–area relationships are actually fairly reliable. From their analyses of North American forest birds they suggest that if such estimates err, they tend to be conservative, i.e. predicting fewer extinctions than actually occur. I am not convinced that we are yet in a position to be so sure in which direction the predictions will err.

None the less, I do not suggest that we should simply discard the theoretical frameworks of island theory, metapopulation theory, etc. They have their relevance to particular systems and a wider and important role in helping us to understand the sorts of effects that might be operative in real-world contexts, and that therefore should be tested for or monitored. As Doak and Mills (1994) have put it, we should face up to the problems rather than use them as an excuse for ignoring the theoretical debates: 'it is incumbent on us to teach such complexity to managers and nonbiologists, rather than attempting to snow them with undefendable over-generalizations'.

It may be politically inept of a natural scientist to challenge the assumptions on which predictions of species losses are based. The problem is a depressingly real one and the pace of species extinctions is undoubtedly accelerating (Whitmore and Sayer 1992; Turner 1996). Yet, as Simberloff *et al.* (1992) observe, ideas derived from island theory, and the so-called SLOSS principles (specifically the benefits of corridors), became embedded in policies of bodies such as the IUCN and World Wildlife Fund without having been validated. Thus, policy makers and legislators have in cases acted upon such uncertain scientific 'principles'. Once released from the bounds of scientific journals, scientific information is often poorly understood. It is therefore important not to offer up general principles and rules where the evidence for them is equivocal and, in cases, contradictory (Shrader-Frechette and McCoy 1993).

The notion that species disassembly (as well as assembly) is fairly structured and predictable, at least in terms of functional ecological types or guilds, is one that has reappeared in several guises in this chapter (e.g. from studies of relaxation, experimental fragmentation, incidence functions, nestedness, and edge effects). In the case of the Amazonian bird fragmentation study, similar structuring was evident both in the initial losses (disassembly) and, in particular circumstances, in the recovery that followed. This reminds us, first, that habitat change does not *have* to be a one-way process, and, secondly, that it might be possible to target conservation efforts fairly directly on to particular sets of vulnerable species. In this sense, 'island' approaches have produced some dividends. It is now a question of learning what they are, and differentiating between those effects which are weak and those which are strong. It seems that we can work out which sorts of species are most threatened and some of the major reasons why they are threatened, and it should therefore be possible to offer some management solutions. These insights have arisen largely from 'island' studies.

For many species, reserves are essential. Very often this is because of specific threats posed by people (or our commensals) and the protection that can be afforded within the reserve. Allied to the reserve strategy, wherever possible, conservationists should argue for greater priority to extensive rather than intensive conservation, i.e. for environmental management policies to encourage the survival of many species outside of the more closely protected reserves systems. In short, to work to prevent reserves becoming more and more like real islands—except in those biogeographical contexts where insuralization is actually beneficial to survival prospects. We have seen in this chapter that issues such as size, shape, and configuration within a landscape are important to reserve success, not just in terms of how many species will be held within a reserve, but also which sets of species. The number of species held in a reserve (or reserve system) is actually less important than to conserve those species which cannot survive outside the remnants (e.g. Newmark 1991). In a recent review Simberloff (1992) comes down in general on the side of large, continuous blocks of habitat. He argues that no existing theory adequately describes the joint effects of loss of area and fragmentation, but that there are empirical observations, such as those to do with predation discussed earlier, to suggest that for many species the combined impact is indeed strongly deleterious: 'Probably any species that has evolved in large, relatively continuous habitat has traits that are maladaptive in small, isolated fragments' (p. 85). Beyond this generalization, the means, rate, and extent to which extinction is delivered, are likely to be idiosyncratic. The theoretical basis for calculating likely regional and global extinctions as a result of the rapid phase of habitat loss currently under way remains highly uncertain. For a careful review of the evidence and basis for calculations see Whitmore and Sayer (1992).

9.11.2. Ecological hierarchies and fragmented landscapes

As with other branches of island theory, approaches to the debates on reserve configuration which take no account of hierarchical interactions between organisms of differing trophic levels are incomplete (Terborgh 1992). For many animal groups, plants (collectively or individually) determine the ability of particular animal species to occupy a habitat island. Successional changes in plant communities may often explain turnover in bird or butterfly communities (Chapter 8; Thomas 1994). Equally important may be the activities of animals as dispersers and pollinators of plants (and as predators), both in contexts of forest maintenance and of recolonization of disturbed areas. Different types of animals disperse different, albeit overlapping, suites of plant species. Studies of cleared areas in the neotropics have suggested that bats may be more significant in seeding open habitat than are day-flying birds. The efficacy of birds, in particular, may also be positively influenced by the availability of some woody cover in an otherwise open area (Estrada *et al.* 1993; Gorchov *et al.* 1993; Guervara and Laborde 1993). Thus, for instance, cheap plantings of small clumps of plants may encourage speedy return of forest around the inocula.

The Krakatau recolonization data demonstrate the significance of birds and bats in transporting seeds between sites, and thus also supports their significance for population interchange between forest patches in the tropics. Maintenance and restoration of forest might thus necessitate protection for vertebrate species, if not in their own right, for their interactions with plants. It is not 'merely' biodiversity which is at stake, as in cases there may be huge economic costs to the loss of pollinator species on which economically useful plants depend (Pannell 1989; Fujita and Tuttle 1991; Cox *et al.* 1992). At the same time, the Krakatau study demonstrated that particular groups of plants, those dispersed by non-volant frugivores and the large-seeded, wind-spread species, are likely to be greatly hampered by patch isolation (Whittaker and Jones 1994*a*,*b*; Whittaker *et al.* 1997). These topics have recently seen a resurgence of research interest, but if we are properly to understand the consequences of fragmentation and develop appropriate responses, much more needs to be done in this field.

It seems likely that that some area effects involve threshold responses and that these operate via hierarchical links within ecosystems (Soulé *et al.* 1988; Terborgh 1992). For instance, below a certain size, an isolated habitat island may be incapable of supporting populations of large predatory vertebrates, with the consequence that higher densities of their prey species may be maintained, with potentially significant consequences further down the food chain. Thus, the densities of medium-sized terrestrial mammals are between 8 and 20 times greater on Barro Colorado Island, which lacks top predators, than in a comparable 'mainland' site in which they occur. Terborgh (1992) suggests that this may be having important selective impacts on plant regeneration. On some smaller islets nearby in Gatun Lake, not only are the (true) predators absent, but so too are those smaller mammals that act as seed-predators, the squirrels, peccaries, agoutis, and pacas which have such high densities on BCI. These small islets have become dominated in the 70 or so years since their isolation by large-seeded tree species. This is plausibly because these plants no longer suffer high rates of attrition at the point of germination and establishment, and have therefore been able to out-compete the smaller-seeded species to a greater extent than usual.

Predators undoubtedly have a crucial role to play in fragmented landscapes, not just in terms of predation within the habitat island, but also within the matrix. Simberloff (1992) notes that in the Amazonian fragmentation study, the bat falcon (*Falco rufigularis*) has repeatedly been seen to chase birds flying over cleared areas from fragment to fragment. It may therefore be unsurprising that most understorey birds will not willingly cross gaps of even 80 m. Similarly, in the USA, the northern spotted owl (*Strix occidentalis caurina*), of metapopulation theory fame, previously was a species which rarely left closed forest. In the newly fragmented landscapes, as many as 80% of yearling males die, apparently from predation by great horned owls (*Bubo virginianus*) and goshawks (*Accipiter gentilis*) as they disperse over cleared areas. This was one consideration that persuaded the interagency federal committee responsible for the management plan to shift their strategy towards larger, more continuous blocks. This brief consideration of predation helps us recall the point that the particular mechanisms responsible for failures to sustain populations or to recolonize habitat islands may thus be poorly predicted by 'species–area' type-approaches, or for that matter by metapopulation models.

There may also be important dynamic features, such as the spatial and temporal oscillations that characterize elephant populations in the Tsavo National Park in Kenya, on a cycle of roughly 50 years. Even with very large reserves, migratory animals may ignore park boundaries and, especially in lean years, may move into areas outside, where they are most unwelcome and perhaps also unsafe. These patterns can provide big management problems, even when a very large area is enclosed as a reserve (Woinarski *et al.* 1992). The elephants illustrate another idea, that of the **keystone species**, meaning a species that is so critical to the functional character of an ecosystem that its removal would cause a chain of alterations (e.g. Soulé 1986; Pimm 1991). Elephants have a key role as grazers. Without them successional changes occur in the vegetation, altering the carrying capacity of the system for other species. High densities of elephant, on the other hand, can lead to overgrazing and widespread destruction of trees. In such cases, conservation management naturally focuses around managing the keystone species.

9.11.3. Climate change and reserve systems

Range shifts driven by significant climatic change have been a feature of the Quaternary period. There is no reason to consider these climatic fluctuations to be at an end, and there is also mounting concern that humans have initiated a phase of rapid climatic warming. Researchers are now focusing on the abilities of plant and animal species to track their required climatic envelope across the now fragmented landscapes (Huntley *et al.* 1995). Key factors are the speed of climatic change, the dispersal abilities of the target species, and the characteristics of the landscapes across which species may have to move, which may range from the optimal to the benign to the hostile. Grappling with such scenarios requires a variety of research tools (Bush 1996), amongst which are the tools of island biogeography reviewed in these pages (Boggs and Murphy 1997).

9.11.4. Concluding remarks

As made clear at the outset of this chapter, this review is not intended as a summary of all important branches of wildlife conservation. However, I have stressed the importance of autecological, distributional, and other forms of data. The student of conservation biology will also need to have a knowledge of such themes and issues as: *ex situ* conservation efforts in zoos, and other dedicated breeding facilities; plant gene banks; species translocation schemes; reserve management practices; CITES and other wildlife legislation; poaching and trade in endangered species; and the philosophy, politics, and economics of conservation (Primack 1993; Shrader-Frechette and McCoy 1993).

It must also be recognized that there can be many aims and purposes for shaping conservation management, ranging from the aesthetic, through the scientific, to the economic. We may wish to conserve systems or species which are 'representative', 'typical', rare, speciose, nice to look at, of recreational value, or provide economic return (Ratcliffe 1977; Shrader-Frechette and McCoy 1993). Such multiplicities of purpose require requisite tools; the island theories have their place in the tool kit, but they should not generally be the first to be reached for.

Conservation requires pragmatic decision-making. As we continue to fragment landscapes, island effects may inform such decision-making, but should not be oversimplified. There is no single message, and no single island effect; indeed insularity may, in at least a minority of cases, bring positive as well as negative effects (Lockwood and Moulton 1994). Island effects may be weak or strong. The implications of insularity vary, depending on such factors as the type(s) of organism involved, the type of landscapes involved, the nature of the environmental dynamics, the biogeographical setting, and the nature of human use and involvement in the system being fragmented. It is unfortunate that the term 'island biogeography theory' has become largely synonymous in a conservation setting with a limited conception of island ecology, stressing the inevitability of stochastically driven trends to equilibrium. Both pure and applied island biogeography are richer than this. This richness of ideas and information needs to be understood and integrated into teaching for both pure and applied purposes.

9.12. Summary

The island analogy can be extended to patches of habitats within continents. The conversion of more-or-less wild habitats to other uses is fragmenting, isolating, and reducing 'wild' areas across much of the globe. The implications of this insularization are examined in this chapter from the scale of individual populations up to whole landscapes.

The minimum viable population (MVP) is the smallest number of individuals required to ensure long-term population persistence. We do not know how big MVPs should be. We do know that MVPs vary from species to species, and that the effective population size is generally smaller than the actual population size, i.e. not all individuals, or all adults, are involved in breeding. Population loss may be due to stochastic demographic and/or genetic effects, or to environmental disturbance and change. Stochastic demographic effects may have been overestimated. Genetic effects are potentially complex, and vary between taxa. The significance of environmental change or catastrophe is probably crucial to population viability, but can be difficult to model. MVPs therefore need to be established separately for different types of species, and management regimes must be responsive to changing circumstances.

The area required by a MVP is termed the minimum viable area, and may be estimated by a knowledge of range size, although this approach assumes no population exchange with other isolates. Another approach to area (or habitat, or isolation) requirements is to use incidence functions, although the patterns revealed may be confounded by multi-causality, and may fail to predict changes that follow within fragmented landscapes.

Where geographically separated groups are interconnected by patterns of extinction and recolonization they constitute a population of populations, or a metapopulation. Here, the idea is that particular patches have their own internal population dynamics, but when they crash to local extinction they are repopulated from another patch within the metapopulation. Such a scenario alters the projections of population viability considerably. In many cases, however, habitat patches appear to describe a source–sink rather than a mutual support system. In this variant, a large patch acts as an effectively permanent population, with smaller sink habitats around it being re-supplied from the source population. The sink habitats may have little relevance to overall persistence-time of the metapopulation. There are few empirical illustrations of metapopulation dynamics; and in the absence of validation, the models should be understood as simplified 'what-if' scenarios.

A heated debate developed around the SLOSS 'principles', which represented an attempt to answer the question of whether it was better to advocate single large or several small reserves of the same overall area, based on the assumptions of the ETIB. Such theory provides no resolution to this debate. In practice, conservationists often have little influence over fragment configuration and, when they do, must consider other biogeographical and practical information. None the less, the implications of fragmentation pattern remain important. What is optimal for one type of organism or landscape may not be so for another. Physical environmental change in both fragment and 'matrix' habitats has not generally been given enough attention in the theoretical literature. Successional change and altered disturbance regimes may have long-term implications for species persistence in altered landscapes. The evidence for 'relaxation', i.e. stochastically driven decline in species number, is surprisingly thin. Species composition and richness are liable to change in fragmented systems: species are indeed lost. However, in relation both to SLOSS and metapopulation scenarios, it is apparent that much turnover may well be deterministic in nature and explicable in relation to 'normal' ecological processes such as succession. Various lines of data suggest strong structure to the disassembly as well as to the assembly of systems. For instance, many systems and species are strongly nested in their distributions, and this can, it seems, be produced in either colonization- or extinction-dominated systems.

It appears possible to predict the types of species that will be in greatest difficulty in fragmented landscapes, and it is these species that require primacy in relation to reserve configuration, population monitoring, and management planning. The consequences of fragmentation can be severe, but they are poorly predicted by classical island theories, which assume stochastic demographic effects to dominate in a system of fragments searching for a new equilibrium in a false sea. The nature of the connectivity of habitat patches may be crucial. In some cases, corridors and short-dispersal distances may be beneficial, whereas in other circumstances isolated reserves may actually provide the best chance of survival for a threatened species. Some species thrive in edge habitats, others are disadvantaged by increased edge effects. The 'filtering' properties of the matrix can be crucial and require careful assessment in conservation planning. Landscape effects may be complex, with the same linear feature acting as corridor, barrier, and lethal hazard for different organisms. Moreover, some species are mobile, seasonally or aseasonally, and their protection may require a more extensive approach to conservation than provided for by any reserve system on its own.

These considerations demand a re-evaluation of the place of island biogeography within conservation biology. Generalizations in this field are fraught with difficulty, although I will venture to say that 'equilibrium island theory' does not provide the basis for meaningful predictions of species loss. Some of the effects traditionally associated with island approaches may actually be fairly weak in general, but other features of insularity may provide powerful and fairly immediate influences on species persistence and system change. The role of hierarchical ecological relationships, e.g. of plants and their dispersers, of vegetation change and altering habitat space for butterflies, of understorey insectivores and their predators, and so forth, may be crucial to the consequences of fragmentation. Such successional and food-web controls may be the most powerful forms of expression of islandness in the altered landscapes.

A line must be drawn somewhere in any text, and in this chapter discussion stops short on alternative approaches to conservation problems, and on themes such as the implication of large-scale climate change under global warming scenarios. Neither is any attempt made to draw out guiding principles for conservation from island studies. The grand hopes for simple unifying principles have proven to be as illusory as the end of the rainbow. It is none the less important to appreciate how these ideas have influenced conservation theory, just as it is important to tackle the ecological complexities of increasing insularity within real landscapes.

10

The human impact on island ecosystems—the lighthouse-keeper's cat and other stories

We cannot discuss the ecology of islands without making a few disparaging comments on goats. These creatures must be the true embodiment of the devil for a plant lover.

(Koopowitz and Kaye 1990, p. 72.)

The biodiversity crisis is nowhere more apparent and in need of urgent attention than on islands. Approximately 90% of all bird extinctions during historic times have occurred on islands...

(Paulay 1994, p. 139.)

10.1. Introduction—the scale of loss

Like the literature on which it is based, much of this book has set to one side the significance of humans as shapers of island biogeography. To do so is unjustified in a contemporary sense, as demonstrated in the previous chapter, and it is equally questionable with respect to the role played by human societies in historical and prehistorical time (Morgan and Woods 1986; Johnson and Stattersfield 1990; Groombridge 1992; Pimm *et al.* 1995). Humans have influenced the ecology of islands in many ways, but the most profound must surely be the extinction of numerous island races and species.

Those collating records of species extinctions commonly take the historic period of island exploration and scientific record to have begun *c.* AD 1600. In the period so defined, far more species of plants and animals are known to have become extinct from islands than from continents: for the taxa listed in Table 10.1, about 80% have been island species. However, we know so little about the losses of invertebrates that it would probably be safer to discount them from the analysis. For the three groups with the best data, the proportion of extinctions from islands varies from 58% for mammals (which are generally lacking from remote islands in the first place), to 80% for molluscs, and 85% for birds. Moreover, of 121 continental species losses cited by Groombridge (1992), about 66% can be classified as aquatic species. To date, therefore, contrary to the impression commonly given, relatively small numbers of terrestrial animal extinctions have been recorded from mainland tropical forest regions such as Amazonia, although of course these are the areas from which large numbers of extinctions are estimated to be occurring or about to occur. In the case of birds, and allowing for the numbers of island versus continental species, island forms have had about a 40 times greater probability of

Table 10.1

Summaries of known animal extinctions on islands and continents since c. *AD 1600 (data from Groombridge 1992; Steadman 1997a)*

Taxon	Number of species	
	Islands	Continents
Birds	97	17–20
Molluscs	151	40
Mammals	34	24
Reptiles	22	1
Amphibians	0	2
Insects	51	10
Total	355	94–97

extinction over this period than have continental species (Johnson and Stattersfield 1990).

Recent palaeontological and archaeological studies have made it abundantly clear that island endemics have been extinguished in significant numbers from islands and archipelagos all over the world, before the explorations of modern collectors brought their biotas to the attention of Western science (Table 10.2). In the case of island birds, at least, the established prehistoric losses greatly outnumber the historical losses. Table 10.2 provides a figure of 158 extinctions of island endemic birds in prehistoric times, but a slightly higher figure, of just over 200 species (recorded as subfossils and which probably vanished due to prehistoric man), is given by Milberg and Tyrberg (1993). There is no doubt that further archaeological and taxonomic work will add to the number of extinctions for the prehistoric period. It should therefore be understood that the year AD 1600 did not mark a key point in island history, and that its use by those compiling the statistics is essentially a matter of convenience. The primary place at which we ought to divide the analysis is the point at which islands have been colonized by humans. Steadman (1997*a*) estimates that rates of natural, pre-human extinction in island birds may be at least two orders of magnitude less than post-human rates. The world's avifauna would have been about 20% richer in species had the islands of the Pacific remained unoccupied by humans. As many as 2000 bird species may have become extinct following human colonization but prior to European contact; the list being dominated by flightless rails, but including moas, petrels, prions, pelicans, ibises, herons, swans, geese, ducks, hawks, eagles, megapodes, kagus, aptornthids, sandpipers, gulls, pigeons, doves, parrots, owls, owlet-nightjars, and many types of passerines. The emphasis on rails stems from observations that virtually every Pacific island that has been examined palaeontologically has been found to have had one or more rails of such reduced flight capabilities that they must have been endemic species. It is conceivable that

Table 10.2

Estimates of actual and threatened losses of island endemic birds (data from Johnson and Stattersfield 1990; Steadman 1997a). The total of 97 extinctions since AD 1600 contrasts with between 17 and 20 lost from continental areas over this period (Table 10.1)

Region	Number of extinctions			Endemism	
	Prehistoric[a]	AD 1600–1899	AD 1900–94	Approximate number of endemics	% endemics threatened
Pacific Ocean	90	28	23	290	38
Indonesia and Borneo	–	0	2	390	22
Indian Ocean	11	30	1	200	33
Philippines	–	0	1	180	19
Caribbean Sea	34	2	1	140	22
New Guinea and Melanesia	10	2	3	500	10
Atlantic Ocean	3	3	1	50	50
Mediterranean Sea	10	0	0	–	–
Total	158	65	32	1750	23

[a]Data are lacking for prehistoric losses from Indonesia and the Philippines. The prehistory category consists of species from prehistoric cultural contexts, and includes only species already described. Many others remain undescribed and doubtless there are more yet to be discovered by the archaeologists; these estimates are thus minimum figures.

more than 800 Pacific islands were once graced with one or more rail species in one of the most dramatic examples of adaptive radiation to have persisted into the Holocene (cf. Trewick 1997). Only a handful of these species remain today.

From current understanding of threatened birds and mammals, IUCN data show that the threat of extinction is switching significantly towards the continents (Groombridge 1992). Yet, even today, one in six plant species grows on oceanic islands and one in three of all known threatened plant species are island endemics. The degree to which the remaining island endemic birds are believed to be threatened is quantified in Table 10.2. These statistics provide measures of the diversity and apparent fragility of island ecosystems and their importance in conservation terms.

10.2. The agencies of destruction

There are four major reasons why island species are reduced by human action: (1) direct predation; (2) the introduction of non-native species; (3) the spread of disease; and (4) habitat degradation or loss. In a global context, the loss of habitat is commonly seen as the greatest problem for biodiversity (Lawton and May 1995), but on particular islands other factors from this list may be more pressing. For instance, on the Galápagos, introduced animals and plants have been described as 'absolutely the most insidious, silent, and dangerous scourge that exists in the archipelago' (*Galapagos Newsletter* 1996, No. 3). Commonly, native endemic species are caught in a pincer action, reduced by habitat loss to a rump, attacked by disease or predation, and facing competition with exotic species. The synergisms between these forces result in the loss of species which conceivably could cope with and adjust to one force alone. Similarly, there may be knock-on effects through food-chains links, such that the loss of prey species might prejudice the survival of native predators. For example, the very large extinct New Zealand eagle, *Harpagornis moorei*, was probably dependent on moas and other large extinct bird species (Milberg and Tyrberg 1993).

10.2.1. Predation by humans

Humans hunt island species both for the pot and for other reasons, such as the societal values attached to brightly coloured feathers. In the case of the extinct Hawaiian mamo (*Drepanis pacifica*), it has been estimated that the famous royal cloak worn by Kamehameha I would have required some 80 000 birds to be killed (Johnson and Stattersfield 1990). Trade in such biological resources has a long history, and was an important feature of the period in which the Polynesians spread across and held sway over the Pacific. Present-day trade in fruit bats (flying foxes) provides an example of how hunting, not primarily for local consumption, but for export, can reduce oceanic island populations to crisis point. The critical role that these bats play as pollinators and as dispersers of plants may mean that their loss will have significant knock-on impacts, i.e. they act in effect as 'keystone' species (Rainey *et al.* 1995; Vitousek *et al.* 1995). Hunting for collections continues to be problematic. Today, the trade is driven mostly by individual collectors, but in the past, hunting for museum collections was common and is, for example, believed to have dealt the final blow to the spectacular New Zealand bird, the huia (*Heteralocha acutirostris*) (Johnson and Stattersfield 1990).

10.2.2. Introduced species

As humans have spread across the islands of the world we have taken with us, either purposefully or inadvertently, a remarkable array of plant and animal species. As often noted, it is the introduction of mammalian predators, such as cats, rats, dogs, and mongooses, which have caused some of the worst problems (Cronk 1989; Groombridge 1992; Benton and Spencer 1995; Keast and Miller 1996). Feral cats are a feature of innumerable islands. They are opportunistic predators, eating what is most easily caught. Cats were introduced to the subantarctic Marion Island in 1949 and at one stage it was estimated that there were about 2000 of them, each taking about 213 birds a year, and thus in total killing well over 400 000 birds per annum, mostly ground-nesting petrels. Given these loses, only by eradicating the cat from the island could the long-term survival of the native bird species be ensured (Leader-Williams and Walton 1989). Cats have been present on another subantarctic island, Macquarie, for a lot longer, and in that case have been responsible for the loss of two endemic bird species. The combination of cats and mongooses on the two largest Fijian islands has resulted in the local extinction of two species of ground-foraging skinks of the genus *Emoia*: they survive only on mongoose-free islands. In the Lesser Antilles, three reptiles have been extirpated (i.e. lost from the particular island but not altogether extinct as a species) from St Lucia within historical time, coincident with the introduction of the mongoose. Domestic dogs can also be devastating, and feral populations have been responsible for the local extinction of land iguanas in the Galápagos. Even invertebrates have led to the extinction of island endemics, notably the losses of all endemic species of *Partula* land snails following the misguided introduction of the carnivorous snail *Euglandina rosea* to the island of Moorea (Williamson 1996). Similarly, exotic ants introduced to the Galápagos and Hawaii (from which they are naturally absent), predate endemic invertebrates and are an increasing cause for concern in both archipelagos. Not only exotic predators have brought about extinctions, however. Browsing animals such as goats and rabbits have also led to losses (Cronk 1989; Johnson and Stattersfield 1990). Examples blamed on the rabbit include the Laysan rail (*Porzanula palmeri*) and Laysan millerbird (*Acrocephalus familiaris*), both of which disappeared during the early twentieth century because rabbits denuded their island.

10.2.3. Disease

The problem of disease is closely associated with exposure to exotic competitors and, indeed, the separation of exotic microbes as a category from other introduced organisms is largely arbitrary. The most striking exemplification of the impact of disease on insular populations has, without doubt, been among the native peoples of islands, particularly on first contact with the Europeans. This factor, in combination with more purposeful persecution, led to the decimation of many island peoples, such that in cases little or no trace of them remain (Kunkel 1976; Watts 1987). That the problem of disease is taken seriously by many conservationists is illustrated by reference to Chapter 9, and the debate concerning translocation of Puerto Rican parrots between a proposed mainland captive breeding pool and the island population.

The most commonly cited example of exotic disease afflicting native island birds is from Hawaii. Hypotheses to explain why so many species of birds became extinct in the period since Cook's landfall in 1778 include habitat destruction, hunting, competition with introduced birds, and introduced predators. Undoubtedly, each has had its role in the degradation of Hawaii's ecosystems. Early work by Warner (1968) on the possible role of disease identified avian malaria (and avian pox), carried by the introduced mosquito *Culex quinquefasciatus*, as an important factor. Warner proposed an 'imaginary line' at about 600 m above sea level (ASL), above which there were no mosquitoes and below which native birds were not found because they had succumbed to the disease. However, later studies have shown that the role played by disease is not as simple as that. Van Riper *et al.* (1986) suggest that avian malaria probably took some time to build up a large enough reservoir and that it did not have a major impact upon the numbers of Hawaiian birds until the 1920s, some time after the extinction pulse of the late nineteenth century. None the less, since then it has had a serious negative impact on the native species. Van Riper *et al.* (1986) report that the malarial parasite *Plasmodium* is found on the big island of Hawaii from sea level to tree line, i.e. to elevations greater than assumed by Warner. The parasite is, however, concentrated in the mid-elevational ranges, in the ecotonal area where introduced bird species and native birds have the greatest distributional overlap. The introduced species have been found to be less susceptible to malaria and thus to act as vectors for the disease. Although a number of native bird species have developed immunogenic and behavioural responses reducing the impact of the parasite, van Riper and colleagues concluded that avian malaria is currently a major limiting factor, restricting both abundance and distribution of the endemic avifauna.

The omao (*Myadestes obscurus*) is one of four surviving species of thrushes on the Hawaiian islands. It occupies no more than 30% of its former range, and while locally abundant in rain forests in particular parts of the big island of Hawaii, it is absent from other areas in which it was once common. Its disappearance from the Kona and Kohala districts during the early part of the twentieth century remains an enigma. A plausible hypothesis is that it failed to develop resistance swiftly in these parts of its range to avian malaria, but in other parts of its range, such as the Puna district, it did develop resistance, thereby allowing its present pattern of coexistence with both the mosquitoes and the malarial parasites in that area (Ralph and Fancy 1994). Given the possibilities of local variations in the development of resistance to disease, it is clearly going to be difficult to be certain about the role of disease in the demise of Hawaii's birds. Added to which, it needs to be remembered that most species were lost prior to European contact, and that a host of other changes have occurred over the past 200 years or so. The pattern of post-European contact losses is bimodal, being concentrated between 1885 and 1900, and 1915 and 1935. The former phase of losses included many species historically confined to higher altitudes (>600 m ASL) and therefore contradicting Warner's (1968) original malarial scenario. The explanation for this phase of losses may relate to another introduced disease, avian pox, or to the impact of habitat modifications and introduced predators. However, the second period does appear to relate to the arrival and spread of malaria. Birds which succumbed during this period were found in the mid-elevational forests in which the highest prevalence of avian malaria was found during studies around 1977–80 (van Riper *et al.* 1986).

Two things emerge. First, disease has played and continues to play a part in the problem, causing range reductions and precipitating extinctions. Secondly, we are dealing with a complex system involving opportunities for the spread of exotics due to habitat alteration and the interactions of several introduced organisms. The exotic birds and mosquitoes act as vectors, and the exotic disease provides the proximal cause of mortality. That introduced birds act as vectors illustrates that important interactions between organisms at the same trophic level are not all obviously 'competitive' in nature.

10.2.4. Habitat degradation and loss

Probably the most renowned example of insular habitat degradation is the case of Easter Island, a depressing experiment in unsustainable exploitation of natural resources, which some have seen as a warning for the rest of us (Bahn and Flenley 1992). Environmental degradation has actually been widespread across the Pacific, and has been linked to both Polynesian and European expansion (Nunn 1990, 1994; Diamond 1991*b*). Habitat changes are hard to separate from other negative influences, but there is no doubt that the reduction in area of particular habitats (notably forest) and the disturbance of that which remains, have been potent forces in the reduction and loss of numerous island endemics (Steadman 1997*a*,*b*). A specific example is the loss of the four-coloured flowerpecker (*Dicaeum quadricolor*) following the almost complete deforestation of the island of Cebu in the Philippines (Johnson and Stattersfield 1990).

A popular image of pre-industrial hunter–gathers and neolithic agriculturalists, is that such peoples lived in harmony with their environment, practising a conservation ethic and avoiding the exploitative destruction ('development') typical of later, industrial societies. This idea is associated with notions of the 'noble savage' and a mythical 'golden age' (Milberg and Tyrberg 1993). Sadly, the evidence for widespread degradation of island ecosystems across the world, prior to contact with modern European societies in the past 400 years or so, is now overwhelming.

The evidence which has accumulated is thus demanding a reappraisal of the significance of direct damage by humans over the past 150 years or so, as compared, for example, with damage caused by the Polynesians (Nunn 1990). This remains, in some parts of the world, a rather sensitive issue and so I will emphasize at the outset of this chapter that there is nothing unique about the degradation brought about on Easter Island, New Zealand, and elsewhere by the Polynesians. It is part of a pattern that has happened to the islands of the Indian Ocean, the Mediterranean, the Atlantic, the Caribbean, and the Subantarctic, and indeed has accompanied the expansion of innumerable human societies throughout the globe, and which has affected land masses from small islands to entire continents (Diamond 1991*b*; Milberg and Tyrberg 1993; Schüle 1993).

Easter Island provides one spectacular example of cultural collapse linked to overexploitation, which included the complete deforestation of the island *before* European contact (below). Prehistoric humans on islands, although dependent on limited animal resources, regularly failed to exploit these in a sustainable way. Several cases where human populations disappeared altogether from Pacific islands may have been due to overexploitation of native animals. It seems likely that the Polynesian taboo system, which prevents overexploitation, was started as a result of the extinctions and misuse of other resources. Apparently the 'ecological balance' between pre-industrial societies and their environment only applied to biota that could not be profitably overexploited with the available technology and that could survive habitat destruction and the introduction of new organisms, such as the rats, dogs, and pigs brought by the Polynesians (Rudge 1989; Bahn and Flenley 1992; Milberg and Tyrberg 1993; Benton and Spencer 1995). In short, the romantic view of the noble savage living in balance with nature is illusory, and where something like it was achieved, it was only accomplished by trial and plenty of error along the way. Pimm *et al.* (1995) estimate that in the Pacific, the first colonists typically exterminated approximately half of the native avifauna. Yet, they also caution that taken at face value some extinctions may be falsely attributed to the first colonists, because intensive collection often began half a century or so after the damage initiated by European discovery. Furthermore, while in many systems it appears that the first colonists had eventually reached some sort of balance with the native faunas and floras, it is clearly the case that more recent colonists have continued the process of disruption such that today the fates of many oceanic island species hang by a thread.

10.3. Trends in the causes of decline

It is easiest to identify trends in the nature of the threat with reference to island birds, for which the most systematic assessments have been carried out. In prehistoric times the majority of losses can be attributed to hunting, followed by the introduction of commensal species. In the recent historical past (last 400 years), it appears that introduced species (not all of them human commensals), followed by hunting, have been the key forces (Table 10.3). Today, the nature of the threat is changing: habitat destruction is cited in respect to over half of the threatened island bird species, and only a fifth now appear to be threatened principally by interactions with exotic species (Johnson and Stattersfield 1990). Many of the potential introductions of predators to once predator-free islands have already occurred and in cases the impacts of exotics are being mediated by protection measures. The legacy from past introductions remains important, however, particularly because they are implicated in the fates of many of the most immediately threatened bird species (the endangered category). For instance, the endemic avifauna of Guam is severely endangered—and probably doomed to extinction—due to nest predation by the introduced brown tree snake (*Boiga irregularis*) (Williamson 1996).

The loss of forest or woodland cover is the biggest issue in terms of habitat change (Fig. 10.1). Johnson and Stattersfield (1990) estimate that about 402 threatened species of

Table 10.3

Principal causes of rarity, decline, and extinction of island endemic bird species (from data in Johnson and Stattersfield 1990)

Threat	Number of species	
	Extinct since *c.* AD 1600	Rare or declining
Habitat destruction	19	206
Limited range	–	165
Introduced species	34	76
Hunting	25	35
Trade	–	16
Human disturbance	1	10
Natural causes	1 or 2	11
Fisheries	–	2
Unknown	41	64

For many species in both 'extinct' and 'rare or declining' categories, more than one cause has been assigned and the values given above reflect this; the overall totals in the dats set were 97 'extinct' and 402 'rare or declining'.

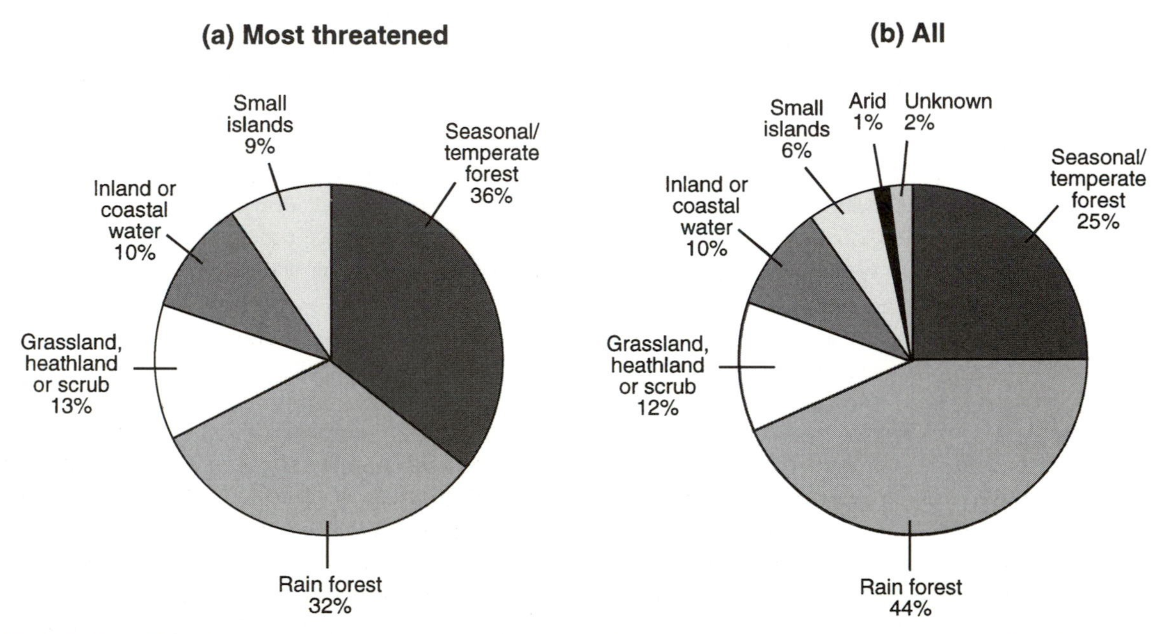

Fig. 10.1. Habitats of threatened island endemic bird species: (a) species listed as either endangered or vulnerable, (b) species listed as endangered, vulnerable, rare, indeterminate, or insufficiently known. Some species have been assigned to more than one habitat in this analysis, hence, for example, the relative importance of the combined forest habitat in part (b) (69%) differs slightly from the estimate for forest species (77%) given in the text (p. 235). Data from Johnson and Stattersfield (1990), according to the then IUCN *Red Data Book* categories: *endangered*—those in danger of extinction and whose survival is unlikely if the causal factors continue operating; *vulnerable*—those likely to move into the endangered category in the near future; *rare*—taxa with small world populations; *indeterminate*—taxa known to belong to one of the foregoing categories, but with insufficient data to say which; *insufficiently known*—taxa suspected but not definitely known to belong to one of the foregoing categories. (The data given in this figure exclude the category *extinct*—taxa no longer known to exist in the wild after repeated searches of their type localities.)

birds are restricted to islands (a further 67 threatened species have both island and continental distributions). Of these 402 species, as many as 77% are forest species, divided between 200 rain-forest (lowland and montane) and 113 seasonal/ temperate-forest species. A recent assessment of threatened birds in insular South-East Asia, found a close relationship between the numbers of bird species considered threatened, and the numbers of extinctions predicted from deforestation using a species–area approach (Brooks *et al.* 1997). While certain caveats must be attached to such crudely derived estimates (cf. Chapter 9; Bodmer *et al.* 1997), there is no doubt that habitat loss is a key driving force no matter which means of assessment is used. The species–area approach may be of most interest in areas where the fit is poorest between its predictions and those made by other means (i.e. in practice, assessments made by reference to both area/fragmentation effects and to other sources of information, such as actual population declines). For instance, the species–area relationship overestimates species extinction threat in the Lesser Sundas, and this was suggested by Brooks *et al.* (1997) to be because the natural vegetation of these islands is deciduous monsoon rather than moist tropical forest. Most endemic birds are found in this deciduous monsoon habitat and they speculate that these birds may adapt better to secondary growth and scrub than do the moist forest species. In contrast, in the Philippines, 78% of the endemic birds are confined to the accessible lowland forests in which deforestation has been concentrated, hence over half of the Philippines' 184 endemic birds are considered threatened.

How these various assessments of risk translate into future patterns of loss will depend on a variety of factors; not least, the protection afforded to particular island habitats and species, and conflicting economic pressures on the resource base of particular islands. It is one thing to identify causes of species extinction, quite another to provide remedies. Solutions to the problems demand political will, legislation, finance, education, and broad public support. None the less, the first step is to assess and articulate the problem.

10.4. A record of passage—patterns of loss across island taxa

These sections are designed to be illustrative rather than globally comprehensive, they provide accounts of how humans have brought about ecological alteration and species extinctions for particular islands and taxa.

10.4.1. Pacific Ocean birds

Birds are typically the most species-rich vertebrates on remote islands, they leave good fossil or subfossil evidence, and have been relatively well studied, often from sites associated with human settlement (i.e. middens). The picture that has emerged of their demise in prehistoric times can thus be more closely related to human activities than is the case for any other taxon—and the oceanic region where the most dramatic and convincing evidence of human involvement has been found is the Pacific. Not only has the Pacific lost large numbers of bird species in the past, but today it holds more threatened bird species than any other oceanic region: approximately 110 species, of which 31 are classified as endangered and 29 vulnerable (the two highest *Red Data Book* categories; Johnson and Stattersfield 1990).

It is believed that the Australian and New Guinea land masses on the Sahul shelf were colonized by people at least 50 000 and 40 000 BP respectively. Voyagers reached New Ireland and the Solomons by 29 000 BP, but then paused for some 25 000 years—islands beyond this zone quite possibly were undetectable to these people (Keast and Miller 1996). Then the ancestors of the Polynesians became the first wave of human colonists to spread across the Pacific, and they did so very rapidly for a society of apparently primitive technology, between 4000 and 1000 BP. So extensive were their explorations that nearly all islands in Oceania (Melanesia, Micronesia, and Polynesia) were inhabited by humans within the prehistoric period (Pimm *et al.* 1995;

Steadman 1997*a*). The human colonists cleared forests, cultivated crops, and raised domesticated animals. Birds provided sources of fat, protein, bones, and feathers for people and so were hunted as well as being indirectly affected by the changes that followed human arrival. European exploration began in the sixteenth century, but their phase of colonization started much later, such that, for instance, the first missionaries arrived in Tahiti in 1795 and in Hawaii in 1779 (Pimm *et al.* 1995).

In the Hawaiian islands, at least 62 endemic species of birds have become extinct since Polynesian colonization some 2000 years BP, and most of these were lost prior to European contact (Steadman 1997*a*). The processes involved are manifold, the role of disease has already been noted, and habitat destruction has also played its part. The commensals introduced by the Polynesians have been supplemented since European arrival by many other exotic species, which have had a powerful impact on the native biota, including black rats, feral dogs, cats, pigs, sheep, horses, cattle, goats, mongooses, and an estimated 3200 species of arthropods (Primack 1993). The extinctions have included flightless geese, ducks, ibises, and rails, and the major part of the radiation of Hawaiian honeycreepers (Olson and James 1982, 1991; James and Olson 1991). Depending upon the assumptions made in calculating the size of the prehistoric avifauna, various estimates are possible. Pimm *et al.* (1995) believe that roughly 125–145 bird species once inhabited the main islands, of which 90–110 are extinct, with another 10 likely to have joined them by *c.* AD 2000. The original avifauna contained only three non-endemic land birds, and all the passerines were endemic, yet in consequence of the enormous losses and of large numbers of species introductions by humans, only about a third of the passerines found on Hawaii today are endemic. Nearly all the birds seen now below 1000 m ASL are aliens. Today, some 48% of the land birds of Hawaii are exotics introduced and released in order to augment the fauna with game birds and birds of beautiful plumage or song. Of 94 species known to have been introduced prior to 1940, 53 species became established at least locally and only 41 failed completely.

The Hawaiian islands serve to illustrate another feature of prehistoric extinctions, namely, how they altered the structure of vertebrate—specifically bird—feeding guilds. All but one species of native predator became extinct (James 1995). Such was the fate of the terrestrial herbivores, with the exception only of the Hawaiian goose (*Branta sandvicensis*), which was rescued from the brink of extinction by an *ex situ* breeding programme initiated by the Wildfowl Trust at Slimbridge in the UK. Losses included the large flightless anseriforms (moa-nalos), which were browsers on understorey vegetation. The terrestrial omnivores all but disappeared too, along with the Hawaiian land crabs. Prehistoric extinctions among insectivores (37%) and arboreal frugivores/omnivores (40%) have been proportional to the losses among passerines generally, but granivores have suffered much worse and nectarivores much better in comparison. With predators disappearing, habitats shifting, and species richness falling during the shake-up following Polynesian settlement, it is likely that some species, even guilds, became both relatively and absolutely more widespread and abundant than under conditions before human contact. James (1995) argues that the differential effects of extinction on vertebrate feeding guilds in prehistoric times may still be affecting plant communities, i.e. that the losses must collectively have implications for the functional characteristics of the remaining ecosystems (cf. section 3.6), but such a legacy is difficult to separate from the other driving forces for change.

New Zealand too, has seen many losses, including at least 44 land birds. The cat referred to in the title of this chapter played its role in the loss of one of the smaller birds, the Stephens Island wren, named *Traversia lyalli* after the lighthouse keeper (Lyall) who discovered it. Stephens Island is a tiny island in Cook Strait, off New Zealand's South Island. The wren was unknown to science until in 1894 the lighthouse-keeper's cat brought in a few specimens it had killed. The exotic predator met the island endemic: end of story. At least, this was the conclusion reached at the time, prompting a correspondent to the Canterbury Press to suggest that in future the Marine Department should see to it that lighthouse keepers sent to such postings should be prohibited from taking any cats with them, 'even if mouse-traps have to be furnished at the cost of the State'. Although cats have been claimed to have by far the worst record in exterminating New Zealand birds (King 1984), I wouldn't want to demonize this particular animal, whether or not it delivered the telling blow. Prior to the arrival of the Polynesians, the wren was actually a widespread mainland resident (Holdaway 1989; Rosenzweig 1995). Its loss from mainland New Zealand was probably due to a combination of habitat changes and the introduction of rats by the Polynesians; the loss from Stephens Island was but the final step to extinction. Interestingly, it is just such small offshore islands that provide the enclaves that may yet save some of New Zealand's remaining endemics from extinction as, cleared of exotic pests, translocated native populations are able to establish and persist (below). Here in microcosm can be found some of the essential ingredients of the diversity disaster of oceanic islands, and a lifeline of hope for the future.

The pre-human avifauna of New Zealand was probably about twice the size of the native avifauna today. The most famous losses were of the giant, flightless moas, the largest of which weighed around 300 kg (about nine times the weight of an emu) (Rudge 1989; Holdaway 1990). The Polynesians discovered New Zealand around AD 950, having voyaged from the area containing the Marquesas, the Society Islands, and the South Cooks, in excess of 3000 km to the east. They brought with them six species of plants and two species of mammals. Polynesian settlement led to the destruction of half of the lowland montane forests, widespread soil erosion, and the loss or reduction of much of the vertebrate fauna. During the phase of their expansion, New Zealand lost its sea lions and sea elephants, and at least 25 bird species, including the 11 species of moa which, in the absence of terrestrial mammals, had evolved to occupy the browser/grazer role normally occupied by animals such as kudu, bushbuck, or deer. The most severe modification occurred between 750 and 500 years ago, when a rapidly increasing human population overexploited animal populations and used fire to clear the land. By about AD 1400, the major New Zealand grazing systems had ceased to exist, their place being taken by unbrowsed systems: the removal of the moas thus had significant effects on the structure and species composition of the vegetation. Holdaway's (1990) analyses of subfossil remains indicate that the pre-human avifauna was dominated by forest species. The high rate of extinction in the avifauna prior to European settlement must have been related to the characteristics of the forest areas removed by the Polynesians: these being shown by palaeoecological data to have been mainly the drier, eastern forests, or those inland areas where drought or severe climate restricted regeneration after clearances. In pre-human times, it was the drier, more structurally diverse forests on more fertile soils which supported the greatest diversities of New Zealand's birds.

Pacific sea-bird populations have also suffered, as Steadman (1997*a*) points out; petrels and shearwaters have seen the greatest loss of species, and many other sea-bird species have had their distribution and population sizes diminished, including albatrosses, storm-petrels, frigatebirds, and boobies. Abbott's booby is a classic example: once widespread in the South Pacific, it now breeds only on Christmas Island in the eastern Indian Ocean.

Easter Island (27°9' S, 109°26' W) is the most isolated piece of inhabited land in the world. When first contacted by Europeans, it presented a fascinating enigma: how could this impoverished treeless island, with its impoverished people, have supported the construction of the remarkable giant statues (*moai*), and how and why had it all gone wrong? The flora of Easter Island consists today of over 200 vascular species, of which only 46 are native. However, formerly the flora was richer, containing several native tree and shrub species. Evidence for a forest cover is provided not just by pollen records, but also by the discovery of extinct endemic achatinellid land snails, which elsewhere in the Pacific appear to be dependent upon forest cover. Palynological studies by Flenley and colleagues demonstrate that the forests of Easter Island persisted for at least 33 000 years (as long as the record goes), during the major climatic shifts of the late Pleistocene and early Holocene. There is now little doubt that deforestation by humans was responsible for the treeless state of the island (Flenley *et al.* 1991; Bahn and Flenley 1992; Flenley 1993). The Polynesians first reached the island about 1500 years BP (AD 400). Their effect on the forest was detectable in the pollen record within 400 years, and within about 1200 years (*c.* AD 1680) the formerly forested island was completely treeless. The remarkable megalithic culture, which had sustained the quarrying, sculpting, transport, and erection of the *moai*, suddenly collapsed. Figure 10.2 sets out a model of the sequence of events which may have led to this collapse, within which a pattern of overpopulation and continuing overexploitation and thus resource depletion appears to be central. This depletion involved not just the loss of a forest cover, but through it the loss of the raw materials for ocean-going canoes, further diminishing resource availability. The loss of trees and other plant species is matched by a more complete loss of native birds than on any comparable island in Oceania (Steadman 1997*a*). Examination of bird bones associated with 600–900-year-old artefacts has revealed that Easter Island once had at least 22 species of seabirds, only seven of which now occur on one or two offshore islets, and just one of which still nests on the main island. Native land birds are known to have included a heron, two rails, two parrots, and an owl. None survive.

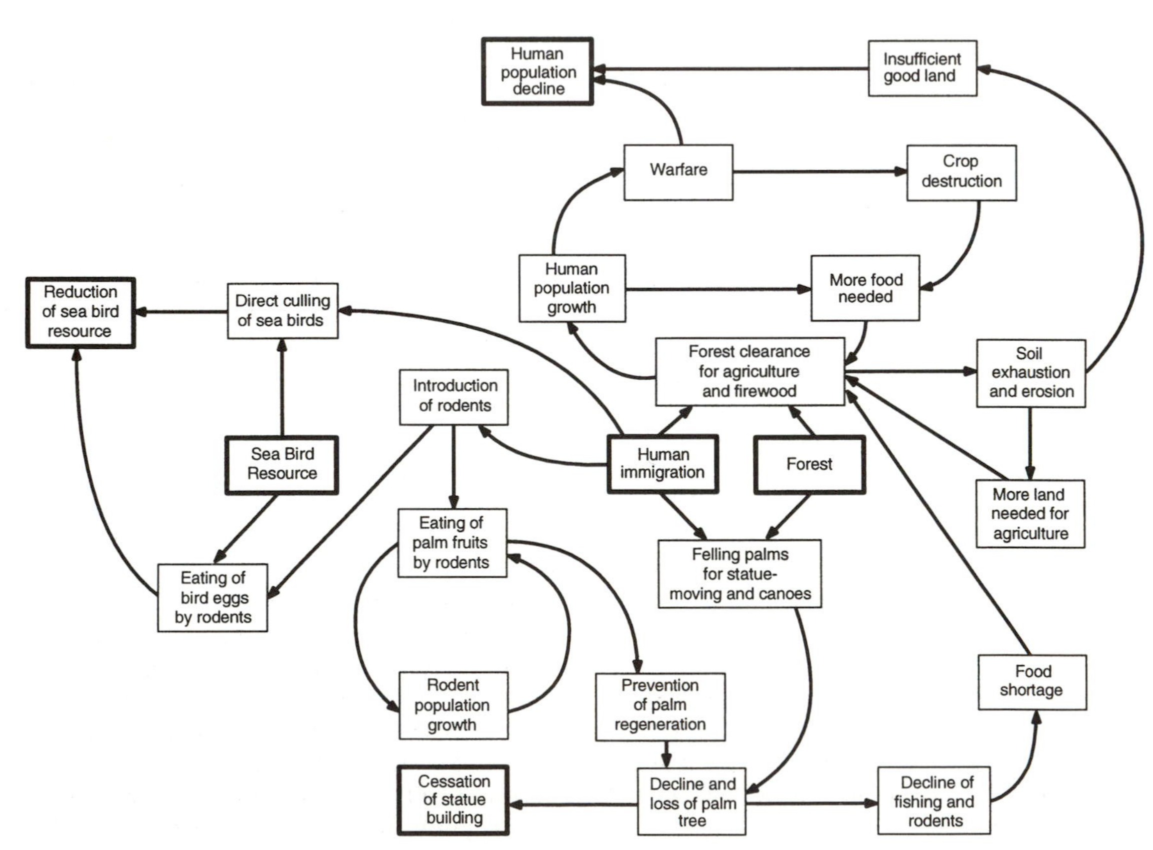

Fig. 10.2. A hypothetical model to indicate the possible course of events on Easter Island as deduced largely from palaeo-ecological study. (Modified from Bahn and Flenley 1992, Fig. 191.)

The islands of the Pitcairn group lie between 23.9 and 24.7° S and 124.7 and 130.7° W in the South Pacific Ocean, and consist of volcanic Pitcairn Island, the raised coralline Henderson Island, and the small coral islands of Ducie and Oeno Atolls. Pitcairn has a special interest, owing to its occupation by the HMS *Bounty* mutineers, but before that it was occupied and then abandoned by the Polynesians. Although the precise timing of the settlement of Pitcairn by the Polynesians remains problematic, it now appears that Henderson Island was settled between AD 800 and 1050 and deserted some time after AD 1450 (Benton and Spencer 1995). During the occupation, marine molluscs, turtles, and birds were heavily predated. Settlement was accompanied by the introduction of cultigens and tree species, and by burning for crop cultivation. A number of endemic bird species and at least 6 out of 22 land-snail species became extinct during this period. It seems likely that sustained occupation was only possible through interaction with Pitcairn and, 400 km to the west, with Mangareva Island.

Weisler (1995) has outlined a prehistory of Mangareva, which suggests a similar pattern of resource exploitation and environmental depletion as Easter Island. Mangareva experienced high human population levels, depleting the resource base to the point where they could no longer support long-distance voyaging, dependent as it was on large ocean-going vessels, skilled crews, food supplies, and exchange commodities. As on Easter Island, the requirement for large trees for canoe manufacture does not appear to have been matched with sustainable tree husbandry! Thus, according to Weisler's reconstructions, the overexploitation of resources by these linked communities resulted in the severing of voyaging linkages and the eventual abandonment of the most marginal locations, including both Henderson and Pitcairn. With its poor soils, sporadic rainfall, lack of permanent sources of potable water, its karstic topography and low diversity of marine and terrestrial life, it is pretty remarkable that Henderson was settled in the first place.

The Henderson petrel (*Pterodrama atrata*), only recently described as a distinct species, breeds exclusively on Henderson Island. A study in 1991/92 showed that its breeding success was greatly reduced by predation by the Pacific rat (*Rattus exulans*) and the species was judged to be threatened (Benton and Spencer 1995). It has been calculated that this petrel has been undergoing a slow decline since the Polynesians introduced the rat some 700 years ago, and it is possible that this may continue until the petrel becomes extinct, although this would seem a remarkably long lag time for a new 'equilibrium' to be reached. Henderson Island has four surviving endemic land birds, including one fruit dove, *Ptilinopus insularis*. In the past, this dove shared the island with at least two other Columbiformes: *Gallicolumba*, a ground dove, and a *Ducula* pigeon. The dove became extinct during Polynesian occupation about the late thirteenth century. Indeed, five of Henderson's nine endemic land birds became extinct as a result of this occupation, as did most of the small ground-nesting seabirds (Wragg 1995). Human predation, as well as predation by (and competition with) the commensal Pacific rat, and habitat alteration by people, each had a role in these changes in the avifauna. The process of change is ongoing, and today a number of the endemic land snails of Pitcairn appear to be restricted to remnants of native vegetation of only about 1 ha or less. These stands are currently under great threat from the spread of invasive exotic plants, which cast a deep shade and create an understorey inimical to the snails (Benton and Spencer 1995).

10.4.2. Indian Ocean birds

As shown in Table 10.2, the islands of the Indian Ocean have also suffered significant losses of birds in prehistoric and historic times. Most notably, on Madagascar, 6–12 species of ratites became extinct, almost certainly as a result of the arrival of the Malagasy people. These losses included the largest bird ever recorded, the giant elephantbird (*Aepyornis maximus*). Other losses include two giant tortoises, pygmy hippo, and at least 14 species of lemur, most of them larger than any surviving species (Diamond 1991*b*; Groombridge 1992).

The islands of Amsterdam and St Paul, some 2500 km from Madagascar in the southern Indian Ocean, and 80 km from each other, are known to have had their own avifauna, but nearly all of the native species were exterminated through fires and the introduction of domesticated and commensal mammals, notably feral cattle (Olson and Jouventin 1996). The extinct avifauna included a species of rail, and a different species of duck on each island. The ducks can be assumed to have been different species because remains of the Amsterdam form, *Anas marecula*, reveal it to have been small and flightless: whatever was on the other island (known only from an historical account) must perforce have been a distinct form.

Without doubt the most celebrated of all extinct island birds, the dodo (*Raphus cucullatus*) was a large, flightless pigeon (Columbiformes) endemic to Mauritius (Livezey 1993). A related species, the solitaire (*Pezophaps solitaria*), occurred on Rodriguez, also in the Mascarene group. The existence of a third species on Réunion is indicated by historical evidence, but has not been backed up by a specimen. The dodo and its relatives were wiped out by the early eighteenth century due to a combination of hunting, habitat destruction, and the negative impacts of introduced vertebrates. Predators implicated include pigs, rats, cats (especially on Rodriguez), and monkeys (Mauritius only). Introduced herbivores responsible for significant destruction of native vegetation were cattle, goats, and deer (Mauritius only). While they proved to be vulnerable to the evolutionarily unpredictable intervention of humankind, these giant, ground-dwelling pigeons were formerly dominant consumers within their respective island ecosystems.

The native forests and woodlands of Mauritius have been reduced in extent by 95% to just 5% of the land area (Safford 1997). This deforestation, combined with the introductions of many exotic animals and plants, has led to the extinction of at least eight other endemic bird species in addition to the dodo and the reduction to critical levels of eight others. Five of the six extant forest-living native passerines are essentially restricted to native vegetation. Moreover, only a small fraction of this habitat is of good quality because of nest predation, reduced food supply, disease, and, in the past, the use of organochlorine pesticides. In 1993, each of these five species was estimated to have populations of the order of only 100–300 pairs. Safford (1997) recommends management of exotic browsers and predators, and weed removal, as part of a broader strategy for conservation on Mauritius. He also suggests that given the impracticality of eliminating predators on the mainland, some of the offshore islets be considered for ecological rehabilitation and translocation programmes. However, he notes that bird populations on small offshore islets would be vulnerable to catastrophes, especially cyclones.

10.4.3. Reptiles

Case *et al.* (1992) review the data for reptilian extinctions worldwide over the past 10 000 years, noting that the great majority have occurred on islands and as a consequence of human impacts. On small islands, humans have raised extinction rates by about an order of magnitude, but rates of loss have been much lower on very large islands. Introduced predators, principally mongooses, rats, cats, and dogs, have been the main agents of human-related extinctions, whereas competition with introduced reptiles appears to have had relatively little impact on native island reptiles. The reptiles most prone to extinction have been those with relatively large body size and endemics with a long history of island isolation.

Case *et al.* (1992) subdivide islands into three categories. First, they discuss islands colonized in prehistory by aboriginal people and then colonized later by Europeans. On such islands, many reptiles are known only as subfossils, having become extinct during the aboriginal period. In New Zealand, for example, three species of lizard known from the Holocene are extinct, and a further nine reptile species are today found only on the off-lying islands. The latter are probably relictual distributions following extinctions from the mainland consequent upon Polynesian settlement. A similar pattern is found in the Caribbean, where several species became extinct during aboriginal occupation on islands such as Hispaniola and Puerto Rico; others have become extinct within the past century. The large herbivorous iguanine *Cyclura* has become extinct on a number of islands in the recent past, and the giant *Cyclura pinguis*, lost from Puerto Rico during the Holocene, survives on the small offshore island of Anegada. Among the lost reptiles of the Caribbean were giant tortoises (*Geochelone* spp.) from the Bahamas, Mono Island, and Curaçao. Giant tortoises also once occurred on Sicily, Malta, and the Balearic islands in the Mediterranean, and on the Mascarene islands. Their disappearance from the latter followed shortly after human contact, but in the Mediterranean the role of humans in the losses is not yet so clearly established (section 3.6; Schüle 1993). The Canary Islands, in the east Atlantic, lost a number of lizard species following their colonization by the aboriginal Guanches between 2000 and 4000 BP. Within the Canaries, fossil lizards are found in association with human artefacts, but lizards were probably not a major item of the diet, and the introduction of commensals, such as rats, dogs, goats, and pigs, were probably of greater significance in the demise of what were once the largest species of laceritid lizards (*Gallotia* spp.) in the world.

The second category identified by Case *et al.* (1992) was of islands with a colonial period but no aboriginal history, on which island endemics survived to be described as living species, very often to meet their demise shortly after. Two spectacular large geckos, *Phelsuma* spp., co-occupied briefly with Europeans on Rodrigues Island in the Indian Ocean in the late seventeenth century, until hunted to extinction by rats and cats. A similar fate befell many endemic reptiles on Mauritius. None of the endemic reptile species of the Galápagos have become extinct yet, but population densities have declined and local extirpations, for instance, of giant tortoises, have occurred. Land iguanas have also been lost from Baltra and James islands, probably due to predation by feral dogs.

Case *et al.*'s (1992) third category was of islands with no permanent human settlement to date. Such islands are generally too bleak or small to support human settlement. For this reason they also tend to support few endemic species. Losses from this category of islands have thus mostly been local rather than global extinctions and of interest principally for the insights they offer into natural turnover, viz. studies of species relaxation from land-bridge islands (cf. Chapter 7).

10.4.4. Caribbean land mammals

Extinctions have dramatically altered the biogeography of West Indian land mammals over the past 20 000 years (Morgan and Woods 1986). Of 76 non-volant mammals, as many as 67 species (88%) have become extinct since the late Pleistocene, including all known primates and Edentata (Table 10.4). Late Pleistocene

Table 10.4

Native West Indian land mammals known from living or fossil records from the past 20 000 years (source: Morgan and Woods 1986)

Order	Total number of species	Living species	Extinct species	Percentage extinction
Non-volant groups				
Rodentia	45	7	38	84
Insectivora	12	2	10	83
Edentata	16	0	16	100
Primates	3	0	3	100
Volant group				
Chiroptera	59	51	8	14
Total	135	60	75	56

extinctions are attributable to climatic change and the postglacial rise in sea level (Chapter 2), but most late Holocene losses are due to humans. Thirty-seven of the losses post-date the arrival of people in the West Indies, which occurred some 4500 years ago. Over this period, species have been lost at the average rate of 1 per 122 years, the figure for the preceding 15 500 years being about 1 per 517 years. In contrast to the data for non-volant mammals, bats appear to have fared relatively well, only 8 of the 59 species having disappeared over the past 20 000 years: they thus appear to have been historically less susceptible to human influence.

Many species of extinct West Indian mammals, especially rodents, are common in archaeological sites, indicating them to have been an important part of the Amerindian diet. In common with other studies cited here, humans brought about extinctions in the West Indies through predation, habitat destruction, and (especially in post-Columbian times) the introduction of exotic species. Knowledge of the precise timings of species disappearances is fragmentary, but it does appear that a number of species disappeared during the Amerindian period, including at least one species of ground sloth, whereas other species undoubtedly survived to post-Columbian times, including several species of rodent.

10.4.5. Island snails

Snail extinctions from islands appear to be largely attributable to two factors, the loss of habitat, and the introduction of exotic species, mostly predatory snails. The endodontoid snails are tiny tropical land snails, only a few millimetres in diameter. Over 600 species have been described from the Pacific, but one in six of these may have become extinct this century (Groombridge 1992). They are mainly ground dwellers in primary forest, and they are threatened by habitat loss and by introduced ants that prey on the eggs. Other important island families are entirely or largely arboreal, e.g. the Partulidae, a restrictively Pacific family of some 120 species, most of which are considered to be threatened. On Hawaii, populations of achatinelline snails have been lost because of overcollecting and habitat modification. They are particularly vulnerable because of very low lifetime fecundity (6–24). In the Caribbean and New Caledonia it is reported that land snails most at risk are those in dry lowland forests which may be lost more rapidly than upland forest to causes such as cattle grazing and other forms of development. In New Zealand, there are many very localized endemic snails that are entirely dependent on native plant associations and which are rapidly being lost. On Madeira, in the eastern Atlantic, it is reported that of 34 species of land snails present in samples of varying ages, nine have become extinct since human settlement in AD 1419, principally, it seems, as a result of habitat loss (Goodfriend *et al.* 1994).

Griffiths *et al.* (1993) have studied the diet of the introduced snail *Euglandina rosea* in Mauritius, where it was introduced to control introduced giant African snails, *Achatina* spp. Unfortunately, as in some other cases of failed biological control schemes, it was found that native snail species were eaten in greater numbers than would be expected from their numbers in the habitat, and that the *Achatina* were not positively identified in the diet of the predator. Some 30% of Mauritian snails are now extinct. As most losses pre-dated the introduction of *Euglandina*, they have been attributed principally to habitat destruction, but the exotic predator now represents one of the major threats to the remaining endemic snail species. On Hawaii, the introduction of *Euglandina* has been implicated in the decline of 44 species of endemic *Achatinella*; and it has been a major factor in the extinction of *Partula* species in French Polynesia, nine species being lost from Moorea alone. Thus far, on Mauritius, primary forests do not appear to have been penetrated by the exotic species, and so habitat conservation remains a key to the conservation of the native species.

10.4.6. Plants in peril

With all the attention given to animals in these pages, it is important to note, for the sake of balance, that the problems also extend to plants. A classic example of the damage to a remote oceanic island brought about by incorporation into European trading routes is provided by St Helena. It is an isolated island of 122 km^2, located in the South Atlantic Ocean. Since discovery by the Portuguese in 1502, the vegetation has been completely transformed by browsing, grazing, erosion, cutting for timber and fuel, the introduction of alien plants, and by clearance for cultivation, plantations, and pasture (Cronk 1989). No native mammals occurred on the island, as is typical of remote oceanic locations. The Portuguese introduced goats, pigs, and cattle, principally to supply homeward-bound carracks from India, and also brought with them rabbits, horses, donkeys, rats, and mice. Within 75 years of their introduction, goats had formed vast herds, devastating the vegetation. The introduction of exotic plants began in earnest in the nineteenth century via British colonial botanic gardens, in part to provide a new plant resource to replace the exhausted native cover. Large numbers of herb and shrub species were introduced, along with timber trees such as *Pinus pisaster* and *Acacia melanoxylon*. By the time the first reliable botanical records were made in the nineteenth century, most of the destruction had already occurred, but some documentary evidence and the relict occurrence of endemic plants in isolated places such as cliffs, provide some indications of the former cover. Of the 46 endemic plants *known* from the island, seven are considered extinct, the rest threatened (Groombridge 1992). Some species are restricted to only a few individuals, and in the early 1980s *Nesiota elliptica* and *Commidendrum rotundifolium* were reduced to just single plants, the latter becoming extinct in the wild in 1986 (Cronk 1989).

Goats have been particularly important in preventing woodland regeneration by grazing and barking saplings. So, as old trees were cut for firewood, areas became converted to a mosaic of anthropogenic formations, notably grassland and *Opuntia* scrub (Fig. 10.3). Erosion has been very rapid, and on decadal time-scales, isolated thunderstorms cause sheet and gully erosion where the vegetation cover has been removed by herbivores. After such erosion little vegetation cover can re-establish. Goats were nearly shot out in the early 1970s and all domestic goats are kept penned by law; however, sheep are still allowed to roam freely, and despite prohibitions, do occasionally cause serious damage. Cronk (1989) considers that a process of rehabilitation is possible, based on trial plantings of native species, but that initially, at least, plantings require fencing.

Fig. 10.3. The dry coastal zone of Punta Teno, Tenerife (Canary Islands) features a high diversity of Macaronesian endemics, including the cactus-like *Euphorbia canariensis* (of similar height to figure); it is also home to the exotic *Opuntia* (foreground)—a true cactus originally introduced as a fodder plant and for the dye industry—which now competes for space with the native plants. Tenerife contains a considerable range of different habitats; compare with Fig. 2.13. (Photo: RJW 1992.)

Turning to one of the showcases of plant evolution, estimates for Hawaii include about 100 extinct taxa, 200 listed as endangered or threatened, and a further 150 recommended for listing (Wiles *et al.* 1996). Numerous introduced plant species have become naturalized, to the point at which the exotic flora outnumbers the endemic flora. Several have become serious pest species (Primack 1993). In the Mariana Islands, Micronesia, the forces involved in the demise of native vegetation include loss of forest to agriculture; war and development; the introduction of alien animals, plants, and plant pathogens; alterations to the fire regime; and declines in animal pollinators and seed dispersers. The impacts on several native plant species have been severe. *Serianthes nelsonii* (Leguminosae) is one of the largest native trees, and is endemic to Rota and Guam. It formerly occurred on volcanic soils but is now restricted to limestone substrates. Just 121 adult trees survive on Rota, and one on Guam. The populations are senescent, with little or no regeneration occurring. The principal threat is from browsing of introduced deer and pigs, plus infestations of herbivorous insects (Wiles *et al.* 1996). On Guam, *S. nelsonii* is only one of at least a dozen species experiencing poor regeneration for these same reasons, and in one case at least, germination problems limit establishment. It appears that the seeds of *Elaeocarpus joga* need to pass through the gut of a bird, yet native frugivores have been wiped out by the introduced brown tree snake (*Boiga irregularis*). Another species, *Pisonia grandis*, is rarely seen with fruit, perhaps because of the loss of native pollinators such as fruit bats or birds. In addition to impacting on seed set and germination, the loss of frugivorous birds and the reduction of the one extant pteropid fruit bat species to a remnant population of a few hundred animals has been found to have had a predictably marked effect on seed dispersal on Guam (Rainey *et al.* 1995). Thus, while habitat removal is again a key driving force of plant reductions and extinctions, a similarly broad array of factors and their synergistic interactions is at work, as discussed more frequently for vertebrates (see Vitousek *et al.* 1995).

In illustration of how differing factors have primacy in different situations, we can turn to Tahiti, where the overriding threat to endemic plants currently takes the form of another plant species. Tahiti is in the Society Islands, Polynesia, and is a large (1045 km^2), high (2241 m) island, of diverse habitats, with a native flora of 467 species, 45% of which are endemic (Meyer and Florence 1996). Some 70% of the endemic plants occur in the montane cloud and sub-alpine forests, which are some of the largest, finest, and most intact to be found throughout the Pacific. The Polynesians arrived *c.* 3500–3000 BP, bringing with them about 30 domestic and 50 other accidental plant introductions. European arrival in the 1760s led to massive resource overuse, and habitat loss, as well as a great flood of exotic plant species—in excess of 1500 now being known for the Society Islands. Many of these have become naturalized, and some have become problem plants. The lowlands and coastal zones were totally transformed by people, and such native forests as remain on Tahiti are endangered by pressures such as housing, agriculture, forestry (e.g. based on exotic *Pinus caribbea*), feral animals, and schemes such as the generation of hydroelectricity. Yet, the highlands remained essentially intact until the arrival of what has become the worst of all exotic escapees, *Miconia calvescens*, a small tree species introduced to the ornamental gardens in 1937. In less than 50 years it had spread to all the mesic habitats between 10 and 1300 m ASL, including the montane cloud forest, often forming pure stands. It now covers over two-thirds of the island of Tahiti, and has spread to the nearby islands of Moorea and Raiatea. It is a potential problem for other islands in the region, and is already recognized as one of the 86 plant pests of Hawaii, having been introduced there by horticulturalists as recently as the early 1970s. It is an effective invader thanks to traits such as: tolerance to a wide range of germination conditions, tolerance of low light levels, fast growth, early reproductive maturity, prolific year-round seed production, and efficient dispersal by small introduced passerines such as the silvereye. In Tahiti alone, 40–50 endemic plant species are considered to be threatened directly, including eight species of *Cyrtandra*, six *Pyschotria*, and two *Fitchia*.

What can be done to halt this decorative invader? One approach is to dig it out. In 3 years (1992, 1993, and 1995), nearly half a million plants were uprooted by hand, and *Miconia* was cleared out of its bridgehead sites on the island of Raiatea. In addition to such plain labour, Meyer and Florence (1996) promote the need for a systematic assessment and inventory of alien and endangered native species, for research, monitoring, and control programmes, and for *ex situ* conservation in the botanic gardens—which still remains only an ornamental garden. Part and parcel of such an integrated approach to conservation, as commonly advocated in other similar studies, is a sustained educational programme.

The problem posed by invasive exotic plants may be first and foremost for native plant species but, if on sufficient scale, can also present a problem for native fauna (Cronk and Fuller 1995). For instance, the now pan-tropical shrub *Lantana camara* forms dense impenetrable stands in parts of the Galápagos, threatening both some endemic plant species and the breeding habitat of an endangered bird, the dark-rumped petrel (*Pterodroma phaeopygia*) (Trillmich 1992). Invasive plants may alter the environment in other, more subtle ways (Cronk and Fuller 1995). The shrub *Myrica faya*, a native of the Canary Islands and the Azores, was introduced to Hawaii in the late nineteenth century, and has since been spread within the National Park by an exotic passerine, the white-eye (*Zosterops japonica*), and by feral pigs. It has a highly effective nitrogen-fixing symbiont, which in part accounts for its ability to outcompete the native flora. By increasing nitrogen inputs relative to native species by 400%, it alters the nutritional balance of the Hawaiian soils. It poses considerable management problems and is predicted to have additional long-term consequences because the altered soil nutrient status is likely to facilitate the spread of other alien species (Vitousek *et al.* 1987; Cuddihy and Stone 1990; Vitousek 1990).

Exotic plants may alter other ecosystem properties, such as fire regimes or hydrological conditions. Examples of each can be drawn from Hawaii. Bunchgrass (*Schizachrium* spp.) increases the flammability of the vegetation, and has prevented regeneration of native species that are not adapted to burning. *Andropogon virginicus* is another exotic grass, considered to be one of the most threatening of aliens. It not only carries fire, but also alters hydrological properties. It occurs in disturbed grassland and scrub on the island of Oahu in areas on red clay soils where native forest vegetation has been replaced by introduced woody and herbaceous plants. Its seasonal pattern of growth and tendency to form dense mats of dead matter are resulting in increasing runoff and accelerated erosion (Cuddihy and Stone 1990).

10.5. How fragile and invasible are island ecosystems?

It has become axiomatic to represent oceanic islands as fragile ecosystems, which, because of their evolution in isolation from continental biotas, are particularly susceptible to the introduction of alien species. Are island systems really particularly fragile or have humans just hit them especially hard? D'Antonio and Dudley (1995) contend that the data support two propositions. First, island ecosystems do typically have a higher representation of alien species in their biota than do mainland systems. Secondly, the severity of the impacts of invasions appears in general to increase with isolation.

Cronk and Fuller (1995) suggest six reasons why oceanic islands may be more susceptible to invasion—(1) species poverty, (2) evolution in isolation, (3) exaggeration of ecological release, (4) early colonization, (5) small scale, and (6) crossroads of intercontinental trade:

1. Species poverty may mean that there is more vacant niche space and less competition from native species. This is plausible, particularly

where humans have disturbed habitats, but it is only one part of the picture.

2. It is often assumed that island species are competitively inferior to continental species due to their long evolutionary isolation. The case of *Myrica faya*, transported from one set of oceanic islands, in Macaronesia, to another, Hawaii, illustrates that the 'continentality' of the land mass from which a species hails may not be of great moment: simply to be from a different biogeographical pool may be sufficient. However, the evolution of flora and fauna in isolation, often without adaptation to grazing, trampling, or predation by land mammals has led to the loss of defensive traits in many oceanic island endemics, expressed in plants as the absence of typical grazing adaptations such as spines, thorns, and pungent leaf chemicals. In the presence of browsers these plants are at a competitive disadvantage and tend to disappear, as was the fate of many endemic plant species on St Helena following the introduction of the goat shortly after the island's discovery in 1502. Similarly, many endemic animals lack a fear of predators, and some birds have (or had!) reduced flight or are flightless, and build their nests on the ground. These are traits that may have no competitive disadvantage when faced with a species of the same trophic level, but which have proven disastrous when faced with terrestrial mammalian predators.

3. Alien species generally arrive on islands without their natural array of pests and diseases, and this provides them with an advantage over native species. A similar ecological release mechanism is proposed as part of the taxon cycle in Chapter 6, and indeed also applies to many introductions to continents.

4. Particularly in the Atlantic, Caribbean, and Indian Oceans, islands were the first landfalls and first colonies of the Europeans, hence they have had a long history of disturbance and of introductions. While this is true, this line of reasoning fails to account for the severity of impact of particular introductions such as the brown tree snake on Guam or *Miconia* on Tahiti: the outcome has been a particular property of the species involved rather than the length of time over which disturbance and introductions have occurred.

5. The small geographical size of islands means that their history is concentrated in a small area, within which there are few physical features large enough to prevent exploitation, disturbance, and introduction on an island-wide basis. They also often sustain surprisingly high human populations (Watts 1987), for example Cronk and Fuller (1995) note that Mauritius, with a population of 530 people/km^2, may be compared culturally and historically with India, which has only about 240 people/km^2.

6. As the crossroads of international trade, particularly in the days of sailing ships, islands have often been used as watering points, resupply posts, and staging posts for trade and for plant transport. As they have often exchanged hands between different colonial powers, they have, moreover, had opportunities to receive introductions from differing networks of nations and biogeographical regions.

The deluge of disturbance and of alien species which has swept across so many oceanic islands goes some way towards explaining the apparent susceptibility of islands to invasion. At least one tropical biologist, who has worked for many years on Hawaii, argues that this is the case.

> Contrary to common opinion, many endemic island species are strong competitors. They would not be eliminated from their niche in the island ecosystem if it were not for the new stress factors introduced by man. The island species evolved with stress factors associated with volcanism, fire in seasonally dry habitats, and occasional hurricanes. The effects of volcanism resulted in superior adaptation of many native species to extreme edaphic conditions existing on volcanic rockland, where soil-water regimes fluctuate almost instantly in direct relation to rainfall.
>
> (Mueller-Dombois 1975, p. 364)

This general position is supported by D'Antonio and Dudley (1995) who conclude that, with exceptions, island habitats are not inherently more easily invaded than are continental habitats, but that, none the less, exotics have had a greater impact on island systems.

Some more formal analysis provides a measure of support for Mueller-Dombois's views. It has been suggested that as a rough rule (the tens rule), about 1 in 10 introductions into exotic territories become established. Williamson (1996) notes that the data for Hawaiian birds indicate a far higher success rate, of over 50% of introduced species establishing. He notes, however, that the majority of introductions have colonized disturbed lowland habitats, and that the figure for species establishing into more or less intact native habitats (some lowland forest, together with upland habitat) falls at between 11 and 17%, which is not significantly out of line with the tens rule. Moreover, even within very disturbed islands, containing large numbers of naturalized (i.e. established) exotic species, those habitats which remain largely undisturbed quite often contain rather few exotics (e.g. Watts 1970, on Barbados; Corlett 1992, on Singapore). It thus appears that the susceptibility of island ecosystems to invasion is above all a reflection of the profound alteration of habitats: although it is equally evident from examples cited in this chapter that some invaders have not required such assistance.

One of the curious features of some invasions, is that invading species sometimes undergo a pattern of increase to a peak of density, followed by a crash, sometimes to extinction. This boom-and-bust pattern appears to have occurred fairly often on islands (Williamson 1996). It has, for instance, been recorded for a number of the passerines introduced to the Hawaiian islands, with time to extinction varying between 1 and 40 years. There may be a number of reasons for these boom-and-bust patterns, but Williamson favours those relating to resource deficiencies as providing the most general explanation.

It is noteworthy from the point of view of island biogeography theory, that the introduction of a single exotic species, such as the brown tree snake (*Boiga irregularis*) on Guam and the shrub *Miconia calvescens* on Tahiti, may cause the local extinction of numerous native island species. Moreover, in the case of *Miconia*, the effect operates within the same taxon, i.e. it is a single species of plant which is implicated in the loss of numerous native plants. This provides an instance in which an equilibrium model involving turnover on a replacement basis, subject only to relaxation effects from area reduction (as discussed in Chapter 9), would be a wholly inappropriate device for predicting species losses.

10.6. Contemporary problems and solutions

10.6.1. Conservation measures

Adsersen (1995) notes the prevalence of the same invasive plants in archipelagos as remote from one another as the Mascarenes, the Galápagos, the Canaries, and the Bahamas. Similarly, many other problems, such as feral animals, habitat loss, and burning, are common to numerous islands (e.g. Cronk 1989; Vitousek *et al.* 1995; Rodriguez-Estrella *et al.* 1996). Practical conservationists are thus dealing with the same problems, but direct contact between them is likely to have been fairly minimal, at least until very recently. There are thus still many benefits to be gained simply from improved information exchange between those involved in conservation management on islands (Adsersen 1995).

The large catalogue of extinctions of native island species is bound to increase. Can anything be done to stem the tide? If we can't save the remaining evolutionary wonders of the showcase islands such as Galápagos and Hawaii—and the signs are not initially encouraging—then the prospect for the many other less famous island forms must be dismal indeed. Yet, it is not a hopeless task. Species can be saved by means of rigorous habitat protection, pest or predator control, and translocation of endangered island species (Franklin and Steadman 1991), provided that political and economic circumstances allow it.

Biological control—a dangerous weapon? Biological control is the term given to pest control by means not of chemical agents but by the introduction into an environment of a species that targets the pest organism without seriously affecting non-target species. Cronk and Fuller (1995) state that it is the long-term goal of conservation managers in Hawaii to achieve the biological control of most of the introduced weeds, but as yet only 21 are controlled biologically, following 70 insect introductions with a 50% success rate. While this form of control can produce some of the best results, it also holds its own attendant risks (above; Williamson 1996). A classic example in earlier times was the introduction of the mongoose as a pest-control agent. The mongoose has been of mixed success in controlling rats but has been extremely effective at devastating many native island bird and reptile species, particularly ground-foraging skinks and snakes (Case *et al.* 1992). So, it is vital in biological control programmes to undertake careful screening and testing to ensure that the control agent does not itself become a pest species.

Translocation and release programmes Translocation and re-release in to the wild have been used as conservation methods with mixed success. Some programmes involve captive-bred populations released into the wild, and other programmes involve the capture of wild animals and their translocation to other sites. The latter may be because of a threat to the species in the area from which they were captured, or because of the desirability of re-establishing a population in an area from which a species has been extirpated. In general, translocated wild animals have been found to establish more successfully than captive-bred animals. Griffith *et al.* (1989) surveyed translocation programmes undertaken in Australia, Canada, Hawaii, mainland USA, and New Zealand over the period 1973–86. They found that only about 46% of release programmes of threatened/endangered species were successful, a far lower percentage than for translocations of native game species, for which the success rate was 86%. They attempted to identify the factors influencing success for 198 bird or mammal translocations from within their survey. The single most important factor in improving the chances of success appears to be the number of animals released. This confirms what might be anticipated from theory. They also found that herbivores were significantly more likely to be successfully translocated than carnivores or omnivores, that it was better to put an animal back into the centre of its historical range than the periphery, and that wild-caught animals fared better than captive-reared animals.

Releases of endangered birds in New Zealand have resulted in many populations being established in new or restored offshore island communities (Lovegrove 1996). By this means the extinction of several species, including the South Island saddleback (*Phiesturnus carunculatus*), have been averted. Conservation of this species has depended on release on to offshore islands because they cannot coexist with carnivores, such as ship rats (*Rattus rattus*), introduced to mainland New Zealand. Clout and Craig (1995) report on the kakapo (*Strigops habroptilus*), a flightless, nocturnal parrot. It was once thought to be effectively extinct, to be temporarily reprieved by the discovery of a new population on Stewart Island. However, they continued to decline, to fewer than 50 individuals, as a result principally of predation by feral cats. All known kakapo (i.e. the entire species) have now been transferred to three predator-free island refuges, where, supported by supplementary feeding, the rate of population decline appears to have slowed but not yet to have been reversed. Problems include the bias of survivors to older birds, a delay in resuming breeding after translocation, and periodic invasions of stoats on one of the islands. Lovegrove (1996), assessing the success of 45 release programmes, concluded that failures occurred where either (1) predators were still present, or arrived after release, or (2) too few birds, with an unbalanced sex ratio, were released. All releases involving more than 15 birds on islands lacking predators were successful, at least in the short term. Other New Zealand studies reportedly show that releases of as few as five birds on predator-free islands are likely to succeed, and, more generally, the review of Griffith *et al.* (1989) indicates that releases of 40 or more birds into good habitat are generally successful.

Clearly, the elimination of exotic predatory mammals is essential. Moreover, as on occasion even individual animals can deal a heavy blow to small populations of threatened species, it may be necessary to monitor islands closely to ensure that predators do not return. Much effort has been channelled into developing effective methods of predator control. Conservationists have been developing a wide range of techniques, often at considerable expense, including the introduction of new rodenticides, and the spreading of bait from helicopters, in order to target particular pest species to greatest effect (e.g. Clout and Craig 1995; Micol and Jouventin 1995). Such schemes are often unaffordable (Herrero 1997).

It is obvious that it is important to translocate populations into suitable habitat, but what this may be is not so obvious, as for many species populations the reasons for their original demise are not fully understood (e.g. Hambler 1994). Given the role of humans in the demise of island species, it is important that translocations take place into areas in which the human influence is managed, or preferably removed. Because the prehistoric human impact was so intense in Polynesia, Franklin and Steadman (1991) argue that the lack of many taxa from uninhabited, forested islands may be due to events that occurred centuries ago, and that with human abandonment and forest regeneration, such islands may once again be able to support the extirpated species. They therefore advocate the translocation of species into such islands, following careful assessment of the resources and problems involved. From their preliminary assessments they argue that low makatea islands, such as the Cooks, hold limited potential, but that there may be greater potential among the 20 or so Tongan islands that are both uninhabited and greater than 2 km^2 in area. Moreover, the fossil record demonstrates the former presence on one such island, 'Eua, of species that still survive in Fiji, Samoa, and other Tongan islands, including the parrots *Vini australis* and *Phyigys soliatrious*. Establishing multiple populations will greatly improve the prospects of a species surviving events such as the loss of a local population in a hurricane. Such initiatives require a sound biogeographical and ecological baseline, together with the political and legal frameworks to support the efforts.

The only wild giant tortoises in the Indian Ocean, *Geochelone gigantea*, survive on Aldabra Atoll. Here, the culling of goats has enabled their population to persist in reasonable numbers. Between 1978 and 1982, 250 tortoises were translocated to Curieuse in the granitic Seychelles (Hambler 1994). As many as 109 of these were probably lost by 1986, and in a detailed survey in 1990 it was found that only 117 animals remained. Poaching probably accounted for the majority of losses, although some of the surviving adults were diseased, and low growth and reproductive rates suggested possible saturation of resources and nutrient limitation (particularly calcium deficiencies). Both feral cats and feral rats are abundant on the island, and probably account for the very low rate of recruitment. To ensure the persistence of the new colony, it became necessary to establish a facility to rear hatchlings until they are over 5 years old and thus relatively safe from predation (if not from poaching). In contrast, a translocation to another of the granitic Seychelles, Frégate Island, has been comparatively successful. The reasons for the difference in outcome on the two islands are as yet uncertain, but it is likely that both habitat differences and differences in predation and poaching are involved. This demonstrates that the success of translocation is not simply a function of initial size of the translocated population. Eradication of the feral predators, and reduction in poaching might be the answer on Curieuse, but it is also possible that the island is basically unsuitable for the long-term persistence of a viable population.

Programmes based on *ex situ* breeding and translocation back into the wild thus have their place in the conservation of island endemics. However, they have to be undertaken as part of a highly organized, managed programme. The haphazard removal of animals from the wild supposedly for captive breeding in zoos may otherwise be counter-productive.

Christian (1993) notes that while several parrots have been taken out of St Vincent and The Grenadines for captive breeding, there have been only three known successes, and to date there is no evidence of any captive-bred parrots being returned to St Vincent for re-release. Thus, not only is it necessary to 'fix' the local environmental problems, but it is also necessary to control the *ex situ* part of the process.

10.6.2. Contemporary problems in the Galápagos

On the Galápagos islands, the main conservation threats include introduced mammals (e.g. goats, pigs, cats, rats, dogs), aggressive alien plants (e.g. *Lantana camara, Rubus* sp.), habitat fragmentation, agricultural encroachment, overexploitation of native woody species, and manset fires (Jackson 1995). The problems of the Galápagos are not solely land-based, and there are now great concerns about exploitative overfishing for overseas markets (e.g. Merlen 1995). The activities of the fishermen extend to the shores too, as they make demands on coastal timber for fuel-wood. Another feature not discussed hitherto, but significant on many islands throughout the world, is the rise of tourist pressures, including that from so-called 'eco-tourists'. In turn, this has encouraged an increase in the resident human population, and there have recently been conflicts between conservationists and other residents concerning the economic development of the islands (Trillmich 1992; Davis *et al.* 1995). It is estimated that the Galápagos generates about half of Ecuador's tourist income. The islands now receive about 48 000 tourists a year, half of which are non-Ecuadorians. Each visitor pays an entrance fee, set at $US40 in 1990, and additional revenue is generated by tours, boats, hotels, and other local services. However, only about 5% of the tourist income remains in the islands, the rest goes to the mainland.

Given the enormous significance of the Galápagos in natural science, it might be anticipated that there would be a matching conservation focus and effort. Indeed, the Galápagos National Park covers 97% of the archipelago, which has also been designated a World Heritage Site and a Biosphere Reserve. Research in the park is co-ordinated by the Charles Darwin Research Station. The management of the park aims to preserve the various important features of interest, while dividing the archipelago into zones of differing intensity of land use. Introduced animals, especially goats, remain the greatest problem, and although feral goats have been eliminated completely from a number of smaller islands—with notable recovery of vegetation—the destruction brought about by pigs and goats on the large island of Santiago has been so great that it is doubted whether a full recovery is now possible (Davis *et al.* 1995). In 1987, a major conservation initiative was launched with laudable short- and long-term aims. The goals included: fencing against feral mammals, inventory of the worst alien plants, development of control and eradication methods, a programme to develop sustainable timber use, and an *ex situ* breeding programme for threatened plants to complement *in situ* efforts. The existence of schemes for labelling, protecting, and conserving these islands are encouraging, but they also hold the danger of promoting a false sense of action and well-being.

In May 1989, seven feral goats were observed on Alcedo volcano on Isabella Island, Galápagos: the first report of goats in the habitat of the most intact race of Galápagos tortoise. By 1997, the goat population was estimated at 75 000, and the habitat of the tortoises and other fauna on Alcedo volcano was reported to be collapsing. It is indicative of the problem facing many conservationists that appeals for donations to finance the goat control programme have had to be issued, for example, in the journal *Conservation Biology*, as the Charles Darwin Foundation lacks the core funding needed (Herrero 1997). The outgoing Director of the Charles Darwin Research Station, Chantal M. Blanton, noted with dismay that the national financial commitment to conservation research in the Galápagos slipped to zero in 1995, while tourist concessions doubled between 1992 and 1996 (*Galapagos Newsletter* 1996). When asked to name the most important problems facing the islands, she listed: first, introduced animals and the lack of a functioning quarantine system; secondly, inadequate political and financial support from government; and thirdly, the politicizing of the National Park Service and of 'technical' posts within the Galápagos in general.

The basic conservation problem of the Galápagos is the ever-accelerating breakdown of the islands' former isolation due to dramatic changes in human population size, activity patterns, and mobility (Trillmich 1992). The real impact of the tourist business occurs not through the direct disturbance to animals by tourists armed with cameras but indirectly via the socio-economic changes on the four inhabited islands, where population is growing rapidly due to the tourist industries. The growth at the end of the 1980s increased pressure on resources, and was accompanied by a fresh influx of aliens—about 100 new plant species arriving in a period of just 5 years—and by a rise in the pet population. A survey in the main settlement on Santa Cruz established that the dog population was about 300, and that they were breeding freely, many unwanted young being released to become feral. They represent a latent threat, as they prey upon young tortoises and tortoise eggs and, for example, in the late 1970s wild dogs attacked a large colony of land iguanas on Santa Cruz, killing over 500 animals (Jackson 1995). While the National Parks Service is working to keep the feral population low, it will be difficult to reduce the number of domestic dogs as long as robberies in the villages increase, creating an incentive for settlers to keep dogs (Trillmich 1992). Feral cats are also problematic. The case articulated by numerous conservationists over more than 25 years for a quarantine system is overwhelming but, as Trillmich argues, this will only be successful if the inhabitants of the islands can be convinced through educational programmes that the tourist attraction, and hence their livelihoods, will ultimately be diminished if the archipelago's biological diversity and uniqueness is eroded too far.

10.6.3. Sustainable development on islands: constraints and remedies

The elements involved in the foregoing case studies are repeated in varying combinations and with varying emphasis for numerous other islands for which conservationist have made management assessments: control exotics, halt or reverse habitat loss, prevent hunting, assist breeding of endangered species, enlist political and legislative support, educate and enthuse the people. It is important, finally, to consider whether there are any special considerations on the human side of the equation which must be included in conservation thinking.

In many respects, oceanic islands are peculiar environments, both biotically (high proportion of endemics, disharmonic, etc.) and abiotically (experiencing wave action from all sides, tending to have small hydrological catchments, peculiar geologies, etc.). Insular peoples have commonly evolved distinctive cultures and retain a strong allegiance to both home and culture (Beller *et al.* 1990). It may be worth noting also that insularity may provide essentially ecological problems for humans. Examples include the extermination of entire cultures, such as the Caribs of the Caribbean, killed off by disease introduced by Europeans; the introduction of malaria into Mauritius in the 1860s, which led to the death of 20% of the capital's population; and a decline of perhaps 50% in the population of Hawaii in the 25 years following the arrival in 1778 of Europeans bringing disease such as syphilis, tuberculosis, and influenza (Cuddihy and Stone 1990). Less directly, an epidemic of swine fever hit São Tomé and Príncipe in 1978, wiping out their entire pig population, and causing an elevenfold increase in meat imports in 3 years. Likewise, one of the most precious resources in respect to attracting outside finance is the possession of ecological interest, to attract international support, or tourist dollars. So, in many respects the problems of the plants and animals of islands, are the problems of people on islands.

Yet, as they include a vast array of climatic, geographic, economic, social, political, and cultural conditions, it is extremely difficult to offer meaningful generalizations as to the problems facing island peoples and how they influence and constrain the options from a conservation management perspective. One approach might be to focus on those islands holding most threatened species. For instance, over 90% of threatened island species are endemic to single geopolitical units and just 11 such units are home to over half the threatened island birds (Groombridge 1992, p. 245). However, even this does not allow much in the way of generalization, as the 11 units concerned are Cuba, Hawaii, Indonesia, the Marquesas, Mauritius, New Zealand, Papua New Guinea, the Philippines, São Tomé and Príncipe, the Seychelles, and the Solomons. This is a pretty mixed bag.

For the human societies and economies of smaller islands (up to about 10 000 km^2) the problems may include the following factors (Beller *et al.* 1990):

1. Lack of water. This provides a particular problem for very small, low, narrow atolls, where salinity conditions the entire land area, but can also affect substantial, high islands such as Tenerife, where the tourist (Fig. 10.4) and horticultural industries have been sustained only on the back of the unsustainable mining of geological reserves of water (cf. Ecker 1976).

2. Susceptibility to rising sea level. For example, the Bimini islanders formally requested advice from the UN on how to plan for their entire land area being below the projected sea level according to greenhouse-warming predictions. It is not simply the loss of area that may be critical, but for small states like the Bahamas, the potential loss of much of their low-lying reservoirs of fresh water (Chapter 2; Stoddart and Walsh 1992).

3. Vulnerability to storm or tectonic damage. Hurricanes and cyclones can badly damage the economy of a small island, forcing the human populations to exploit biological resources themselves damaged by such events. Certain types of islands are particularly prone to volcanic and tectonic hazards and to related phenomena such as tsunamis (Chapter 2). Such phenomena can deal a savage blow to the sustainable human use of an island, as evident from the devastation inflicted on the island of Montserrat by eruptions in 1997.

4. Climatic variation. The agricultural base of a small island may be hard hit by atypical weather conditions. The El Niño event of 1983 caused immense disruption to marine and terrestrial habitats, both proximal and distal to the tropics where the unusually high water temperatures were recorded. Central Pacific seabird communities had a dramatic reproduction failure and briefly abandoned atolls such as Christmas Island. Island life is clearly vulnerable to short-term weather events as well as long-term climatic change (Stoddart and Walsh 1992). Because of the limited options provided by the discrete area of a small island, agricultural/fishing economies are similarly constrained.

5. Narrow agricultural base. The constraints of the often narrow agricultural base of island societies is clearly demonstrated by the history of Barbados, which has been dominated until modern times by the boom–bust fortunes of the sugar industry since colonization of the islands by Europeans in 1492 (Watts 1987). When sugar was first established as the export crop, the island's ecology was devastated in just 20 years, at the end of which time 80% of the land area was cultivated for sugar cane. Many indigenous species were lost in this first, dramatic conversion. Subsequently, changes in the global market for sugar, and occasional pest problems, have resulted in very significant fluctuations in the area of cane, and in the security and wealth of the island. The narrow environmental range provided by such small size, and the small land area, means that it is difficult for the local economy to diversify, and hard to produce exportable quantities of more than a limited range of agricultural crops. The boom-and-bust cycle of periodic export booms, subsequent deflation and resource exhaustion, and chronic emigration appears to be a feature common to many small island histories (Beller *et al.* 1990).

6. Suitability of land for agriculture. Very much tied in to the previous point, is that there may be few crops suitable for the island soils and climate. The Bahamian island of Andros, for instance, is characterized by a viciously dissected coral rag substrate. Mechanized agriculture is not feasible and the land essentially does not provide a living.

7. Dependence on the seas. Dependence on the seas is all very well when the fish are

plentiful. Problems arise not only from anomalies in climate or ocean currents, but also from the activities of foreign fishing fleets, over which local peoples may have limited sway (cf. Merlen 1995).

8. Small or fragmented states. Tied in to all of the foregoing is the political geography of small or fragmented states. In the case of countries such as the Bahamas, or Jamaica, with a shallow natural resource base, a fluctuation in the environment (e.g. a hurricane) can deal a significant blow to the resources of the state.

In his review of resource development on Pacific islands, Hamnett (1990) identifies five major trends within the region: (1) increasing pressure on the land and freshwater resources; (2) intensification of agriculture; (3) loss of native forest resources; (4) river and stream siltation and the loss of aquatic resources; and (5) increasing use of coastal areas and the degradation of harbours and lagoon environments. We might wish to add others to this list, notably the increasing interest from the tourist, seeking the idyllic desert island on which to holiday, but expecting the facilities of a modern society—the hotels, roads, fresh water, and other amenities—as well, perhaps, as the exotic wildlife. This sixth trend is powerfully evident in the Caribbean, which, in the period after the Second World War, has witnessed rapid urbanization, a significant increase in tourism, and marine degradation from the construction of transport (air and sea) and petroleum facilities and hotels (Beller *et al.* 1990). Many of the islands of the Mediterranean, and certain Atlantic islands, notably the Canaries, have experienced very similar patterns of expansion in the tourist industries. Tourist pressures are also extremely significant on Hawaii and the Galápagos (Cuddihy and Stone 1990; above). This is just one way in which, in common with the problems facing native plants and animals, exogenously driven forces (social, economic, and technological) are affecting and disrupting many island cultures, and eroding traditional skills.

The task of confronting these pressures is not a simple one. Some of the themes involved in sustainable development planning for small islands are denoted in Table 10.5. In many ways the ecological management goals are the easiest part of the equation to specify, while managing the societal, economic, and political issues is very difficult (Beller *et al.* 1990). In illustration, concepts of ownership among island nations can be both extremely varied and complex. In Papua New Guinea, the Solomon Islands, and Vanuatu, ownership is defined by oral traditions, land is typically owned by family groups rather than individuals, and there may

Fig. 10.4. A tourist town on Tenerife, Canary Islands. The small fishing town of Puerto de la Cruz has been swallowed by the development of the tourist industry. Whilst tourism has brought large economic inputs to Tenerife it has also contributed to pressures on the environment. (Photo: RJW 1994.)

Table 10.5

Sustainable development options for small islands should aim at increasing self-reliance, and can be categorized under six basic headings (after Hess 1990)

Categories	Examples
Resource preservation	Conservation zones, multiple-use options, control of hunting
Resource restoration	Replanting, re-introductions, alien herbivore removal
Resource enhancement	Freshwater re-use, sea-water desalination
Sustainable resource development	Small-scale diversified, closely managed forms of resource-based enterprises (agriculture, fisheries, tourism)
Provision of human services	Alternative energy generation sources and distribution systems, waste disposal
Non-resource-dependent development options	Financial services (tax havens), light industries processing imported materials, rental of fishing rights

be secondary and tertiary levels of 'ownership' providing rights of use but not implying actual ownership of the land (Keast and Miller 1996). Success in such circumstances will depend on how well insular resource planning and management can provide solutions tailored to local circumstances and cultures, and whether they carry with them the support of the island communities. The involvement of major users, farmers, fishermen, charter operators, etc., is one element in this. The mobilization of non-governmental organizations (NGOs) in island problems is another element, whereby energies and expertise exist that could be harnessed provided interest can be raised. Perhaps above all, it is necessary to foster a general public awareness of the environmental and ecological constraints and problems, and the significance to the island societies themselves of the ecological 'goods' at stake (Christian 1993).

The importance of public awareness and support can be illustrated by reference to parrot conservation in the West Indies. Since the arrival of Columbus at the end of the fifteenth century, 14 species and two genera of parrots have been lost. The remaining nine endemic members of the genus *Amazona* are each confined to individual islands. The case of the Puerto Rican parrot, *Amazona vittata*, and its conservation was outlined in Chapter 9. Within the Lesser Antilles, seven endemic parrots have become extinct in historical times, with just four remaining. They are now threatened by deforestation, predation, illegal hunting and collecting, competition with exotics, and natural disasters (Christian 1993; Christian *et al.* 1996). Impacts on the parrots resulting from habitat alteration include the loss of nesting cavities, food, and shelter. Given these pressures, management intervention is essential for their survival.

Measures for parrot conservation in the Lesser Antilles consist of environmental education, habitat protection, enforcement of appropriate legislation, and enhancement of wild breeding and captive breeding programmes. Expensive programmes, such as those adopted for the conservation of the Puerto Rican parrot (Chapter 9), are not feasible in the Lesser Antilles due to shortage of funds. Despite this, effective targeting of conservation resources has delivered some degree of success. None the less, more needs to be done; first, to protect habitats; secondly, to discontinue the export (legal or illegal) of live parrots; and thirdly, to involve local people as much as possible in both the rationale of and the employ of the management programmes. Christian (1993) has described the efforts made by the government of St Vincent and The Grenadines to ensure the long-term survival of the Vincentian parrot, *Amazona guildingii*, which has been reduced to about 450 birds. Aided by financial support from bodies such as the World Wildlife Fund, the RARE Centre for Tropical Bird Conservation, and notably the local St Vincent Brewery, the Forestry Department has been able to bring the conservation message to the local population and enlist their support. A significant amount of the island's limited resources have been channelled into activities and programmes that support and complement parrot conservation. Thus this small island nation has already made a big contribution and shown its commitment to biodiversity conservation. Christian stresses the importance of continuing to demonstrate the links between habitat protection for the parrot, and the opportunities for eco-tourism, extraction of minor forest products, and of soil and watershed protection. He cautions that failure to demonstrate such direct linkages would risk the loss of public co-operation and support so vital for conservation.

This final section has served merely to introduce a few of the topics of sustainable development of islands and to present, briefly, the case that islands have many problems different in kind from those of larger land masses (for a fuller account see Beller *et al.* 1990). There is in this a parallel with the underlying ecological or nature conservation problems of oceanic islands. In short, small or remote oceanic islands are ecologically special, they constitute evolutionary treasure-houses, threatened by contact with other regions and especially the continents. The nature of the ecological problems involved is now well understood and the pattern of continuing degradation and species loss is all too obvious and, in the general sense, predictable. Dealing with these problems requires a sensitivity to, and respect for, the island condition of the human societies that occupy them. There are remedies, but they require solutions tailored to the islands and not simply exported from the continents.

10.7. Summary

As a result of human action, island birds have been 40 times more likely to become extinct within the past 400 years than continental birds, while altogether about 80% of documented animal species extinctions in this period have been of island species. Moreover, the losses caused by humans in prehistoric times exceed those of the past 400 years. This is most clearly established for birds, but applies to other taxa and to islands across the world. The scale of environmental degradation on certain Pacific islands, exemplified by Easter Island and Henderson Island, has been shown to have been so great as to have led to complete cultural collapse and, in the latter case, human abandonment of the island.

The major causes of local and global extinction have been: (1) predation by humans, (2) the introduction of exotics, (3) the spread of disease, and (4) habitat loss, of which habitat loss and the introduction of alien species currently give greatest cause for concern. The introduction of exotic species, in cases a single exotic species, can cause the local extinction of numerous native island species. The shrub *Miconia calvescens* on Tahiti and the brown tree snake (*Boiga irregularis*) on Guam provide two classic examples. However, case details provided in this chapter from the Pacific, Indian Ocean, Caribbean, and Atlantic illustrate that while some losses can indeed be attributed to a sole major cause, many species losses have resulted from synergisms between several causes, such as habitat alteration allowing the invasion of exotic plants, which in turn are spread by exotic birds, which in turn carry exotic diseases. While it is commonly perceived by conservation biologists that the big problem for continental organisms is the increasing insularization of their habitats, the problem for oceanic island biotas today is the increasing breakdown of the insularization of theirs.

While oceanic islands have a high representation of exotic species, it does not follow that island habitats, if undisturbed, are necessarily more invasible than continental habitats. However, exotic invaders have had a significant impact on many island ecosystems, and this susceptibility to exotic impact can be related both to peculiarities of human use of islands and to evolutionary features of the island biotas, especially to their evolution in the absence of terrestrial mammalian browsers and predators.

Island species continue to be lost, yet species can be saved, and are being saved, by means of rigorous habitat protection, pest and predator control, the control of hunting, and by breeding, translocation, and re-release programmes. Integrated programmes involving the re-insularization of populations, for instance by transferring them to small offshore islands from which predators have been eradicated, offer hope for the persistence of many threatened island endemics (getting rid of the lighthouse-keeper's cat and other commensals is thus often a key part of the solution). Many practical problems face island conservationists, and they are illustrated here by reference to the Galápagos, wherein the political, financial, and socio-economic problems are shown to be the root of current concerns. The strategy required for effective conservation of such island systems must be tailored to meet the particular features of island environments and their human societies. To be successful, conservation goals must be part of broader management for the sustainable development of island ecosystems, some elements of which are briefly reviewed in the present chapter. It is stressed that it is crucial to enlist political and legislative support, and to build educational programmes to engender the interest and support of island peoples.

Further reading

Publication details of volumes that appeared too late for inclusion in the text are given in full, the remainder are to be found in the reference list. Much important material has been published in the journal literature, but I have restricted this section to books.

Island environments

Nunn (1994) provides an excellent review of island environments and origins; Menard (1986) a beautifully illustrated account. Darwin's (1842) account of coral reefs and atolls retains value, and for those who cannot easily get hold of it, Ridley (1994) provides in a lightly edited excerpt the core of his theory, together with some other important passages on islands and other topics from a variety of Darwin's books. See also Williamson (1981) and Keast and Miller (1996).

Biodiversity hot-spots

A reading of Wallace's classic *Island life* (1880, 1895, 1902), is instructive in showing how much of the foundations of this subject were established by the end of the nineteenth century. See Groombridge (1992) and Davis *et al.* (1994, 1995) for recent estimates of the relative significance of islands in global biodiversity. For the biogeography of island biodiversity see Carlquist (1974), Whitmore (1987), and Keast and Miller (1996).

Speciation and the island condition; arrival and change; emergent models of island evolution

Darwin's *Origin of species*, and the Galápagos chapter of the *Voyage of the Beagle* (1845 edition) are both essential historical reading. Otte and Endler (1989) provide a wide-ranging series of advanced-level papers illustrating the great diversity of concepts, viewpoints, and empirical information on evolution. For an excellent general text on evolution, see Ridley (1996). Wagner and Funk (1995) focus on Hawaii, bringing together the comparative analyses of different taxa using modern cladistic techniques. Useful material can also be found in Williamson (1981), Myers and Giller (1988), Rosenzweig (1995), Keast and Miller (1996), and Vitousek *et al.* (1995). Finally, a significantly revised version of Clarke and Grant (1996) has been issued in book form: Grant, P. R. (ed.) (1998) *Evolution on islands.* Oxford University Press, Oxford, providing a valuable set of papers on microevolution, species formation, and adaptive radiations on islands.

Species numbers games—community assembly and dynamics

The obvious place to start is MacArthur and Wilson (1967). I also recommend reading some of Lack's work (e.g. 1976), as his views are often misunderstood or misrepresented. Williamson (1981) and Brown and Gibson (1983) provide clear, if now somewhat dated, reviews of these topics; while Shrader-Frechette and McCoy (1993) provide an insightful critique of equilibrium theory. Papers in Cody and Diamond (1975) and Strong *et al.* (1984) cover island assembly issues. Thornton (1996) provides a useful synthesis of the Krakatau re-colonization story.

Island theory and conservation

Numerous books have been published in recent years on the theme of conservation biology. The following provide a range of differing approaches: Harris (1984), realism allied to ETIB theory; Shafer (1990), a general review of the 'island' approach; Primack (1993), a wide-ranging conservation text, moving well beyond the confines of island approaches; Shrader-Frechette and McCoy (1993), a thought-provoking critique; Hanski, I. and Gilpin, M. E. (ed.) (1997) (*Metapopulation biology: ecology, genetics, and evolution.* Academic Press, San Diego), advances in metapopulation thinking; Laurance, W. F. and Bierregaard, R. O., Jr (ed.) (1997) (*Tropical forest remnants: ecology, management, and conservation of fragmented communities.* The University of Chicago Press, Chicago), a substantial body of papers on the themes in the title. Schwartz, M. W. (1997) (*Conservation in highly fragmented landscapes.* Chapman & Hall, New York), conservation in the American Midwest.

The human impact on island ecosystems

For accounts of the impacts of humans on oceanic islands see: Watts (1987), on the West Indies; Cuddihy and Stone (1990), on Hawaii; Diamond (1991*b*), for the global picture; Bahn and Flenley (1992), on the tragedies of Easter Island; Mitchell, A. (1989) (*A fragile Island paradise: nature and man in the Pacific.* Collins, London), written in the style of a travel documentary; Quammen, D. (1996) (*The song of the dodo; island biogeography in an age of extinctions.* Hutchinson, London), also written in documentary style, this volume also tackles numerous debates in island evolution and ecology, replete with biographical sketches of key figures in island biogeography. For insights on the human dimension and ingredients for sustainable development, see Beller *et al.* (1990) and Vitousek *et al.* (1995).

References

Abbott, I. (1983). The meaning of z in species/area regressions and the study of species turnover in island biogeography. *Oikos*, **41**, 385–90.

Abbott, I. and Black, R. (1980). Changes in species composition of floras on islets near Perth, Western Australia. *Journal of Biogeography*, 7, 399–410.

Abbott, I. and Grant, P. R. (1976). Nonequilibrial bird faunas on islands. *The American Naturalist*, **110**, 507–28.

Adler, G. H. (1992). Endemism in birds of tropical Pacific islands. *Evolutionary Ecology*, **6**, 296–306.

Adler, G. H. (1994). Avifaunal diversity and endemism on tropical Indian Ocean islands. *Journal of Biogeography*, **21**, 85–95.

Adler, G. H. and Dudley, R. (1994). Butterfly biogeography and endemism on tropical Pacific Islands. *Biological Journal of the Linnean Society*, **52**, 151–62.

Adler, G. H. and Levins, R. (1994). The island syndrome in rodent populations. *The Quarterly Review of Biology*, **69**, 473–90.

Adler, G. H. and Seamon, J. O. (1991). Distribution and abundance of a tropical rodent, the spiny rat, on Islands in Panama. *Journal of Tropical Ecology*, 7, 349–60.

Adler, G. H. and Wilson, M. L. (1985). Small mammals on Massachusetts islands: the use of probability functions in clarifying biogeographic relationships. *Oecologia*, **66**, 178–86.

Adler, G. H. and Wilson, M. L. (1989). Insular distributions of voles and shrews: the rescue effect and implications for species co-occurrence patterns. *Coenoses*, **4**, 69–72.

Adler, G. H., Wilson, M. L., and DeRosa, M. J. (1986). Influence of island area and isolation on population characteristics of *Peromyscus leucopus*. *Journal of Mammalogy*, **67**, 406–9.

Adler, G. H., Austin, C. A., and Dudley, R. (1995). Dispersal and speciation of skinks among archipelagos in the tropical Pacific Ocean. *Evolutionary Ecology*, **9**, 529–41.

Adsersen, H. (1995). Research on islands: classic, recent, and prospective approaches. In *Islands: biological diversity and ecosystem function* (ed. P. M. Vitousek, L. L. Loope, and H. Adsersen), Ecological Studies 115, pp. 7–21. Springer-Verlag, Berlin.

Alonso, J. C., Alonso, J. A., and Munoz-Pulido, R. (1994). Mitigation of bird collisions with transmission lines through groundwire marking. *Biological Conservation*, **67**, 129–34.

Armstrong, P. (1982). Rabbits (*Oryctolagus cuniculus*) on islands: a case study of successful colonization. *Journal of Biogeography*, **9**, 353–62.

Arnold, E. N. (1979). Indian Ocean giant tortoises: their systematics and island adaptations. *Philosophical Transactions of the Royal Society of London*, **B 286**, 127–45.

Ashmole, N. P. and Ashmole, M. J. (1988). Insect dispersal on Tenerife, Canary Islands: high altitude fallout and seaward drift. *Arctic and Alpine Research*, **20**, 1–12.

Asquith, A. (1995). Evolution of *Sarona* (Heroptera, Miridae): Speciation on geographic and ecological islands. In *Hawaiian Biogeography: evolution on a hot spot archipelago* (ed. W. L. Wagner and V. A. Funk), pp. 90–120. Smithsonian Institution Press, Washington.

Atmar, A. and Patterson, B. D. (1993). The measure of order and disorder in the distribution of species in fragmented habitat. *Oecologia*, **96**, 373–82.

Ayers, J. M., Bodmer, R. E., and Mittermeier, R. A. (1991). Financial considerations of reserve design in countries with high primate diversity. *Conservation Biology*, **5**, 109–14.

Backer, C. A. (1929). *The problem of Krakatao as seen by a Botanist.* Published by the Author, Surabaya.

Bahn, P. and Flenley, J. R. (1992). *Easter Island, Earth island.* Thames and Hudson, London.

Barrett, S. C. H. (1989). Mating system evolution and speciation in heterostylous plants. In *Speciation and its consequences* (ed. D. Otte and J. A. Endler), pp. 257–83. Sinauer, Sunderland, Massachusetts.

Barrett, S. C. H. (1996). The reproductive biology and genetics of island plants. *Philosophical Transactions of the Royal Society of London*, **B 351**, 725–33.

Barton, N. H. (1989). Founder effect speciation. In *Speciation and its consequences* (ed. D. Otte and J. A. Endler), pp. 229–56. Sinauer, Sunderland, Massachusetts.

Barton, N. H. and Charlesworth, B. (1984). Genetic revolutions, founder effects, and speciation. *Annual Review of Ecology and Systematics*, **15**, 133–64.

Bawa, K. S. (1980). Evolution of dioecy in flowering plants. *Annual Review of Ecology and Systematics*, **11**, 15–19.

Bawa, K. S. (1982). Outcrossing and incidence of dioecism in island floras. *American Naturalist*, **119**, 866–71.

Begon, M., Harper, J. L., and Townsend, C. R. (1986). *Ecology*. Blackwell, Oxford.

Begon, M., Harper, J. L., and Townsend, C. R. (1990). *Ecology* (2nd edn). Blackwell, Oxford.

Bell, M. and Walker, M. J. C. (1992). *Late Quaternary environmental change: physical and human perspectives*. Longman, Harlow.

Beller, W., d'Ayala, P., and Hein, P. (ed.) (1990). *Sustainable development and environmental management of small islands*, Vol. 5, Man and the Biosphere Series. UNESCO/Parthenon Publishing, Paris.

Benton, T. and Spencer, T. (ed.) (1995). *The Pitcairn Islands: biogeography, ecology and prehistory*. Academic Press, London.

Berry, R. J. (1986). Genetics of insular populations of mammals with particular reference to differentiation and founder effects in British small mammals. *Biological Journal of the Linnean Society*, **28**, 205–30.

Berry, R. J. (1992). The significance of island biotas. *Biological Journal of the Linnean Society*, **46**, 3–12.

Berry, R. J., Berry, A. J., Anderson, T. J. C., and Scriven, P. (1992). The house mice of Faray, Orkney. *Journal of Zoology (London)*, **228**, 233–46.

Bevanger, K. (1996). Estimates and population consequences of tetraonid mortality caused by collisions with high tension power lines in Norway. *Journal of Applied Ecology*, **32**, 745–53.

Bibby, C. J. (1995). Recent past and future extinctions in birds. In *Extinction rates* (ed. J. H. Lawton and R. M. May), pp. 98–110. OUP, Oxford.

Blake, J. G. (1991). Nested subsets and the distribution of birds on isolated woodlots. *Conservation Biology*, **5**, 58–66.

Bodmer, R. E., Eisenberg, J. F., and Redford, K. H. (1997). Hunting and the likelihood of extinction of Amazonian mammals. *Conservation Biology*, **11**, 460–6.

Boecklen, W. J. (1986). Effects of habitat heterogeneity on the species–area relationships of forest birds. *Journal of Biogeography*, **13**, 59–68.

Boecklen, W. J. and Gotelli, N. J. (1984). Island biogeographic theory and conservation practice: species–area or specious–area relationships? *Biological Conservation*, **29**, 63–80.

Boggs, C. L. and Murphy, D. D. (1997). Community composition in mountain ecosystems: climatic determinants of montane butterfly distributions. *Global Ecology and Biogeography Letters*, **6**, 39–48.

Borgen, L. (1979). Karyology of the Canarian flora. In *Plants and islands* (ed. D. Bramwell), pp. 329–46. Academic Press, London.

Bramwell, D. (ed.) (1979). *Plants and islands*. Academic Press, London.

Bramwell, D. and Bramwell, Z. I. (1974). *Wild flowers of the Canary Islands*. Stanley Thornes, Cheltenham.

Brooks, T. and Balmford, A. (1996). Atlantic forest extinctions. *Nature*, **380**, 115.

Brooks, T. M., Pimm, S. L., and Collar, N. J. (1997). Deforestation predicts the number of threatened birds in insular Southeast Asia. *Conservation Biology*, **11**, 382–94.

Brothers, T. S. and Spingarn, A. (1992). Forest fragmentation and alien plant invasion of Central Indiana old-growth forests. *Conservation Biology*, **6**, 91–100.

Brown, J. H. (1971). Mammals on mountaintops: nonequilibrium insular biogeography. *The American Naturalist*, **105**, 467–78.

Brown, J. H. (1981). Two decades of homage to Santa Rosalia: toward a general theory of diversity. *American Zoologist*, **21**, 877–88.

Brown, J. H. (1986). Two decades of interaction between the MacArthur–Wilson model and the complexities of mammalian distribution. *The Biological Journal of the Linnean Society*, **28**, 231–51.

Brown, J. H. and Gibson, A. C. (1983). *Biogeography*. Mosby, St Louis.

Brown, J. H. and Kodric-Brown, A. (1977). Turnover rates in insular biogeography: effect of immigration on extinction. *Ecology*, **58**, 445–9.

Brown, J. H. and Lomolino, M. V. (1989). Independent discovery of the equilibrium theory of island biogeography. *Ecology*, **70**, 1954–7.

Brown, J. W. (1987). The peninsular effect in Baja California: an entomological assessment. *Journal of Biogeography*, **14**, 359–66.

Brown, J. W. and Opler, P. A. (1990). Patterns of butterfly species density in peninsular Florida. *Journal of Biogeography*, **17**, 615–22.

Brown, K. S., Jr and Brown, G. G. (1992). Habitat alteration and species loss in Brazilian forests. In *Tropical deforestation and species extinction* (ed. T. C. Whitmore and J. A. Sayer), pp. 119–42. Chapman & Hall, London.

Brown, M. and Dinsmore, J. J. (1988). Habitat islands and the equilibrium theory of island biogeography: testing some predictions. *Oecologia*, **75**, 426–9.

Brown, W. L., Jr and Wilson, E. O. (1956). Character displacement. *Systematic Zoology*, **7**, 49–64.

Browne, J. and Neve, M. (1989). Introduction to the abridged version of Charles Darwin's 1839 *Voyage of the Beagle*. Penguin Classics, London.

Bruijnzeel, L. A., Waterloo, M. J., Proctor, J., Kuiters, A. T., and Kotterink, B. (1993). Hydrological observations in montane rain forests on Gunung Silam, Sabah, Malaysia, with special reference to the '*Massenerhebung*' effect. *Journal of Ecology*, **81**, 145–67.

Buckley, R. C. (1981). Scale-dependent equilibrium on highly heterogeneous islands: plant geography of northern Great Barrier Reef sand cays and shingle islets. *Australian Journal of Ecology*, **6**, 143–8.

Buckley, R. C. (1982). The Habitat-unit model of island biogeography. *Journal of Biogeography*, **9**, 339–44.

Buckley, R. C. (1985). Distinguishing the effects of area and habitat type on island plant species richness by separating floristic elements and substrate types and controlling for island isolation. *Journal of Biogeography*, **12**, 527–35.

Buckley, R. C. and Knedlhans, S. B. (1986). Beachcomber biogeography: interception of dispersing propagules by islands. *Journal of Biogeography*, **13**, 68–70.

Burkey, T. V. (1993). Edge effects in seed and egg predation at two neotropical rainforest sites. *Biological Conservation*, **66**, 139–43.

Burkhardt, F. (ed.) (1996). *Charles Darwin's Letters: a selection 1825–1859*. Cambridge University Press, Cambridge.

Bush, M. B. (1994). Amazonian speciation—a necessarily complex model. *Journal of Biogeography*, **21**, 5–17.

Bush, M. B. (1996). Amazonian conservation in a changing world. *Biological Conservation*, **76**, 219–28.

Bush, M. B. and Whittaker, R. J. (1991). Krakatau: colonization patterns and hierarchies. *Journal of Biogeography*, **18**, 341–56.

Bush, M. B. and Whittaker, R. J. (1993). Non-equilibration in island theory of Krakatau. *Journal of Biogeography*, **20**, 453–8.

Bush, M. B., Whittaker, R. J., and Partomihardjo, T. (1992). Forest development on Rakata, Panjang and Sertung: contemporary dynamics (1979–1989). *Geojournal*, **28**, 185–99.

Buskirk, R. E. (1985). Zoogeographic patterns and tectonic history of Jamaica and the northern Caribbean. *Journal of Biogeography*, **12**, 445–61.

Cameron, R. A. D., Cook, L. M., and Hallows, J. D. (1996). Land snails on Porto Santo: adaptive and non-adaptive radiation. *Philosophical Transactions of the Royal Society of London*, **B 351**, 309–27.

Carlquist, S. (1965). *Island life: a natural history of the islands of the world*. Natural History Press, New York.

Carlquist, S. (1974). *Island biology*. Columbia University Press, New York.

Carlquist, S. (1995). Introduction. In *Hawaiian Biogeography: evolution on a hot spot archipelago* (ed. W. L. Wagner and V. A. Funk), pp. 1–13. Smithsonian Institution Press, Washington.

Carr, G. D. *et al.* (1989). Adaptive radiation of the Hawaiian silversword alliance (Compositae—Madiinae): a comparison with Hawaiian picture-winged *Drosophila*. In *Genetics, speciation and the founder principle* (ed. L. Y. Giddings, K. Y. Kaneshiro, and W. W. Anderson), pp. 79–95. Oxford University Press, New York.

Carson, H. L. (1983). Chromosomal sequences and interisland colonizations in Hawaiian Drosophilidae. *Genetics*, **103**, 465–82.

Carson, H. L. (1992). Genetic change after colonization. *Geojournal*, **28**, 297–302.

Carson, H. L., Hardy, D. E., Spieth, H. T., and Stone, W. S. (1970). The evolutionary biology of the Hawaiian Drosophilidae. In *Essays in evolution and genetics in honor of Theodosius Dobzhansky* (ed. M. K. Hecht and W. C. Steere), pp. 437–543. Appleton-Century-Crofts, New York.

Carson, H. L., Lockwood, J. P., and Craddock, E. M. (1990). Extinction and recolonization of local populations on a growing shield volcano. *Proceedings of the National Academy of Sciences, USA*, **87**, 7055–7.

Case, T. J. and Cody, M. L. (1987). Testing theories of island biogeography. *American Scientist*, **75**, 402–11.

Case, T. J., Bolger, D. T., and Richman, A. D. (1992). Reptilian extinctions: the last ten thousand years. In *Conservation Biology: the theory and practice of nature conservation and management* (ed. P. L. Fielder and S. K. Jain), pp. 91–125. Chapman & Hall, New York.

Caswell, H. (1978). Predator-mediated coexistence: a non-equilibrium model. *The American Naturalist*, **112**, 127–54.

Caswell, H. (1989). Life-history strategies. In *Ecological concepts: the contribution of ecology to an understanding of the natural world* (ed. J. M. Cherrett), pp. 285–307. Blackwell Scientific Publications, Oxford.

Christian, C. S. (1993). The challenge of parrot conservation in St Vincent and the Grenadines. *Journal of Biogeography*, **20**, 463–9.

Christian, C. S., Lacher, T. E., Jr, Zamore, M. P., Potts, T. D., and Burnett, G. W. (1996). Parrot conservation in the Lesser Antilles with some comparison to the Puerto Rican efforts. *Biological Conservation*, 77, 159–67.

Christiansen, M. B. and Pitter, E. (1997). Species loss in a forest bird community near Lagoa Santa in Southeastern Brazil. *Biological Conservation*, **80**, 23–32.

Clarke, B. and Grant, P. R. (ed.) (1996). Evolution on islands. *Philosophical Transactions of the Royal Society of London*, **B 351**, 723–854.

Clarke, B. Johnson, M. S., and Murray, J. (1996). Clines in the genetic distance between two species of island land snails: how 'molecular leakage' can mislead us about speciation. *Philosophical Transactions of the Royal Society of London*, **B 351**, 773–84.

Clout, M. N. and Craig, J. L. (1995). The conservation of critically endangered flightless birds in New Zealand. *The Ibis*, **137**, S181–90.

Cody, M. L. and Diamond, J. M. (ed.) (1975). *Ecology and evolution of communities*. Harvard University Press, Cambridge, Massachusetts, USA.

Cody, M. L. and Overton, J. McC. (1996) Short-term evolution of reduced dispersal in island plant populations. *Journal of Ecology*, **84**, 53–61.

Colinvaux, P. A. (1972). Climate and the Galápagos Islands. *Nature*, **219**, 590–4.

Colwell, R. K. and Winkler, D. W. (1984). A null model for null models in biogeography. In *Ecological communities: conceptual issues and the evidence* (ed. D. R. Strong, Jr, D. Simberloff, L. G. Abele, and A. B. Thistle), pp. 344–59. Princeton University Press, Princeton, New Jersey.

Compton, S. G., Ross, S. J., and Thornton, I. W. B. (1994). Pollinator limitation of fig tree reproduction on the island of Anak Krakatau (Indonesia). *Biotropica*, **26**, 180–6.

Conant, S. (1988). Geographic variation in the Laysan finch (*Telespyza cantans*). *Evolutionary Ecology*, **2**, 270–82.

Connell, J. H. (1980). Diversity and the coevolution of competitors, or the ghost of competition past. *Oikos*, **35**, 131–8.

Connor, E. F. and McCoy, E. D. (1979). The statistics and biology of the species–area relationship. *The American Naturalist*, **113**, 791–833.

Connor, E. F. and Simberloff, D. (1979). The assembly of species communities: chance or competition? *Ecology*, **60**, 1132–40.

Connor, E. F. and Simberloff, D. (1983). Interspecific competition and species co-occurrence patterns on islands: null models and the evaluation of evidence. *Oikos*, **41**, 455–65.

Connor, E. F., McCoy, E. D., and Cosby, B. J. (1983). Model discrimination and expected slope values in species–area studies. *The American Naturalist*, **122**, 789–96.

Cook, R. R. (1995). The relationship between nested subsets, habitat subdivision, and species diversity. *Oecologia*, **101**, 204–10.

Cook, R. R. and Quinn, J. F. (1995). The influence of colonization in nested species subsets. *Oecologia*, **102**, 413–25.

Corlett, R. T. (1992). The ecological transformation of Singapore, 1819–1990. *Journal of Biogeography*, **19**, 411–20.

Cox, C. B. and Moore, P. D. (1993). *Biogeography: an ecological and evolutionary approach* (5th edn). Blackwell Scientific Publications, Oxford.

Cox, G. W. and Ricklefs, R. E. (1977). Species diversity, ecological release, and community structuring in Caribbean land bird faunas. *Oikos*, **28**, 113–22.

Cox, P. A., Elmqvist, T., Pierson, E. D., and Rainey, W. E. (1992). Flying foxes as pollinators and seed dispersers in Pacific Island ecosystems. In *Pacific Island Flying Foxes: Proceedings of an International Conservation Conference*, Biological Report 90 (23), July 1992 (ed. D. E. Wilson and G. L. Graham), pp. 18–23. US Department of the Interior, Fish and Wildlife Service, Washington.

Crawford, D. J., Stuessy, T. F., Haines, D. W., Cosner, M. B. O., Silva, M., and Lopez, P. (1992). Allozyme diversity within and divergence among four species of *Robinsonia* (Asteraceae: Senecioneae), a genus endemic to the Juan Fernandez islands, Chile. *American Journal of Botany*, **79**, 962–6.

Cronk, Q. C. B. (1987). History of the endemic flora of St. Helena: a relictual series. *New Phytologist*, **105**, 509–20.

Cronk, Q. C. B. (1989). The past and present vegetation of St. Helena. *Journal of Biogeography*, **16**, 47–64.

Cronk, Q. C. B. (1990). History of the endemic flora of St. Helena: late Miocene '*Trochetiopsis*-like' pollen from St. Helena and the origin of *Trochetiopsis*. *New Phytologist*, **114**, 159–65.

Cronk, Q. C. B. (1992). Relict floras of Atlantic islands: patterns assessed. *Biological Journal of the Linnean Society*, **46**, 91–103.

Cronk, Q. C. B. and Fuller, J. L. (1995) *Plant invaders*. Chapman & Hall, London.

Crowell, K. L. (1986). A comparison of relict versus equilibrium models for insular mammals of the Gulf of Maine. *Biological Journal of the Linnean Society*, **28**, 37–64.

Cuddihy, L. W. and Stone, C. P. (1990). *Alteration of native Hawaiian vegetation: effects of humans, their activities and introductions*. University of Hawaii

Cooperative National Park Resources Studies Unit, 3190 Maile Way, Honolulu, Hawaii.

Cutler, A. (1991). Nested fauna and extinction in fragmented habitats. *Conservation Biology*, **5**, 496–505.

Dammerman, K. W. (1948). The fauna of Krakatau (1883–1933). *Verhandelingen der Koninklijke Nederlandse Akademie Van Wetenschappen Afdeling Natuurkunde (II)*, **44**, 1–594. N. V. Noord-Hollandsche Uitgevers Maatschappij, Amsterdam.

Damuth, J. (1993). Cope's rule, the island rule and the scaling of mammalian population density. *Nature*, **365**, 748–50.

D'Antonio, C. M. and Dudley, T. L. (1995). Biological invasions as agents of change on islands versus mainlands. In *Islands: biological diversity and ecosystem function* (ed. P. M. Vitousek, L. L. Loope, and H. Adsersen), Ecological Studies, **115**, pp. 103–21. Springer-Verlag, Berlin.

Darlington, P. J. (1957). *Zoogeography: the geographical distribution of animals*. Wiley, New York.

Darwin, C. (1842). *The structure and distribution of coral reefs. Being the first part of the geology of the voyage of the Beagle, under the command of Capt. Fitzroy, R. N., during the years 1832–36*. Smith, Elder, and Company, London.

Darwin, C. (1845). *Voyage of the Beagle*. (First published 1839. Abridged version edited by J. Browne and M. Neve (1989). Penguin Books, London.)

Darwin, C. (1859). *On the origin of species by means of natural selection*. J. Murray, London. (Page numbers cited here are from an edition published by Avenel, New York, in 1979 under the title *The origin of species*.)

Davis, S. D., Heywood, V. H., and Hamilton, A. C. (ed.) (1994). *Centres of plant diversity: a guide and strategy for their conservation. Volume 1: Europe, Africa, South West Asia and the Middle East*. WWF and IUCN, Cambridge.

Davis, S. D., Heywood, V. H., and Hamilton, A. C. (ed.) (1995). *Centres of plant diversity: a guide and strategy for their conservation. Volume 2: Asia, Australasia and the Pacific*. WWF and IUCN, Cambridge.

Dawson, D. (1994). Are habitat corridors conduits for animals and plants in a fragmented landscape? A review of the scientific evidence. *English Nature Research Reports*, No. 94.

Decker, R. W. and Decker, B. B. (1991). *Mountains of fire: the nature of volcanoes*. Cambridge University Press, Cambridge.

Delcourt, H. R. and Delcourt, P. A. (1988). Quaternary landscape ecology: relevant scales in space and time. *Landscape Ecology*, **2**, 23–44.

Delcourt, H. R. and Delcourt, P. A. (1991). *Quaternary Ecology: A paleoecological perspective*. Chapman & Hall, London.

Deshaye, J. and Morisset, P. (1988). Floristic richness, area, and habitat diversity in a hemiarctic archipelago. *Journal of Biogeography*, **15**, 721–8.

Diamond, J. M. (1969). Avifaunal equilibria and species turnover rates on the Channel Islands of California. *Proceedings of the National Academy of Sciences, USA*, **64**, 57–63.

Diamond, J. M. (1972). Biogeographical kinetics: estimation of relaxation times for avifaunas of southwest Pacific islands. *Proceedings of the National Academy of Sciences, USA*, **69**, 3199–203.

Diamond, J. M. (1974). Colonization of exploded volcanic islands by birds: the supertramp strategy. *Science*, **184**, 803–6.

Diamond, J. M. (1975*a*). Assembly of species communities. In *Ecology and evolution of communities* (ed. M. L. Cody and J. M. Diamond), pp. 342–444. Harvard University Press, Cambridge, Massachusetts, USA.

Diamond, J. M. (1975*b*). The island dilemma: lessons of modern biogeographic studies for the design of natural preserves. *Biological Conservation*, **7**, 129–46.

Diamond, J. M. (1977). Continental and insular speciation in Pacific land birds. *Systematic Zoology*, **26**, 263–8.

Diamond, J. M. (1984). 'Normal' extinctions of isolated populations. In *Extinctions* (ed. M. H. Nitecki), pp. 191–246. The University of Chicago Press, Chicago.

Diamond, J. M. (1986). The design of a nature reserve system for Indonesian New Guinea. In *Conservation biology: the science of scarcity and diversity* (ed. M. Soulé), pp. 485–503. Sinauer Associates, Sunderland, Massachusetts.

Diamond, J. M. (1991*a*). A new species of rail from the Solomon islands and convergent evolution of insular flightlessness. *The Auk*, **108**, 461–70.

Diamond, J. M. (1991*b*). *The rise and fall of the third chimpanzee*. Vintage, London.

Diamond, J. M. (1996). Daisy gives an evolutionary answer. *Nature*, **380**, 103–4.

Diamond, J. M. and Gilpin, M. E. (1982). Examination of the 'null' model of Connor and Simberloff for species co-occurrences on islands. *Oecologia*, **52**, 64–74.

Diamond, J. M. and Gilpin, M. E. (1983). Biogeographical umbilici and the origin of the Philippine avifauna. *Oikos*, **41**, 307–21.

Diamond, J. M. and May, R. M. (1977). Species turnover rates on islands: dependence on census interval. *Science*, **197**, 266–70.

Diamond, J. M. and May, R. M. (1981). Island Biogeography and the Design of Nature Reserves. In *Theoretical Ecology* (2nd edn) (ed. R. M. May), pp. 228–52. Blackwell, Oxford.

Diamond, J. M. and May, R. M. (1985). Conservation biology: a discipline with a time limit. *Nature*, **317**, 111–12.

Diamond, J. and Pimm, S. (1993). Survival times of bird populations: a reply. *The American Naturalist*, **142**, 1030–5.

Diamond, J. M., Pimm, S. L., Gilpin, M. E., and LeCroy, M. (1989). Rapid evolution of character displacement in Myzomelid Honeyeaters. *The American Naturalist*, **134**, 675–708.

Diehl, S. and Bush, G. L. (1989). The role of habitat preference in adaptation and speciation. In *Speciation and its consequences* (ed. D. Otte and J. A. Endler), pp. 345–65. Sinauer, Sunderland, Massachusetts.

Doak, D. F. and Mills, L. S. (1994). A useful role for theory in conservation. *Ecology*, **75**, 615–26.

Docters van Leeuwen, W. M. (1936). Krakatau 1883–1933. *Annales du Jardin botanique de Buitenzorg*, **46–47**, 1–506.

Dunn, C. P. and Loehle, C. (1988). Species–area parameter estimation: testing the null model of lack of relationship. *Journal of Biogeography*, **15**, 721–8.

Dunstan, C. E. and Fox, B. J. (1996). The effects of fragmentation and disturbance of rainforest on ground-dwelling small mammals on the Robertson Plateau, New South Wales, Australia. *Journal of Biogeography*, **23**, 187–201.

Ecker, A. (1976). Groundwater behaviour in Tenerife, volcanic island (Canary Islands, Spain). *Journal of Hydrology*, **28**, 73–86.

Ehrendorfer, E. (1979). Reproductive biology in island plants. In *Plants and islands* (ed. D. Bramwell), pp. 293–306. Academic Press, London.

Elisens, W. J. (1992). Genetic divergence in *Galvezia* (Scrophulariaceae): evolutionary and biogeographic relationships among South American and Galápagos species. *American Journal of Botany*, **79**, 198–206.

Elmqvist, T., Rainey, W. E., Pierson, E. D., and Cox, P. A. (1994). Effects of tropical cyclones Ofa and Val on the structure of a Samoan lowland rain forest. *Biotropica*, **26**, 384–91.

Ernst, A. (1908). *The new flora of the volcanic island of Krakatau* (translated into English by A. C. Seward). Cambridge University Press, Cambridge.

Erwin, R. M., Hatfield, J. S., and Winters, T. J. (1995). The value and vulnerability of small estuarine islands for conserving metapopulations of breeding waterbirds. *Biological Conservation*, **71**, 187–91.

Estrada, A., Coates-Estrada, R., and Meritt, D. (1993). Bat species richness and abundance in tropical rain forest fragments and in agricultural habitats at Los Tuxtlas, Mexico. *Ecography*, **16**, 309–18.

Fahrig, L. and Merriam, G. (1994). Conservation of fragmented populations. *Conservation Biology*, **8**, 50–9.

Fernandopullé, D. (1976). Climate characteristics of the Canary Islands. In *Biogeography and ecology in the Canary Islands* (ed. G. Kunkel), pp. 185–206. W. Junk, The Hague.

Fiedler, P. L. and Jain, S. K. (ed.) (1992). *Conservation Biology: the theory and practice of nature conservation, preservation and management.* Chapman & Hall, London.

Fisher, R. A., Corbet, A. S., and Williams, C. B. (1943). The relation between the number of species and the number of individuals in a random sample of an animal population. *Journal of Animal Ecology*, **12**, 42–58.

Flenley, J. R. (1993). The palaeoecology of Easter Island, and its ecological disaster. In *Easter Island studies. Contributions to the history of Rapanui in memory of William T. Mulloy* (ed. S. R. Fisher), pp. 27–45. Oxbow, Oxford.

Flenley, J. R., King, S. M., Jackson, J., Chew, C., Teller, J. T., and Prentice, M. E. (1991). The Late Quaternary vegetational and climatic history of Easter Island. *Journal of Quaternary Science*, **6**, 85–115.

Forsman, A. (1991). Adaptive variation in head size in *Vipera berus* L. populations. *Biological Journal of the Linnean Society*, **43**, 281–96.

Fosberg, F. R. (1948). Derivation of the flora of the Hawaiian Islands. In *Insects of Hawaii*, Vol. 1 (ed. E. C. Zimmerman), pp. 107–19. University Press, Hawaii, Honolulu.

Foster, J. B. (1964). Evolution of mammals on islands. *Nature*, **202**, 234–5.

Franklin, J. and Steadman, D. W. (1991). The potential for conservation of Polynesian birds through habitat mapping and species translocation. *Conservation Biology*, **5**, 506–21.

Fujita, M. S. and Tuttle, M. D. (1991). Flying foxes (Chiroptera: Pteropodidae): threatened animals of key ecological and economic importance. *Conservation Biology*, **5**, 455–63.

Galapagos Newsletter Autumn 1996, No. 3. Galapagos Conservation Trust, 18 Curzon Street, London.

Gardner, A. S. (1986). The biogeography of the lizards of the Seychelles Islands. *Journal of Biogeography*, **13**, 237–53.

Gaston, K. J. (1991). Estimates of the near-imponderable: a reply to Erwin. *Conservation Biology*, **5**, 564–6.

Gibbs, H. L. and Grant, P. R. (1987). Ecological consequences of an exceptionally strong El Niño event on Darwin's finches. *Ecology*, **68**, 1735–46.

Gibson, C. W. D. (1986). Management history in relation to changes in the flora of different habitats on an Oxfordshire estate, England. *Biological Conservation*, **38**, 217–32.

Gilbert, F. S. (1980). The equilibrium theory of island biogeography, Fact or Fiction? *Journal of Biogeography*, **7**, 209–35.

Gilpin, M. E. and Diamond, J. M. (1982). Factors contributing to non-randomness in species co-occurrences on islands. *Oecologia*, **52**, 75–84.

Gittenberger, E. (1991). What about non-adaptive radiation? *Biological Journal of the Linnean Society*, **43**, 263–72.

Godwin, H. (1975). *The history of the British flora: a factual basis for phytogeography* (2nd edn). Cambridge University Press, Cambridge.

Goodfriend, G. A., Cameron, R. A. D., and Cook, L. M. (1994). Fossil evidence of recent human impact on the land snail fauna of Madeira. *Journal of Biogeography*, **21**, 309–20.

Gorchov, D. L., Cornejo, F., Ascorra, C., and Jaramillo, M. (1993). The role of seed dispersal in the natural regeneration of rain forest after strip-cutting in the Peruvian Amazon. *Vegetatio*, **107/108**, 339–49.

Gotelli, N. J. (1991). Metapopulation models: the rescue effect, the propagule rain, and the core-satellite hypothesis. *The American Naturalist*, **138**, 768–76.

Grainger, A. (1993). Rates of deforestation in the humid tropics: estimates and measurements. *The Geographical Journal*, **159**, 33–44.

Grant, B. R. and Grant, P. R. (1996). High survival of Darwin's finch hybrids: effects of beak morphology and diets. *Ecology*, **77**, 500–9.

Grant, P. R. (1981). Speciation and the adaptive radiation of Darwin's finches. *American Scientist*, **69**, 653–63.

Grant, P. R. (1984). The endemic land birds. In *Galapagos* (ed. R. Perry), Key Environments Series, pp. 175–89. IUCN/Pergamon Press, Oxford.

Grant, P. R. (1986). Interspecific competition in fluctuating environments. In *Community Ecology* (ed. J. M. Diamond and T. J. Case), pp. 173–91. Harper & Row, New York.

Grant, P. R. (1994). Population variation and hybridization: comparison of finches from two archipelagos. *Evolutionary Ecology*, **8**, 598–617.

Grant, P. R. and Grant, B. R. (1989). Sympatric speciation and Darwin's finches. In *Speciation and its consequences* (ed. D. Otte and J. A. Endler), pp. 433–57. Sinauer, Sunderland, Massachusetts.

Grant, P. R. and Grant, B. R. (1994). Phenotypic and genetic effects of hybridization in Darwin's finches. *Evolution*, **48**, 297–316.

Grant, P. R. and Grant, B. R. (1996). Speciation and hybridization in island birds. *Philosophical Transactions of the Royal Society of London*, **B 351**, 765–72.

Graves, G. R. and Gotelli, N. J. (1983). Neotropical land-bridge avifaunas: new approaches to null hypotheses in biogeography. *Oikos*, **41**, 322–33.

Grayson, D. L. and Livingston, S. D. (1993). Missing mammals on Great Basin mountains: Holocene extinctions and inadequate knowledge. *Conservation Biology*, **7**, 527–32.

Greenslade, P. J. M. (1968). Island patterns in the Solomon Islands bird fauna. *Evolution*, **22**, 751–61.

Griffith, B., Scott, J. M., Carpenter, J. W. and Reed, C. (1989). Translocation as a species conservation tool: status and strategy. *Science*, **245**, 477–80.

Griffiths, O., Cook, A., and Wells, S. M. (1993). The diet of introduced carnivorous snail *Euglandina rosea* in Mauritius and its implications for threatened island gastropod faunas. *Journal of Zoology (London)*, **229**, 78–89.

Groombridge, B. (ed.) (1992). *Global biodiversity: status of the Earth's living resources.* (A report compiled by the World Conservation Monitoring Centre.) Chapman & Hall, London.

Guervara, S. and Laborde, J. (1993). Monitoring seed dispersal at isolated standing trees in tropical pastures: consequences for local species availability. *Vegetatio*, **107/108**, 319–38.

Haas, A. C., Dunski, J. F., and Maxson, L. R. (1995). Divergent lineages within the *Bufo margaritifera* Complex (Amphibia: Anura; Bufonidae) revealed by albumin immunology. *Biotropica*, **27**, 238–49.

Haila, Y. (1990). Towards an ecological definition of an island: a northwest European perspective. *Journal of Biogeography*, **17**, 561–8.

Haila, Y. and Hanski, I. K. (1993). Birds breeding on small British islands and extinction risks. *The American Naturalist*, **142**, 1025–9.

Haila, Y. and Järvinen, O. (1983). Land bird communities on a Finnish island: species impoverishment and abundance patterns. *Oikos*, **41**, 255–73.

Hair, J. B. (1966). Biosystematics of the New Zealand flora 1945–64. *New Zealand Journal of Botany*, **4**, 559–95.

Hambler, C. (1994). Giant tortoise *Geochelone gigantea* translocation to Curieuse Island (Seychelles): Success or failure? *Biological Conservation*, **69**, 293–9.

Hamnett, M. P. (1990) Pacific island resource development and environmental management. In *Sustainable development and environmental management of small islands*. Vol. 5, Man and the Biosphere Series (ed. W. Beller, P. d'Ayala, and P. Hein), pp. 227–57. UNESCO/Parthenon Publishing, Paris.

Hanley, K. A., Bolger, D. T., and Case, T. J. (1994). Comparative ecology of sexual and asexual gecko species (*Lepidodactylus*) in French Polynesia. *Evolutionary Ecology*, **8**, 438–54.

Hanski, I. (1986). Population dynamics of shrews on small islands accord with the equilibrium model. *Biological Journal of the Linnean Society*, **28**, 23–36.

Hanski, I. (1992). Inferences from ecological incidence functions. *The American Naturalist*, **138**, 657–62.

Hanski, I. (1996). Metapopulation ecology. In *Population dynamics in ecological space and time* (ed. O. E. Rhodes, Jr, R. K. Chesser, and M. H. Smith), pp. 13–43. Chicago University Press, Chicago.

Hanski, I., Moilanen, A., and Gyllenberg, M. (1996). Minimum viable metapopulation size. *The American Naturalist*, **147**, 527–41.

Harris, L. D. (1984). *The fragmented forest: island biogeography theory and the preservation of biotic diversity*. The University of Chicago Press, Chicago.

Harrison, S. (1994). Metapopulations and conservation. In *Large-scale ecology and conservation biology* (ed. P. J. Edwards, R. M. May, and N. R. Webb), pp. 111–28. Blackwell Scientific Publications, Oxford.

Harrison, S., Murphy, D. D., and Ehrlich, P. R. (1988). Distribution of the Bay Checkerspot Butterfly, *Euphydryas editha bayensis*: evidence for a metapopulation model. *The American Naturalist*, **132**, 360–82.

Hass, C. A., Dunski, J. F., Maxson, L. R., and Hoogmoed, M. S. (1995). Divergent lineages within the *Bufo margaritifera* Complex (Amphibia: Anura: Bufonidae) revealed by albumin immunology. *Biotropica*, **27**, 238–49.

Heads, M. J. (1990). Mesozoic tectonics and the deconstruction of biogeography: a new model of Australasian biology. *Journal of Biogeography*, **17**, 223–5.

Heaney, L. R. (1986). Biogeography of mammals in SE Asia: estimates of rates of colonization, extinction and speciation. *Biological Journal of the Linnean Society*, **28**, 127–65.

Helenurm, K. and Ganders, F. R. (1985). Adaptive radiation and genetic differentiation in Hawaiian *Bidens*. *Evolution*, **39**, 753–65.

Herrero, S. (1997). Galapagos tortoises threatened. *Conservation Biology*, **11**, 305.

Hess, A. L. (1990). Overview: sustainable development and environmental management of small islands. In *Sustainable development and environmental management of small islands*. Vol. 5, Man and the Biosphere Series (ed. W. Beller, P. d'Ayala, and P. Hein), pp. 3–14. UNESCO/Parthenon Publishing, Paris.

Hess, G. R. (1996). Linking extinction to connectivity and habitat destruction in metapopulation models. *The American Naturalist*, **148**, 226–36.

Hinsley, S. A., Bellamy, P. E., Newton, I., and Sparks, T. H. (1994). Factors influencing the presence of individual breeding bird species in woodland fragments. *English Nature Research Reports*, No. 99.

Hobbs, R. J. and Huenneke, L. F. (1992). Disturbance, diversity, and invasion: implications for conservation. *Conservation Biology*, **6**, 324–37.

Holdaway, R. N. (1989). New Zealand's pre-human avifauna and its vulnerability. In Rudge, M. R. (ed.), Moas, man and climate in the ecological history of New Zealand. *New Zealand Journal of Ecology*, **12** (Suppl.), 11–25.

Holdaway, R. N. (1990). Changes in the diversity of New Zealand forest birds. *New Zealand Journal of Zoology*, **17**, 309–21.

Holloway, J. D. (1996). The Lepidoptera of Norfolk Island, actual and potential, their origins and dynamics. In *The origin and evolution of Pacific Island Biotas, New Guinea to Eastern Polynesia: patterns and processes* (ed. A. Keast, and S. E. Miller), pp. 123–51. SPB Academic Publishing, Amsterdam.

Hommel, P. W. F. (1990). Ujung Kulon: landscape survey and land evaluation as a habitat for the Javan rhinoceros. *ITC Journal*, **1990–1**, 1–15.

Hoogerwerf, A. (1953). Notes on the vertebrate fauna of the Krakatau Islands, with special reference to the birds. *Treubia*, **22**, 319–48.

Huggett, R. J. (1995). *Geoecology: an evolutionary approach.* Routledge, London.

Hughes, L. *et al.* (1994). Predicting dispersal spectra: a minimal set of hypotheses based on plant attributes. *Journal of Ecology*, **82**, 933–50.

Humphreys, W. F. and Kitchener, D. J. (1982). The effect of habitat utilization on species–area curves: implications for optimal reserve area. *Journal of Biogeography*, **9**, 381–96.

Humphries, C. J. (1979). Endemism and evolution in Macaronesia. In *Plants and islands* (ed. D. Bramwell), pp. 171–99. Academic Press, London.

Hunt, G. L., Jr and Hunt, M. W. (1974). Trophic levels and turnover rates: the avifauna of Santa Barbara island, California. *The Condor*, **76**, 363–9.

Huntley, B., Berry, P. M., Cramer, W., and McDonald, A. P. (1995). Modelling present and potential future ranges of some European higher plants using climate response surfaces. *Journal of Biogeography*, **22**, 967–1001.

Hutchinson, G. E. (1957). Concluding remarks. *Cold Spring Harbor Symposia on Quantitative Biology*, **22**, 415–27.

Hutchinson, G. E. (1959). Homage to Santa Rosalia: or why are there so many kinds of animals? *The American Naturalist*, **93**, 145–59.

Itow, S. (1988). Species diversity of mainland- and island forests in the Pacific area. *Vegetatio*, **77**, 193–200.

Jackson, E. D., Silver, E. A., and Dalrymple, G. B. (1972). Hawaiian-Emperor chain and its relation to Cenozoic circumpacific tectonics. *Geological Society of America, Bulletin*, **83**, 601–18.

Jackson, M. H. (1995). *Galápagos: a natural history.* University of Calgary Press, Calgary.

James, H. F. (1995). Prehistoric extinctions and ecological changes on Oceanic islands. In *Islands: biological diversity and ecosystem function* (ed. P. M. Vitousek, L. L. Loope, and H. Adsersen), Ecological Studies 115, pp. 87–102. Springer-Verlag, Berlin.

James, H. F. and Olson, S. L. (1991). Descriptions of thirty-two new species of birds from the Hawaiian Islands: Part II. Passeriformes. *Ornithological Monographs*, **46**, 1–88.

Janzen, D. H. (1983). No park is an island: increase in interference from outside as park size increases. *Oikos*, **41**, 402–10.

Järvinen, O. and Haila, Y. (1984) Assembly of land bird communities on Northern Islands: a quantitative analysis of insular impoverishment. In *Ecological communities: conceptual issues and the evidence* (ed. D. R. Strong, Jr, D. Simberloff, L. G. Abele, and A. B. Thistle), pp. 138–47. Princeton University Press, Princeton, New Jersey.

Johnson, D. L. (1980). Problems in the land vertebrate zoogeography of certain islands and the swimming power of elephants. *Journal of Biogeography*, **7**, 383–98.

Johnson, T. H. and Stattersfield, A. J. (1990). A global review of island endemic birds. *The Ibis*, **132**, 167–80.

Juvig, J. O. and Austring, A. O. (1979). The Hawaiian avifauna: biogeographical theory in evolutionary time. *Journal of Biogeography*, **6**, 205–24.

Kadmon, R. (1995). Nested species subsets and geographic isolation: a case study. *Ecology*, **76**, 458–65.

Kaneshiro, K. Y. (1989). The dynamics of sexual selection and founder effects in species formation. In *Genetics, speciation and the founder principle* (ed. L. V. Giddings, K. Y. Kaneshiro, and W. W. Anderson), pp. 279–96. Oxford University Press, New York.

Kaneshiro, K. Y. (1995). Evolution, speciation, and the genetic structure of island populations. In *Islands: biological diversity and ecosystem function* (ed. P. M. Vitousek, L. L. Loope, and H. Adsersen), Ecological Studies 115, pp. 22–33. Springer-Verlag, Berlin.

Kaneshiro, K. Y., Gillespie, R. G., and Carson, H. L. (1995). Chromosomes and male genitalia of Hawaiian *Drosophila*: tools for interpreting phylogeny and geography. In *Hawaiian biogeography: evolution on a hot spot archipelago* (ed. W. L. Wagner and V. A. Funk), pp. 57–71. Smithsonian Institution Press, Washington.

Kapos, V. (1989). Effects of isolation on the water status of forest patches in the Brazilian Amazon. *Journal of Tropical Ecology*, **5**, 173–85.

Keast, A. and Miller, S. E. (ed.) (1996). *The origin and evolution of Pacific Island Biotas, New Guinea to Eastern Polynesia: patterns and processes.* SPB Academic Publishing, Amsterdam.

Kellman, M. (1996). Redefining roles: plant community reorganization and species preservation in fragmented systems. *Global Ecology and Biogeography Letters*, **5**, 111–16.

Kelly, B. J., Wilson, J. B., and Mark, A. F. (1989). Causes of the species–area relation: a study of islands in Lake Manapouri, New Zealand, *Journal of Ecology*, **77**, 1021–8.

Kim, S.-C., Crawford, D. J., Francisco-Ortega, J., and Santos-Guerra, A. (1996). A common origin for woody *Sonchus* and five related genera in the Macaronesian islands: Molecular evidence for extensive radiation. *Proceedings of the National Academy of Sciences, USA*, **93**, 7743–8.

Kindvall, O. and Ahlén, I. (1992). Geometrical factors and metapopulation dynamics of the Bush Cricket, *Metrioptera bicolor* Philippi (Orthoptera: Tettigoniidae). *Conservation Biology*, **6**, 520–9.

King, C. (1984). *Immigrant killers*. Oxford University Press, Auckland.

Klein, N. K. and Brown, W. M. (1994). Intraspecific molecular phylogeny in the Yellow Warbler (*Dendroica petechia*), and implications for avian biogeography in the West Indies. *Evolution*, **48**, 1914–32.

Kohn, D. D. and Walsh, D. M. (1994). Plant species richness—the effect of island size and habitat diversity. *Journal of Ecology*, **82**, 367–77.

Koopowitz, H. and Kaye, H. (1990). *Plant extinction: a global crisis* (2nd edn). Christopher Helm, London.

Korn, H. (1994). Genetic, demographic, spatial, environmental and catastrophic effects on the survival probability of small populations of mammals. In *Minimum animal populations* (ed. H. Remmert), pp. 39–49. Ecological Studies 106. Springer-Verlag, Berlin.

Kunkel, G. (ed.) (1976). *Biogeography and ecology of the Canary Islands*. W. Junk, The Hague.

Lack, D. (1947). *Darwin's finches: an essay on the general biological theory of evolution*. Cambridge University Press, Cambridge.

Lack, D. (1969). The numbers of bird species on islands. *Bird Study*, **16**, 193–209.

Lack, D. (1970). The endemic ducks of remote islands. *Wildfowl*, **21**, 5–10.

Lack, D. (1976). *Island Biology illustrated by the land birds of Jamaica*. Blackwell Scientific Publications, Oxford.

Lacy, R. C. (1992). The effects of inbreeding on isolated populations: are minimum viable population sizes predictable? In *Conservation Biology: the theory and practice of nature conservation and management* (ed. P. L. Fiedler and S. K. Jain), pp. 276–320. Chapman & Hall, New York.

Law, R. and Watkinson, A. R. (1989). Competition. In *Ecological concepts: the contribution of ecology to an understanding of the natural world* (ed. J. M. Cherrett), pp. 243–84. Blackwell Scientific Publications, Oxford.

Lawlor, T. E. (1986). Comparative biogeography of mammals on islands. *Biological Journal of the Linnean Society*, **28**, 99–125.

Lawton, J. H. and May, R. M. (ed.) (1995). *Extinction rates*. Oxford University Press, Oxford.

Leader-Williams, N. and Walton, D. (1989). The isle and the pussycat. *New Scientist*, **121** (1651), 11 February, 48–51.

Leberg, P. L. (1991). Influence of fragmentation and bottlenecks on genetic divergence of wild turkey populations. *Conservation Biology*, **5**, 522–30.

Leuschner, C. (1996). Timberline and alpine vegetation on the tropical and warm-temperate oceanic islands of the world: elevation, structure, and floristics. *Vegetatio*, **123**, 193–206.

Lister, A. M. (1993). Mammoths in miniature. *Nature*, **362**, 288–9.

Livezey, B. C. (1993). An ecomorphological review of the dodo (*Raphus cucullatus*) and solitaire (*Pezophaps solitaria*), flightless Columbiformes of the Mascarene Islands. *Journal of Zoology, London*, **230**, 247–92.

Lockwood, J. L. and Moulton, M. P. (1994). Ecomorphological pattern in Bermuda birds: the influence of competition and implications for nature preserves. *Evolutionary Ecology*, **8**, 53–60.

Lomolino, M. V. (1984*a*). Immigrant selection, predation, and the distribution of *Microtus pennsylvanicus* and *Blarina brevicauda* on islands. *The American Naturalist*, **123**, 468–83.

Lomolino, M. V. (1984*b*). Mammalian island biogeography: effects of area, isolation, and vagility. *Oecologia*, **61**, 376–82.

Lomolino, M. V. (1985). Body size of mammals on islands: the island rule reexamined. *The American Naturalist*, **125**, 310–16.

Lomolino, M. V. (1986). Mammalian community structure on islands: the importance of immigration, extinction and interactive effects. *Biological Journal of the Linnean Society*, **28**, 1–21.

Lomolino, M. V. (1989). Interpretations and comparisons of constants in the species–area relationship: an additional caution. *The American Naturalist*, **133**, 277–80.

Lomolino, M. V. (1990). The target area hypothesis: the influence of island area on immigration rates of non-volant mammals. *Oikos*, **57**, 297–300.

Lomolino, M. V. (1996). Investigating causality of nestedness of insular communities: selective immigrations or extinctions? *Journal of Biogeography*, **23**, 699–703.

Lomolino, M. V. and Davis, R. (1997). Biogeographic scale and biodiversity of mountain forest mammals of western North America. *Global Ecology and Biogeography Letters*, **6**, 57–76.

Lomolino, M. V., Brown, J. H., and Davis, R. (1989). Island biogeography of montane forest mammals in the American southwest. *Ecology*, **70**, 180–94.

Losos, J. B. (1990). A phylogenetic analysis of character displacement in Caribbean *Anolis* lizards. *Evolution*, **44**, 558–69.

Losos, J. B. (1994). Integrative approaches to evolutionary ecology: *Anolis* lizards as model systems. *Annual Review of Ecology and Systematics*, **25**, 467–93.

Losos, J. B. (1996). Phylogenetic perspectives on community ecology. *Ecology*, **77**, 1344–54.

Losos, J. B., Warheit, K. I., and Schoener, T. W (1997). Adaptive differentiation following experimental island colonization in *Anolis* lizards. *Nature*, **387**, 70–3.

Lovegrove, T. G. (1996). Island releases of Saddlebacks *Phiesturnus carunculatus* in New Zealand. *Biological Conservation*, **77**, 151–7.

Lovejoy, T. E. *et al.* (1986). Edge and other effects of isolation on Amazon forest fragments. In *Conservation biology: the science of scarcity and diversity* (ed. M. Soulé), pp. 257–85. Sinauer Associates Inc., Sunderland, Massachusetts.

Ludwig, D. (1996). The distribution of population survival times. *The American Naturalist*, **147**, 506–26.

Lugo, A. E. (1988). Ecological aspects of catastrophes in Caribbean Islands. *Acta Cientifica*, **2**, 24–31.

Lynch, J. D. and Johnson, N. V. (1974). Turnover and equilibria in insular avifaunas, with special reference to the California Channel Islands. *The Condor*, **76**, 370–84.

Mabberley, D. J. (1979). Pachycaul plants and islands. In *Plants and islands* (ed. D. Bramwell), pp. 259–77. Academic Press, London.

MacArthur, R. H. and Wilson, E. O. (1963). An equilibrium theory of insular zoogeography. *Evolution*, **17**, 373–87.

MacArthur, R. H. and Wilson, E. O. (1967). *The theory of island biogeography.* Princeton University Press, Princeton.

McDonald, M. A. and Smith, M. H. (1990). Speciation, heterochrony, and genetic variation in Hispaniolan palm-tanagers. *The Auk*, **107**, 707–17.

McGuinness, K. A. (1984). Equations and explanations in the study of species–area curves. *Biological Reviews*, **59**, 423–40.

Madsen, T., Stille, B., and Shine, R. (1996). Inbreeding depression in an isolated population of adders *Vipera berus*. *Biological Conservation*, **75**, 113–18.

Mallet, J. (1995). A species definition for the modern synthesis. *Trends in Ecology and Evolution*, **10**, 294–9.

Mangel, M. and Tier, C. (1994). Four facts every conservation biologist should know about persistence. *Ecology*, **75**, 607–14.

Martin, J.-L., Gaston, A. J., and Hitier, S. (1995). The effect of island size and isolation on old growth forest habitat and bird diversity in Gwaii Haanas (Queen Charlotte Islands, Canada). *Oikos*, **72**, 115–31.

Martin, T. E. (1981). Species–area slopes and coefficients: a caution on their interpretation. *The American Naturalist*, **188**, 823–37.

Mawdsley, N. A., Compton, S. G., and Whittaker, R. J. (1998). Population persistence, pollination mutualisms, and figs in fragmented tropical landscapes. *Conservation Biology*, in press.

Mayr, E. (1942). *Systematics and the origin of species.* Columbia University Press, New York.

Mayr, E. (1954). Change of genetic environment and evolution. In *Evolution as a process* (ed. J. S. Huxley, A. C. Hardy, and E. B. Ford), pp. 156–80. Allen and Unwin, London.

Means, D. B. and Simberloff, D. (1987). The peninsula effect: habitat-correlated species decline in Florida's herpetofauna. *Journal of Biogeography*, **14**, 551–68.

Menard, H. W. (1986). *Islands*. Scientific American Library, New York.

Menges, E. S. (1992). Stochastic modeling of extinction in plant populations. In *Conservation Biology: the theory and practice of nature conservation, preservation and management* (ed. P. L. Fiedler and S. K. Jain), pp. 253–75. Chapman & Hall, London.

Merlen, G. (1995) Use and misuse of the seas around the Galápagos Archipelago. *Oryx*, **29**, 99–106.

Meyer, J.-Y. and Florence, J. (1996). Tahiti's native flora endangered by the invasion of *Miconia calvescens* DC. (Melastomataceae). *Journal of Biogeography*, **23**, 775–81.

Micol, T. and Jouventin, P. (1995). Restoration of Amsterdam Island, South Indian Ocean, following control of feral cattle. *Biological Conservation*, **73**, 199–206.

Mielke, H. W. (1989). *Patterns of life: biogeography of a changing world.* Unwin Hyman, Boston.

Milberg, P. and Lamont, B. B. (1995). Fire enhances weed invasion of roadside vegetation in Southwestern Australia. *Biological Conservation*, **73**, 45–9.

Milberg, P. and Tyrberg, T. (1993). Naive birds and noble savages—a review of man-caused prehistoric extinctions of island birds. *Ecography*, **16**, 229–50.

Miskelly, C. M. (1990). Effects of the 1982–83 El Niño event on two endemic landbirds on the Snares Islands, New Zealand. *Emu*, **90**, 24–7.

Moore, D. M. (1979). The origins of temperate island floras. In *Plants and islands* (ed. D. Bramwell), pp. 69–86. Academic Press, London.

Morgan, G. S. and Woods, C. A. (1986). Extinction and zoogeography of West Indian land mammals. *Biological Journal of the Linnean Society*, **28**, 167–203.

Morin, M. P. (1992). The breeding biology of an endangered Hawaiian Honeycreeper, the Laysan Finch. *The Condor*, **94**, 646–67.

Morrison, L. W. (1997). The insular biogeography of small Bahamian cays. *Journal of Ecology*, **85**, 441–54.

Mueller-Dombois, D. (1975). Some aspects of island ecosystem analysis. In *Tropical ecological systems: trends in terrestrial and aquatic research* (ed. F. B. Golley and E. Medina), pp. 353–66. Springer-Verlag, New York.

Myers, A. A. (1991). How did Hawaii accumulate its biota? A test from the Amphipoda. *Global Ecology and Biogeography Letters*, **1**, 24–9.

Myers, A. A. and Giller, P. S. (ed.) (1988). *Analytical biogeography*. Chapman & Hall, London.

Newmark, W. D. (1991). Tropical forest fragmentation and the local extinction of understorey birds in the Eastern Usambara Mountains, Tanzania. *Conservation Biology*, **5**, 67–78.

Nilsson, I. N. and Nilsson, S. G. (1985). Experimental estimates of census efficiency and pseudo-turnover on islands: error trend and between-observer variation when recording vascular plants. *Journal of Ecology*, **73**, 65–70.

Nores, M. (1995). Insular biogeography of birds on mountain-tops in north western Argentina. *Journal of Biogeography*, **22**, 61–70.

Nunn, P. D. (1990). Recent environmental changes on Pacific Islands. *Geographical Journal*, **156**, 125–40.

Nunn, P. D. (1994). *Oceanic islands*. Blackwell, Oxford.

O'Brien, E. M. (1993). Climatic gradients in woody plant species richness: towards an explanation based on an analysis of southern Africa's woody flora. *Journal of Biogeography*, **20**, 181–98.

Olson, S. L. and James, H. F. (1982). Fossil birds from the Hawaiian Islands: evidence for wholesale extinction by man before western contact. *Science*, **217**, 633–5.

Olson, S. L. and James, H. F. (1991). Descriptions of thirty-two new species of birds from the Hawaiian Islands: Part I. Non-passeriformes. *Ornithological Monographs*, **45**, 1–88.

Olson, S. L. and Jouventin, P. (1996). A new species of small flightless duck from Amsterdam Island, Southern Indian Ocean (Anatidae: *Anas*). *The Condor*, **98**, 1–9.

Otte, D. (1989). Speciation in Hawaiian crickets. In *Speciation and its consequences* (ed. D. Otte and J. A. Endler), pp. 482–526. Sinauer, Sunderland, Massachusetts.

Otte, D. and Endler, J. A. (ed.) (1989). *Speciation and its consequences*. Sinauer, Sunderland, Massachusetts.

Paine, R. T. (1985). Re-establishment of an insular winter wren population following a severe freeze. *The Condor*, **87**, 558–9.

Pannell, C. (1989). The role of animals in natural regeneration and the management of equatorial rain forests for conservation and timber production. *Commonwealth Forestry Review*, **68**, 309–13.

Partomihardjo, T., Mirmanto, E., and Whittaker, R. J. (1992). Anak Krakatau's vegetation and flora circa 1991, with observations on a decade of development and change. *Geojournal*, **28**, 233–48.

Paton, P. W. C. (1994). The effect of edge on avian nest success: how strong is the evidence? *Conservation Biology*, **8**, 17–26.

Patterson, B. D. (1990). On the temporal development of nested subset patterns of species composition. *Oikos*, **59**, 330–42.

Patterson, B. D. and Atmar, W. (1986). Nested subsets and the structure of insular mammalian faunas and archipelagos. *Biological Journal of the Linnean Society*, **28**, 65–82.

Paulay, G. (1994). Biodiversity on oceanic islands: its origin and extinction. *American Zoologist*, **34**, 134–44.

Peltonen, A. and Hanski, A. (1991). Patterns of island occupancy explained by colonization and extinction rates in shrews. *Ecology*, **72**, 1698–708.

Perry, R. (ed.) (1984). *Galapagos*, Key Environments Series. IUCN/Pergamon Press, Oxford.

Pickett, S. T. A. and Thompson, J. N. (1978). Patch dynamics and the design of nature reserves. *Biological Conservation*, **13**, 27–37.

Pickett, S. T. A. and White, P. S. (ed.) (1985). *The ecology of natural disturbance and patch dynamics*. Academic Press, Orlando, Florida.

Pickett, S. T. A., Parker, V. T., and Fiedler, P. L. (1992). The new paradigm in ecology: implications for conservation biology above the species level. In *Conservation biology: the theory and practice of nature conservation and management* (ed. P. L. Fiedler and S. K. Jain), pp. 91–125. Chapman & Hall, New York.

Pierson, E. D., Elmqvist, T., Rainey, W. E., and Cox, P. A. (1996). Effects of tropical cyclonic storms on flying fox populations on the South

Pacific islands of Samoa. *Conservation Biology*, **10**, 438–51.

Pimm, S. L. (1991). *The balance of nature? Ecological issues in the conservation of species and communities*. The University of Chicago Press, Chicago.

Pimm. S. L. and Askins, R. A. (1995). Forest losses predict bird extinctions in eastern North America. *Proceedings of the National Academy of Sciences, USA*, **92**, 9343–7.

Pimm, S. L., Jones, H. L., and Diamond, J. (1988). On the risk of extinction. *The American Naturalist*, **132**, 757–85.

Pimm, S. L., Moulton, M. P., and Justice, L. J. (1995). Bird extinctions in the central Pacific. In *Extinction rates* (ed. J. H. Lawton and R. M. May), pp. 75–87. Oxford University Press, Oxford.

Pole, M. (1994). The New Zealand flora—entirely long-distance dispersal? *Journal of Biogeography*, **21**, 625–35.

Porter, D. M. (1979). Endemism and evolution in Galapagos islands vascular plants. In *Plants and islands* (ed. D. Bramwell), pp. 225–56. Academic Press, London.

Porter, D. M. (1984). Endemism and evolution in terrestrial plants. In *Galapagos*, Key Environments Series (ed. R. Perry), pp. 85–99. IUCN/Pergamon Press, Oxford.

Power, D. M. (1972). Numbers of bird species on the California islands. *Evolution*, **26**, 451–63.

Pratt, H. D., Bruner, P. L., and Berrett, D. G. (1987). *A field guide to the Birds of Hawaii and the Tropical Pacific*. Princeton University Press, Princeton, New Jersey.

Pregill, G. K. (1986). Body size of insular lizards: a pattern of Holocene dwarfism. *Evolution* **40**, 997–1008.

Pregill, G. K. and Olson, S. L. (1981). Zoogeography of West Indian vertebrates in relation to Pleistocene climate cycles. *Annual Review of Ecology and Systematics*, **12**, 75–98.

Preston, F. W. (1948). The commonness, and rarity, of species. *Ecology*, **48**, 254–83.

Preston, F. W. (1962). The canonical distribution of commonness and rarity. *Ecology*, **43**, part I, pp. 185–215; part II, pp. 410–32.

Price, O., Woinarski, J. C. Z., Liddle, D. L., and Russell-Smith, J. (1995). Patterns of species composition and reserve design for a fragmented estate: monsoon rainforests in the Northern Territory, Australia. *Biological Conservation*, **74**, 9–19.

Primack, R. B. (1993). *Essentials of conservation biology*. Sinauer, Sunderland, Massachusetts.

Primack, R. B. and Miao, S. L. (1992). Dispersal can limit local plant distribution. *Conservation Biology*, **6**, 513–19.

Pulliam, H. R. (1996). Sources and sinks: empirical evidence and population consequences. In *Population dynamics in ecological space and time* (ed. O. E. Rhodes, Jr, R. K. Chesser, and M. H. Smith), pp. 45–69. Chicago University Press, Chicago.

Quinn, J. F. and Harrison, S. P. (1988). Effects of habitat fragmentation and isolation on species richness: evidence from biogeographic patterns. *Oecologia*, **75**, 132–40.

Rafe, R. W., Usher, M. B., and Jefferson, R. G. (1985). Birds on reserves: the influence of area and habitat on species richness. *Journal of Applied Ecology*, **22**, 327–35.

Rainey, W. E., Pierson, E. D., Elmqvist, T., and Cox, P. A. (1995). The role of flying foxes (Pteropodidae) in oceanic island ecosystems of the Pacific. *Symposium of the Zoological Society of London*, **67**, 47–62.

Ralph, C. J. and Fancy, S. G. (1994). Demography and movements of the Omao (*Myadestes obscurus*). *The Condor*, **96**, 503–11.

Rassmann, K., Trillmich, F., and Tautz, D. (1997). Hybridization between the Galápagos land and marine iguana (*Conlophus subcristatus* and *Amblyrhynchus cristatus*) on Plaza Sur. *Journal of Zoology (London)*, **242**, 729–39.

Ratcliffe, D. A. (ed.) (1977). *A nature conservation review*. Cambridge University Press, Cambridge.

Rawlinson, P. A., Zann, R. A., van Balen, S., and Thornton, I. W. B. (1992). Colonization of the Krakatau islands by vertebrates. *Geojournal*, **28**, 225–31.

Reijnen, R., Foppen, R., and Meeuwsen, H. (1996). The effects of traffic on the density of breeding birds in Dutch agricultural grasslands. *Biological Conservation*, **75**, 255–60.

Renvoize, S. A. (1979). The origins of Indian Ocean island floras. In *Plants and islands* (ed. D. Bramwell), pp. 107–29. Academic Press, London.

Rey, J. R. (1984). Experimental tests of island biogeographic theory. In *Ecological communities: conceptual issues and the evidence* (ed. D. R. Strong, Jr, D. Simberloff, L. G. Abele, and A. B. Thistle), pp. 101–12. Princeton University Press, Princeton, New Jersey.

Rey, J. R. (1985). Insular ecology of salt marsh arthropods: species level patterns. *Journal of Biogeography*, **12**, 97–107.

Reyment, R. A. (1983). Palaeontological aspects of island biogeography: colonization and evolution of mammals on Mediterranean islands. *Oikos*, **41**, 299–306.

Ricklefs, R. E. (1989). Speciation and diversity: the integration of local and regional processes. In *Speciation and its consequences* (ed. D. Otte and J. A. Endler), pp. 599–622. Sinauer, Sunderland, Massachusetts.

Ricklefs, R. E. and Cox, G. W. (1972). Taxon cycles in the West Indian avifauna. *The American Naturalist*, **106**, 195–219.

Ricklefs, R. E. and Cox, G. W. (1978). Stage of taxon cycle, habitat distribution, and population density in the avifauna of the West Indies. *The American Naturalist*, **112**, 875–95.

Ridley, H. N. (1930). *The dispersal of plants throughout the world.* Reeve, Ashford, England.

Ridley, M. (ed.) (1994). *A Darwin selection.* Fontana Press, London.

Ridley, M. (1996). *Evolution* (2nd edn). Blackwell Science, Cambridge, Massachusetts.

Roberts, A. and Stone, L. (1990). Island-sharing by archipelago species. *Oecologia*, **83**, 560–7.

Rodriguez-Estrella, R., Leon de la Luz, J. L., Breceda, A., Castellanos, A., Cancino, J., and Llinas, J. (1996). Status, density and habitat relationships of the endemic terrestrial birds of Socorro island, Revillagigedo Islands, Mexico. *Biological Conservation*, **76**, 195–202.

Roff, D. A. (1991). The evolution of flightlessness in insects. *Ecological Monographs*, **60**, 389–421.

Roff, D. A. (1994). The evolution of flightlessness: is history important? *Evolutionary Ecology*, **8**, 639–57.

Rosenzweig, M. L. (1978). Competitive speciation. *Biological Journal of the Linnean Society*, **10**, 275–89.

Rosenzweig, M. L. (1995). *Species diversity in space and time.* Cambridge University Press, Cambridge.

Rosenzweig, M. L. and Clark, C. W. (1994). Island extinction rates from regular censuses. *Conservation Biology*, **8**, 491–4.

Roughgarden, J. (1989). The structure and assembly of communities. In *Perspectives in Ecological Theory* (ed. J. Roughgarden, R. M. May, and S. A. Levin), pp. 203–26. Princeton University Press, Princeton.

Roughgarden, J. and Pacala, S. (1989). Taxon cycle among *Anolis* lizard populations: review of evidence. In *Speciation and its consequences* (ed. D. Otte and J. A. Endler), pp. 403–32. Sinauer, Sunderland, Massachusetts.

Rudge, M. R. (ed.) (1989). Moas, man and climate in the ecological history of New Zealand. *New Zealand Journal of Ecology*, **12** (Suppl.), 11–25.

Rummel, J. D. and Roughgarden, J. D. (1985). A theory of faunal buildup for competition communities. *Evolution*, **39**, 1009–33.

Runhaar, J., van Gool, C. R., and Groen, C. L. G. (1996). Impact of hydrological changes on nature conservation areas in the Netherlands. *Biological Conservation*, **76**, 269–76.

Russell, G. J., Diamond, J. M., Pimm, S. L., and Reed, T. M. (1995). A century of turnover: community dynamics at three timescales. *Journal of Animal Ecology*, **64**, 628–41.

Russell, R. W., Carpenter, F. L., Hixon, M. A., and Paton, D. C. (1994). The impact of variation in stopover habitat quality on migrant rufous hummingbirds. *Conservation Biology*, **8**, 483–90.

Safford, R. J. (1997). Distribution studies on the forest-living native passerines of Mauritius. *Biological Conservation*, **80**, 189–98.

Sakai, A. K., Wagner, W. L., Ferguson, D. M., and Herbst, D. R. (1995*a*). Origins of dioecy in the Hawaiian flora. *Ecology*, **76**, 2517–29.

Sakai, A. K., Wagner, W. L., Ferguson, D. M., and Herbst, D. R. (1995*b*). Biogeographical and ecological correlates of dioecy in the Hawaiian flora. *Ecology*, **76**, 2530–43.

Sarre, S. and Dearn, J. M. (1991). Morphological variation and fluctuating asymmetry among insular populations of the Sleepy Lizard, *Trachydosaurus rugosus* Gray (Squamata: Scincidea). *Australian Journal of Zoology*, **39**, 91–104.

Sarre, S., Schwaner, T. D., and Georges, A. (1990). Genetic variation among insular populations of the Sleepy Lizard, *Trachydosaurus rugosus* Gray (Squamata: Scincidea). *Australian Journal of Zoology*, **38**, 603–16.

Sauer, J. D. (1969). Oceanic island and biogeographic theory: a review. *The Geographical Review*, **59**, 582–93.

Sauer, J. D. (1990). Allopatric speciation: deduced but not detected. *Journal of Biogeography*, **17**, 1–3.

Saunders, D. A. and Hobbs, R. J. (1989). Corridors for conservation. *New Scientist*, **121** (1648), 28 January, 63–8.

Saunders, D. A., Hobbs, R. J., and Margules, C. R. (1991). Biological consequences of ecosystem fragmentation: a review. *Conservation Biology*, **5**, 18–32.

Scatena, F. N. and Larsen, M. C. (1991). Physical aspects of Hurricane Hugo in Puerto Rico. *Biotropica*, **23**, 317–23.

Schluter, D. (1988). Character displacement and the adaptive divergence of finches on islands and continents. *The American Naturalist*, **131**, 799–824.

Schmitt, S. and Whittaker, R. J. (1998). Disturbance and succession on the Krakatau Islands, Indonesia. In *Dynamics of tropical communities*. British Ecological Society Symposium, Vol. 37 (ed. D. M. Newbery, H. N. T. Prins, and N. D. Brown), pp. 515–48. Blackwell Science, Oxford.

Schoener, T. W. (1975). Presence and absence of habitat shift in some widespread lizard species. *Ecological Monographs*, **45**, 233–58.

Schoener, T. W. (1983). Rate of species turnover decreases from lower to higher organisms: a review of the data. *Oikos*, **41**, 372–7.

Schoener, T. W. (1986). Patterns in terrestrial vertebrate versus arthropod communities: do systematic differences in regularity exist? In *Community ecology* (ed. J. M. Diamond and T. J. Case), pp. 556–86. Harper & Row, New York.

Schoener, T. W. (1988). Testing for non-randomness in sizes and habitats of West Indian lizards: choice of species pool affects conclusions from null models. *Evolutionary Ecology*, **2**, 1–26.

Schoener, T. W. (1989*a*). The ecological niche In *Ecological concepts: the contribution of ecology to an understanding of the natural world* (ed. J. M. Cherrett), pp. 79–113. Blackwell Scientific Publications, Oxford.

Schoener, T. W. (1989*b*). Food webs from the small to the large. *Ecology*, **70**, 1559–89.

Schoener, T. W. and Gorman, G. C. (1968). Some niche differences in three Lesser Antillean lizards of the genus *Anolis*. *Ecology*, **49**, 819–30.

Schoener, T. W. and Schoener, A. (1983). Distribution of vertebrates on some very small islands. I. Occurrence sequences of individual species. *Journal of Animal Ecology*, **52**, 209–35.

Schoener, T. W. and Spiller, D. A. (1987). High population persistence in a system with high turnover. *Nature*, **330**, 474–7.

Schüle, W. (1993). Mammals, vegetation and the initial human settlement of the Mediterranean islands: a palaeoecological approach. *Journal of Biogeography*, **20**, 399–411.

Schwaner, T. D. and Sarre, S. D. (1988). Body size of Tiger Snakes in Southern Australia, with particular reference to *Notechis ater serventyi* (Elapidae) on Chappell Island. *Journal of Herpetology*, **22**, 24–33.

Scott, T. A. (1994). Irruptive dispersal of black-shouldered kites to a coastal island. *The Condor*, **96**, 197–200.

Sfenthourakis, S. (1996). The species–area relationship of terrestrial isopods (Isopoda; Oniscidea) from the Aegean archipelago (Greece): a comparative study. *Global Ecology and Biogeography Letters*, **5**, 149–57.

Shafer, C. L. (1990). *Nature reserves: island theory and conservation practice*. Smithsonian Institution Press, Washington.

Shennan, I. (1983) Flandrian and late Devensian sea-level changes and crustal movements in England and Wales. In *Shorelines and isostacy* (ed. D. E. Smith and A. G. Dawson), pp. 255–84. Academic Press, London.

Short, J. and Turner, B. (1994). A test of the vegetation mosaic hypothesis: a hypothesis to explain the decline and extinction of Australian mammals. *Conservation Biology*, **8**, 439–49.

Shrader-Frechette, K. S. and McCoy, E. D. (1993). *Method in ecology: strategies for conservation*. Cambridge University Press, Cambridge.

Simberloff, D. (1976). Species turnover and equilibrium island biogeography. *Science*, **194**, 572–8.

Simberloff, D. (1978). Using island biogeographic distributions to determine if colonization is stochastic. *The American Naturalist*, **112**, 713–26.

Simberloff, D. (1983). When is an island community in equilibrium? *Science*, **220**, 1275–7.

Simberloff, D. (1992). Do species–area curves predict extinction in fragmented forests? In *Tropical deforestation and species extinction* (ed. T. C. Whitmore and J. A. Sayer), pp. 119–42. Chapman & Hall, London.

Simberloff, D. and Levin, B. (1985). Predictable sequences of species loss with decreasing island area—land birds in two archipelagoes. *New Zealand Journal of Ecology*, **8**, 11–20.

Simberloff, D. and Martin, J.-L. (1991). Nestedness of insular avifaunas: simple summary statistics masking complex species patterns. *Ornis Fennica*, **68**, 187–92.

Simberloff, D. and Wilson, E. O. (1969). Experimental zoogeography of islands. The colonisation of empty islands. *Ecology*, **50**, 278–96.

Simberloff, D. and Wilson, E. O. (1970). Experimental zoogeography of islands. A two year record of colonization. *Ecology*, **51**, 934–7.

Simberloff, D., Farr, J. A., Cox, J., and Mehlman, D. W. (1992). Movement corridors: conservation bargains or poor investments? *Conservation Biology*, **6**, 493–504.

Simkin, T. (1984). Geology of Galápagos Islands. In *Galapagos*, Key Environments Series (ed. R. Perry), pp. 15–41. IUCN/Pergamon Press, Oxford.

Simpson, B. B. (1974). Glacial migration of plants: island biogeographical evidence. *Science*, **185**, 698–700.

Sohmer, S. H. and Gustafson, R. (1993). *Plants and flowers of Hawaii* (3rd printing). University of Hawaii Press, Honolulu.

Soulé, M. (ed.) (1986). *Conservation biology: the science of scarcity and diversity.* Sinauer Associates Inc., Sunderland, Massachusetts.

Soulé, M. E., Bolger, D. T., Alberts, A. C., Wright, J., Sorice, M., and Hill, S. (1988). Reconstructed dynamics of rapid extinctions of Chaparral-requiring birds in urban habitat islands. *Conservation Biology*, **2**, 75–92.

Spellerberg, I. F. and Gaywood, M. J. (1993). Linear features: linear habitats and wildlife corridors. *English Nature Research Reports*, No. 60.

Spencer-Smith, D., Ramos, S. J., McKenzie, F., Munroe, E., and Miller, L. D. (1988). Biogeographical affinities of the butterflies of a 'forgotten' island: Mona (Puerto Rico). *Bulletin of the Allyn Museum*, No. 121, pp. 1–35.

Spencer-Smith, D., Miller, L. D., and Miller, J. Y. (1994). *The butterflies of the West Indies and South Florida.* Oxford University Press, Oxford.

Spiller, D. A. and Schoener, T. W. (1995). Long-term variation in the effect of lizards on spider density is linked to rainfall. *Oecologia*, **103**, 133–9.

Stace, C. A. (1989). Dispersal versus vicariance—no contest! *Journal of Biogeography*, **16**, 201–2.

Steadman, D. W. (1997*a*). Human-caused extinctions of birds. In *Biodiversity II: understanding and protecting our biological resources* (ed. M. L. Reaka-Kudla, W. E. Wilson, and W. O. Wilson), pp. 139–61. Joseph Henry Press, Washington DC.

Steadman, D. W. (1997*b*). The historic biogeography and community ecology of Polynesian pigeons and doves. *Journal of Biogeography*, **24**, 737–53.

Steers, J. A. and Stoddart, D. R. (1977). The origin of fringing reefs, barrier reefs and atolls. In *Biology and geology of coral reefs* (ed. O. A. Jones and R. Endean), pp. 21–57. Academic Press, New York.

Stoddart, D. R. and Walsh, R. P. D. (1992). Environmental variability and environmental extremes as factors in the island ecosystem. *Atoll Research Bulletin*, No. 356.

Stone, L. and Roberts, A. (1990). The checkerboard score and species distributions. *Oecologia*, **85**, 74–9.

Stone, L. and Roberts, A. (1992). Competitive exclusion, or species aggregation? *Oecologia*, **91**, 419–24.

Stouffer, P. C. and Bierregaard, R. O., Jr (1995). Use of Amazonian forest fragments by understory insectivorous birds. *Ecology*, **76**, 2429–45.

Stouffer, P. C. and Bierregaard, R. O., Jr (1996). Effects of forest fragmentation on understorey hummingbirds in Amazonian Brazil. *Conservation Biology*, **9**, 1085–94.

Strong, D. R., Jr, Simberloff, D., Abele, L. G., and Thistle, A. B. (ed.) (1984). *Ecological communities: conceptual issues and the evidence.* Princeton University Press, Princeton, New Jersey.

Stuessy, T. F., Crawford, D. J., and Marticorena, C. (1990). Patterns of phylogeny in the endemic vascular flora of the Juan Fernandez Islands, Chile. *Systematic Botany*, **15**, 338–46.

Sugihara, G. (1981). $S=CA^z$, $z=1/4$: a reply to Connor and McCoy. *The American Naturalist*, **117**, 790–3.

Sunding, P. (1979). Origins of the Macaronesian Flora. In *Plants and islands* (ed. D. Bramwell), pp. 13–40. Academic Press, London.

Tarr, C. L. and Fleischer, R. C. (1995). Evolutionary relationships of the Hawaiian Honeycreepers (Aves, Drepanidae). In *Hawaiian biogeography: evolution on a hot spot archipelago* (ed. W. L. Wagner and V. A. Funk), pp. 147–59. Smithsonian Institution Press, Washington.

Tauber, C. A. and Tauber, M. J. (1989). Sympatric speciation in insects: perception and perspective. In *Speciation and its consequences* (ed. D. Otte and J. A. Endler), pp. 307–44. Sinauer, Sunderland, Massachusetts.

Taylor, B. W. (1957) Plant succession on recent volcanoes in Papua. *Journal of Ecology*, **45**, 233–43.

Taylor, R. J. (1987). The geometry of colonization: 1. Islands. *Oikos*, **48**, 225–31.

Tellería, J. L. and Santos, T. (1995). Effects of forest fragmentation on a guild of wintering passerines: the role of habitat selection. *Biological Conservation*, **71**, 61–7.

Terborgh, J. (1992). Maintenance of diversity in tropical forests. *Biotropica*, **24**, 283–92.

Thomas, C. D. (1994). Extinction, colonization, and metapopulations: environmental tracking by rare species. *Conservation Biology*, **8**, 373–8.

Thomas, C. D. and Harrison, S. (1992). Spatial dynamics of a patchily distributed butterfly species. *Journal of Animal Ecology*, **61**, 437–46.

Thomas, C. D., Singer, M. C., and Boughton, D. A. (1996). Catastrophic extinction of population sources in a butterfly metapopulation. *The American Naturalist*, **148**, 957–75.

Thomson, J. D. and Barrett, S. C. H. (1981). Selection for outcrossing, sexual selection, and the evolution of dioecy in plants. *The American Naturalist*, **118**, 443–9.

Thornton, I. W. B. (1992). K. W. Dammerman—fore-runner of island equilibrium theory. *Global Ecology and Biogeography Letters*, **2**, 145–8.

Thornton, I. W. B. (1996). *Krakatau—the destruction and reassembly of an island ecosystem*. Harvard University Press, Cambridge, Massachusetts.

Thornton, I. W. B. and Walsh, D. (1992). Photographic evidence of rate of development of plant cover on the emergent island Anak Krakatau from 1971 to 1991 and implications for the effect of volcanism. *Geojournal*, **28**, 249–59.

Thornton, I. W. B., Zann, R. A., and Stephenson, D. G. (1990). Colonization of the Krakatau islands by land birds, and the approach to an equilibrium number of species. *Philosophical Transactions of the Royal Society of London*, **B 328**, 55–93.

Thornton, I. W. B., Ward, S. A., Zann, R. A., and New, T. R. (1992). Anak Krakatau—a colonization model within a colonization model? *Geojournal*, **28**, 271–86.

Thornton, I. W. B., Zann, R. A., and van Balen, S. (1993). Colonization of Rakata (Krakatau Is.) by non-migrant land birds from 1883 to 1992 and implications for the value of island equilibrium theory. *Journal of Biogeography*, **20**, 441–52.

Thornton, I. W. B., Partomihardjo, T., and Yukawa, J. (1994). Observations on the effects, up to July 1993, of the current eruptive episode of Anak Krakatau. *Global Ecology and Biogeography Letters*, **4**, 88–94.

Thorpe, R. S. and Malhotra, A. (1996). Molecular and morphological evolution within small islands. *Philosophical Transactions of the Royal Society of London*, **351**, 815–22.

Thorpe, R. S., McGregor, D. P., Cumming, A. M., and Jordan, W. C. (1994). DNA evolution and colonization sequence of island lizards in relation to geological history: mtDNA RFLP, cytochrome B, cytochrome oxidase, 12s RRNA sequence and nuclear RAPD analysis. *Evolution*, **48**, 230–40.

Tidemann, C. D., Kitchener, D. J., Zann, R. A., and Thornton, I. W. B. (1990). Recolonization of the Krakatau islands and adjacent areas of West Java, Indonesia, by bats (Chiroptera) 1883–1986. *Philosophical Transactions of the Royal Society of London*, **B 328**, 121–30.

Toft, C. A. and Schoener, T. W. (1983). Abundance and diversity of orb spiders on 106 Bahamian islands: biogeography at an intermediate trophic level. *Oikos*, **41**, 411–26.

Tracy, C. R. and George. L. (1992). On the determinants of extinction. *The American Naturalist*, **139**, 102–22.

Trewick, S. A. (1997). Flightlessness and phylogeny amongst endemic rails (Aves: Rallidae) of the New Zealand region. *Philosophical Transactions of the Royal Society of London*, **B 352**, 429–56.

Trillmich, F. (1992). Conservation problems on Galápagos: the showcase of evolution in danger. *Naturwissenschaften*, **79**, 1–6.

Tudge, C. (1991). Time to save rhinoceroses. *New Scientist*, **131** (1788), 30–5.

Turner, I. M. (1996). Species loss in fragments of tropical rain forest: a review of the evidence. *Journal of Applied Ecology*, **33**, 200–9.

van Balgooy, M. M. J., Hovenkamp, P. H., and van Welzen, P. C. (1996). Phytogeography of the Pacific—floristic and historical distribution patterns in plants. In *The origin and evolution of Pacific island biotas, New Guinea to Eastern Polynesia: patterns and processes* (ed. A. Keast and S. E. Miller), pp. 191–213. SPB Academic Publishing, Amsterdam.

van Riper, C. III, van Riper, S. G., Goff, M. L., and Laird, M. (1986). The epizootiology and ecological significance of Malaria in Hawaiian land birds. *Ecological Monographs*, **56**, 327–44.

van Steenis, C. G. G. J. (1972) *The mountain flora of Java*. E. J. Brill, Leiden.

Vartanyan, S. L., Garutt, V. E., and Sher, A. V. (1993). Holocene dwarf mammoths from Wrangel Island in the Siberian Arctic. *Nature*, **362**, 337–40.

Villa, F., Rossi, O., and Sartore, F. (1992). Understanding the role of chronic environmental disturbance in the context of island biogeographic theory. *Environmental Management*, **16**, 653–66.

Vincek, V. *et al.* (1997). How large was the founding population of Darwin's finches? *Proceedings of the Royal Society of London*, **B 264**, 111–18.

Vitousek, P. M. (1990). Biological invasion and ecosystem processes: towards an integration of population biology and ecosystem studies. *Oikos*, **57**, 7–13.

Vitousek, P. M., Walker, L. R., Whiteaker, L. D., Mueller-Dombois, D., and Matson, P. A. (1987). Biological invasion by *Myrica faya* alters ecosystem development in Hawaii. *Science*, **238**, 802–4.

Vitousek, P. M., Loope, L. L., and Adsersen, H. (ed.) (1995). *Islands: biological diversity and ecosystem function*, Ecological Studies 115. Springer-Verlag, Berlin.

Wagner, W. L. and Funk, V. A. (ed.) (1995). *Hawaiian biogeography: evolution on a hot spot archipelago*. Smithsonian Institution Press, Washington.

Waide, R. B. (1991). Summary of the responses of animal populations to hurricanes in the Caribbean. *Biotropica*, **23**, 508–12.

Walker, L. R., Lodge, D. J., Brokaw, N. V. L., and Waide, R. B. (1991*a*). An introduction to hurricanes in the Caribbean. *Biotropica*, **23**, 313–16.

Walker, L. R., Brokaw, N. V. L., Lodge, D. J., and Waide, R. B. (ed.) (1991*b*). Special issue: ecosystem, plant, and animal responses to hurricanes in the Caribbean. *Biotropica*, **23**, 313–521.

Walker, L. R., Silver, W. L., Willig, M. R., and Zimmerman, J. K. (ed.) (1996). Special Issue: long term responses of Caribbean ecosystems to disturbance. *Biotropica*, **28**, 414–613.

Wallace, A. R. (1902). *Island life* (3rd edn). Macmillan, London.

Warner, R. E. (1968). The role of introduced diseases in the extinction of the endemic Hawaiian avifauna. *The Condor*, **70**, 101–20.

Waterloo, M. J., Schelleken, J., Bruijnzeel, L. A., Vugts, H. F., Assenberg, P. N., and Rawaqa, T. T. (1997). Chemistry of bulk precipitation in southwestern Viti Levu, Fiji. *Journal of Tropical Ecology*, **13**, 427–47.

Watts, D. (1970). Persistence and change in the vegetation of oceanic islands: an example from Barbados, West Indies. *Canadian Geographer*, **14**, 91–109.

Watts, D. (1987). *The West Indies: patterns of development, culture and environmental change since 1492.* Cambridge University Press, Cambridge.

Weaver, M. and Kellman, M. (1981). The effects of forest fragmentation on woodlot tree biotas in Southern Ontario. *Journal of Biogeography*, **8**, 199–210.

Weiher, E. and Keddy, P. A. (1995). Assembly rules, null models, and trait dispersion: new questions from old patterns. *Oikos*, **74**, 159–64.

Weins, J. (1984). On understanding a non-equilibrium world: myth and reality in community patterns and processes. In *Ecological communities: conceptual issues and the evidence* (ed. D. R. Strong, Jr, D. Simberloff, L. G. Abele, and A. B. Thistle), pp. 439–57. Princeton University Press, Princeton, New Jersey.

Weisler, M. I. (1995). Henderson Island prehistory: colonization and extinction on a remote Polynesian island. *Biological Journal of the Linnean Society*, **56**, 377–404.

Werner, T. K. and Sherry, T. W. (1987). Behavioural feeding specialization in *Pinaroloxias inornata*, the 'Darwin's Finch' of Cocos Island, Costa Rica. *Proceedings of the National Academy of Sciences, USA*, **84**, 5506–10.

Westerbergh, A. and Saura, A. (1994). Genetic differentiation in endemic *Silene* (Caryophyllaceae) on the Hawaiian islands. *American Journal of Botany*, **81**, 1487–93.

White, G. B. (1981). Semispecies, sibling species and superspecies. In *The evolving biosphere (chance, change and challenge)* (ed. P. L. Forey), pp. 21–8. British Museum (Natural History) and Cambridge University Press, Cambridge.

Whitehead, D. R. and Jones, C. E. (1969). Small islands and the equilibrium theory of insular biogeography. *Evolution*, **23**, 171–9.

Whitmore, T. C. (1984). *Tropical rain forests of the Far East* (2nd edn). Clarendon Press, Oxford.

Whitmore, T. C. (ed.) (1987). *Biogeographical evolution of the Malay Archipelago*, Oxford Monographs on Biogeography No. 4. Oxford University Press, Oxford.

Whitmore, T. C. and Sayer, J. A. (ed.) (1992). *Tropical deforestation and species extinction.* Chapman & Hall, London.

Whittaker, R. J. (1992). Stochasticism and determinism in island ecology. *Journal of Biogeography*, **19**, 587–91.

Whittaker, R. J. (1995). Disturbed island ecology. *Trends in Ecology and Evolution*, **10**, 421–5.

Whittaker, R. J. and Bush, M. B. (1993). Anak Krakatau and old Krakatau: a reply. *Geojournal*, **29**, 417–20.

Whittaker, R. J. and Jones, S. H. (1994*a*). Structure in re-building insular ecosystems: an empirically derived model. *Oikos*, **69**, 524–9.

Whittaker, R. J. and Jones, S. H. (1994*b*). The role of frugivorous bats and birds in the rebuilding of a tropical forest ecosystem, Krakatau, Indonesia. *Journal of Biogeography*, **21**, 689–702.

Whittaker, R. J., Bush, M. B., and Richards, K. (1989). Plant recolonization and vegetation succession on the Krakatau Islands, Indonesia. *Ecological Monographs*, **59**, 59–123.

Whittaker, R. J., Bush, M. B., Partomihardjo, T., Asquith, N. M., and Richards, K. (1992*a*). Ecological aspects of plant colonisation of the Krakatau Islands. *Geojournal*, **28**, 201–11.

Whittaker, R. J., Walden, J., and Hill, J. (1992*b*). Post-1883 ash fall on Panjang and Sertung and its ecological impact. *Geojournal*, **28**, 153–71.

Whittaker, R. J., Partomihardjo, T., and Riswan, S. (1995). Surface and buried seed banks from Krakatau, Indonesia: implications for the sterilization hypothesis. *Biotropica*, **27**, 346–54.

Whittaker, R. J., Jones, S. H., and Partomihardjo, T. (1997). The re-building of an isolated rain forest assemblage: how disharmonic is the flora of Krakatau? *Biodiversity and Conservation*, **6**, 1671–96.

Whittaker, R. J., Schmitt, S., Jones, S. H., Partomihardjo, T., and Bush, M. B. (1998). Stand biomass and turnover of permanent forest plots on Krakatau, 1989–95. *Biotropica*, **30**, in press.

Wilcove, D. S., McLellan, C. H., and Dobson, A. P. (1986). Habitat fragmentation in the temperate zone. In *Conservation biology: the science of scarcity and diversity* (ed. M. Soulé), pp. 237–56. Sinauer Associates Inc., Sunderland, Massachusetts.

Wiles, G. J., Schreiner, I. H., Nafus, D., Jurgensen, L. K., and Manglona, J. C. (1996). The status, biology, and conservation of *Serianthes nelsonii* (Fabaceae), an endangered Micronesian tree. *Biological Conservation*, **76**, 229–39.

Wilkinson, D. M. (1993). Equilibrium island biogeography: its independent invention and the marketing of scientific theories. *Global Ecology and Biogeography Letters*, **3**, 65–6.

Williams, E. E. (1972). The origin of faunas. Evolution of lizard congeners in a complex island fauna: A trial analysis. *Evolutionary Biology*, **6**, 47–88.

Williams, M. R. (1996). Species–area curves: the need to include zeroes. *Global Ecology and Biogeography Letters*, **5**, 91–3.

Williamson, M. H. (1981). *Islands populations.* Oxford University Press, Oxford.

Williamson, M. H. (1983). The land-bird community of Skokholm: ordination and turnover. *Oikos*, **41**, 378–84.

Williamson, M. H. (1984). Sir Joseph Hooker's lecture on insular floras. *Biological Journal of the Linnean Society*, **22**, 55–77.

Williamson, M. H. (1988). Relationship of species number to area, distance and other variables. In *Analytical Biogeography, an integrated approach to the study of animal and plant distributions* (ed. A. A. Myers and P. S. Giller), pp. 91–115. Chapman & Hall, London.

Williamson, M. H. (1989*a*). The MacArthur and Wilson theory today: true but trivial. *Journal of Biogeography*, **16**, 3–4.

Williamson, M. H. (1989*b*). Natural extinction on islands. *Philosophical Transactions of the Royal Society of London*, **B 325**, 457–68.

Williamson, M. H. (1996). *Biological invasions.* Population and community biology series 15, Chapman & Hall, London.

Wilson, D. E. and Graham, G. L. (ed.) (1992). *Pacific island flying foxes: proceedings of an international conservation conference.* US Fish and Wildlife Service, Biological Report 90 (23), Washington, DC.

Wilson, E. O. (1959). Adaptive shift and dispersal in a tropical ant fauna. *Evolution*, **13**, 122–44.

Wilson, E. O. (1961). The nature of the taxon cycle in the Melanesian ant fauna. *The American Naturalist*, **95**, 169–93.

Wilson, E. O. (1969). The species equilibrium. *Brookhaven Symposia in Biology*, **22**, 38–47.

Wilson, E. O. (1995). *Naturalist.* Allen Lane, The Penguin Press, London.

Wilson, E. O. and Willis, E. O. (1975). Applied biogeography. In *Ecology and evolution of communities* (ed. M. L. Cody and J. M. Diamond), pp. 522–34. Harvard University Press, Cambridge, Massachusetts, USA.

Wilson, J. B. and Roxburgh, S. H. (1994). A demonstration of guild-based assembly rules for a plant community, and the determination of intrinsic guilds. *Oikos*, **69**, 267–76.

Wilson, J. B. and Whittaker, R. J. (1995). Assembly rules demonstrated in a saltmarsh community. *Journal of Ecology*, **83**, 801–7.

Wilson, J. T. (1963). A possible origin of the Hawaiian islands. *Canadian Journal of Physics*, **41**, 863–70.

Wilson, M. H. *et al.* (1994). Puerto Rican parrots and potential limitations of the metapopulation approach to species conservation. *Conservation Biology*, **8**, 114–23.

Woinarski, J. C. Z., Whitehead, P. J., Bowman, D. M. J. S., and Russell-Smith, J. (1992). Conservation of mobile species in a variable environment: the problem of reserve design in the Northern Territory, Australia. *Global Ecology and Biogeography Letters*, **2**, 1–10.

Worthen, W. B. (1996). Community composition and nested-subset analyses: basic descriptors for community ecology. *Oikos*, **76**, 417–26.

Wragg, G. M. (1995). The fossil birds of Henderson Island, Pitcairn Group: natural turnover and human impact, a synopsis. *Biological Journal of the Linnean Society*, **56**, 405–14.

Wright, D. H. (1983). Species–energy theory: an extension of species–area theory. *Oikos*, **41**, 496–506.

Wright, D. H. and Reeves, J. J. (1992). On the meaning and measurement of nestedness of species assemblages. *Oecologia*, **92**, 416–28.

Wright, D. H., Currie, D. J., and Maurer, B. A. (1993). Energy supply and patterns of species richness on local and regional scales. In *Species diversity in ecological communities: historical and geographical perspectives* (ed. R. Ricklefs and D. Schluter), pp. 66–74. University of Chicago Press, Chicago.

Wyatt-Smith, J. (1953). The vegetation of Jarak Island, Straits of Malacca. *Journal of Ecology*, **41**, 207–25.

Wylie, J. L. and Currie, D. J. (1993*a*). Species–energy theory and patterns of species richness: I. Patterns of bird, angiosperm, and mammal richness on islands. *Biological Conservation*, **63**, 137–44.

Wylie, J. L. and Currie, D. J. (1993*b*). Species–energy theory and patterns of species richness: II. Predicting mammal species richness on isolated nature reserves. *Biological Conservation*, **63**, 145–8.

Young, T. P. (1994). Natural die-offs of large mammals: implications for conservation. *Conservation Biology*, **8**, 410–18.

Zann, R. A. and Darjono (1992). The birds of Anak Krakatau: the assembly of an avian community. *Geojournal*, **28**, 261–70.

Zann, R. A., Male, E. B., and Darjono (1990). Bird colonization of Anak Krakatau, an emergent volcanic island. *Philosophical Transactions of the Royal Society of London*, **B 328**, 95–121.

Zimmerman, B. L. and Bierregaard, R. O., Jr (1986). Relevance of the equilibrium theory of island biogeography and species–area relations to conservation with a case from Amazonia. *Journal of Biogeography*, **13**, 133–43.

Index